READING the EARTH

LANDFORMS IN THE MAKING

Jerome Wyckoff

Adastra West, Inc.
Publishers

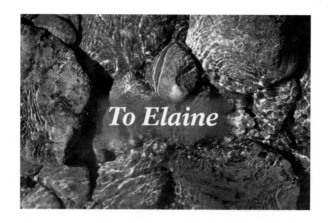

To Elaine

Donald R. Coates, Ph.D.
Professor of Geology Emeritus
Binghamton University
Technical Adviser

Jerome Wyckoff
Design and Typesetting

Maja Britton
Production Management
Illustration Processing
Cover Design

Produced and Published by
Adastra West, Inc.
P.O. Box 874, Mahwah, NJ 07430

Other acknowledgments appear on page 336, which is considered to be an extension of the present page for copyright purposes. Most credits appear on page 336; others are given with illustrations on certain pages in the text.

Cover: Hickman Natural Bridge, Capitol Reef National Park, Utah

ISBN # 0-9674075-0-8
Library of Congress Catalog Card Number: 99-95582

Pre-press by Raimond Graphics, Inc., Englewood Cliffs, New Jersey
Printed in the U.S.A. by World Color Press, Inc.

Foreword

NEVER in history has humankind held such a friendly, inquiring interest in nature as we have today. This interest is focused mainly on the world of animals and plants, and on the conditions under which they are born, live, and die. As living things ourselves, we can hardly question the priority of this interest. Yet there remains a broad aspect of nature which is much neglected: the physical features that make up Earth scenery. Mountains and plains, river valleys and seacoasts, limestone caverns and desert dunes and volcanoes – such features are recognized, often rhapsodized about, but how they are created, undergo natural change, and eventually are obliterated or recycled is understood by few. In fact, these inanimate things are well worth more than a passing acquaintance, not only for geologists and for physical geographers but by all observers of nature who like to understand.

Landforms and the processes that make them have more than purely scientific and esthetic significance. In multitudes of ways, what happens on and in Earth's crust continually affects human beings and, indeed, all other living things. Earthquakes and landslides, advances of glaciers and flooding by rivers, gullying in fields and erosion of beaches, subsidence of land and eruptions of volcanoes – these are but a few of the events that create landforms and are of importance to all inhabitants of the planet. With some understanding of such events we can better appreciate them, somewhat control them, and, at least, learn to live with them.

This book is an introduction to landforms, mainly as scenery. Here are illustrations and descriptions of hundreds of features, many of them scenic, in localities throughout the world. Features shown and described are those most likely to be seen and wondered about by ordinary observers. A deeper understanding of geomorphology – that branch of geology which investigates the forms of Earth's crust – can be sought in textbooks; meanwhile this book can be an informal guide in the field and an armchair companion as well, raising memories of what has been seen in the past, showing features perhaps to be encountered in the future, and telling in everyday language how they originate and evolve. This book can be, indeed, a window on "art" in nature, introducing forms which from any viewpoint deserve understanding and appreciation.

Almost two centuries ago James Hutton, that observant farmer who became the father of geology in England, poetically described Earth's surface as "the ruin of former worlds." So it is – and it is also the scene of wondrous worlds in the making. To understand it, even if only rudimentarily, is one of the rewards of residence on this planet.

– J.W

Contents

Contents

1
Introducing Landforms

WE TAKE Earth scenes around us for granted, almost as if mountain and valley, sea cliff and desert, cavern and glacier have always been as they are now and always will be. Earth seems an almost changeless backdrop against which the drama of living things is played. In reality, Earth changes continuously. Great plates that form the crust push up, down, and sidewise. Mountains rise, only to be leveled by time and the elements. The sea advances over the land, and then withdraws. Volcanic eruptions flood the land with seething lava for millennia, then become dormant. Glaciers spread over vast regions, and melt away. Meanwhile, land is shaped by agents of change into myriad forms. Some changes occur in an instant; others take millions of years. All testify that this planet is a place of fascinating transformations.

Looking at the land, we moderns consider it inanimate, and we deal with it in practical (if not always wise) ways. Yet, though aware of its nature as physical (whether or not divinely created), we can still feel, when we view the majesty of Earth scenes, some of the wonder felt by our ancestors long ago as they contemplated a world little understood.

Basic in recognition and understanding of Earth scenery is the awareness that its great variety of features do not come in clear-cut species, as do people, sparrows, and quartz crystals. Every species of landform, as diverse as fault-block mountains, valley glaciers, and sea stacks, occurs in so many variations that no description, no picture, can fully represent all members of the species. Landforms that may resemble one another superficially – cinder cones, kames, and haystack hills are examples – often are totally different in origin and constitution. Thus landform identification usually requires some knowledge of land-forming processes as well as care in observation. As it happens, most Earth scenery has been more or less eroded, and thus much evidence of its origins has been removed. Further, the foundations of many landforms lie beneath the surface, more or less hidden. Geologists, like Sherlock Holmes, often must search for the tiniest clues to understand the genesis of features they study.

But amateurs need not despair. Each species of landform often has distinguishing characteristics that tend to repeat from specimen to specimen. These clues are useful. When the observer has some familiarity with common rock types, a fundamental knowledge of processes that make scenery, and an awareness of the kind of terrane he is on (such as volcanic, karst, or glaciated), tentative identification is achievable. Beyond is the challenge to expand one's knowledge and perhaps to develop, in time, professional capability in all-around understanding.

Any degree of understanding that one brings to the observation of landforms is helpful and rewarding. Even for professionals there is continually the pleasure of encountering something new, because so few landforms – if any – are totally alike. Understanding, incidentally, not only is a matter of technical knowledge but involves an awareness of beauty.

This book is for ordinary observers; it is organized according to how such observers are likely to encounter and become curious about landforms. Early pages sketch out the basic processes and conditions involved in the shaping of Earth's crust – plate movements, climatic change, tectonic and igneous activity, erosion. Next we look at the nature of rocks that form the crust and at typical forms in which they appear. Further chapters tell in more detail about the processes that shape the crust and the kinds of landscapes that result. Throughout, the focus is more on specific kinds of landforms than on the highly complex systems that control landform evolution. We seek the degree of understanding that is possible without resort to lengthy study of formal texts.

Reading these chapters in order is recommended: thus a somewhat structured understanding can be developed. Any part of this book, however, can be read independently according to the reader's interest at the moment. For the benefit of such readers, and because similar rocks and geologic processes keep appearing in different environments, in changing relationships, some overlapping and repetition

Processes in Earth Scenery: At Lower Ausable Lake, in New York's Adirondacks, the rock was made by the generation, movement, and solidification of molten materials far below the surface, the land was elevated by an upheaval of Earth's crust, the rock was uncovered by erosion, the valley was cut by weathering, gravity movements, and erosion by running water and a glacier, and the lake formed as debris from the glaciation dammed the stream in the valley. Thus landforms may be the product of many processes.

occur from chapter to chapter. Incidentally, since information about a particular topic may occur at two or more locations, the index, which is highly detailed, should be consulted regularly.

Technical terms are defined to the extent necessary where first used in the text. Definitions are to be found also, in context, on pages containing the principal information about an indexed subject; such pages are underlined in the Index. The term "terrain" is used for any piece of land, and "terrane" for any such piece with regard to its geological nature. Measurements are given in terms of the British-American system or the metric system according to the source from which the measurement was obtained.

On page 337 are listed some publications that provide interesting, readable, authoritative information beyond this book's scope.

A Natural Phenomenon: During most of human history, mountains were regarded as abodes of gods and even as gods themselves. Today we understand them as parts of the same nature of which we humans are members. These mountains are the Sierra Nevada as viewed westward over the Owens Valley from near Lone Pine, California. They are distinctive in their abrupt rise and in the sharp sculpturing of their granite masses by thaw-freeze, gravity movements, streams, and glaciers. In contrast are the low Alabama Hills (*foreground*), which are lava masses shaped by erosion.

Our
2 Changing
Planet

IN THE grand array of Earth's landforms, not one exists in isolation, unrelated to other landforms and to the system of processes that made them and are changing them through geologic time. Every landform results from a long train of events extending back to the beginnings of Earth and, indeed, the beginnings of the solar system, of our Milky Way galaxy, and of the universe. The events by which landforms are created and changed are a system in which, as in the ecosystem, all phenomena are related, directly or indirectly, and in which each event results from, and causes, other events. When one strand of a spider's web is touched, the entire web moves; so it is with the web of processes under which all landforms, like all living things, are created and transformed. As we observe, hear, and read about landforms, we can think of them as members of an ever-active, ever-changing community – the community whose members together form the crust of Earth.

Ancestral Earth

As Earth has changed through ages, so have humankind's views of it. For primitive people, landforms were solid realities that constantly and directly affected their lives; thus cliffs were forbidding obstacles to foraging, caves were shelters against rain, wind, and the cold. Landforms were also abodes or embodiments of spirits, to be revered and placated. In ancient Greece, gods lived on the highest peaks, and caves served as homes of other mythical beings and as entrances to the underworld. From the Middle Ages almost into the twentieth century, a common belief was that Earth was created divinely only a few thousand years ago, its interior is the fiery realm of Satan, and eye-catching landforms such as deep canyons, folded rock strata, and rugged profiles result from the violence of Noah's Flood. Contrary to such ideas, called "catastrophism," today's scientific view is that Earth was formed (whether or not by a divine hand) by natural forces billions of years ago and that landforms are expressions, or results, of natural processes operating on and in the planet ever since.

Astronomical and geological evidence indicates that the solar system's planets formed originally as gravitational forces caused clustering and coalescence of materials drifting in space. By perhaps 4.5 billion years ago one cluster had become Earth, whose surface was cooling enough to begin forming patches of crust. The planet's interior was extremely hot and molten because of pressure (due to the planet's own gravity) and radioactivity. By convection, molten material surged up through breaks in the crust, remelting parts of

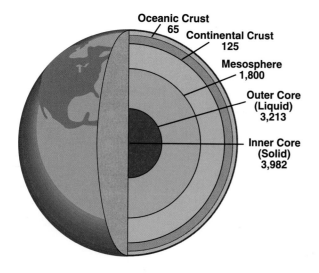

Oceanic Crust
65   **Continental Crust**
125

Mesosphere
1,800

Outer Core
(Liquid)
3,213

Inner Core
(Solid)
3,982

Earth's Interior: Our planet, structured like an onion, consists of concentric layers around a core. This drawing shows approximate distances in miles from Earth's surface to the inner boundaries of the successive layers. The drawing is not to scale.

it and cooling at the surface to form new crust. In the turmoil, heavier elements such as iron and nickel sank toward the planet's center while compounds of lighter elements, notably silicon and aluminum, became portions of crust.

In time, degassing of materials that erupted from the interior of the young planet created the hydrosphere, or ocean, and the atmosphere. Molten materials, continuing to rise through fissures in the crust, piled up on the ocean floor to form more rock, which here and there emerged above the waves and became the first lands. These were weathered and

eroded during millions of years. Rock debris that accumulated on the ocean bottom consolidated to form new rock, and in some areas was raised by movements of crust to become more new land. By about 4 billion years ago, if not somewhat earlier, continents and seas had formed, and the planet's inner structures may have been about what they are today.

Anatomy of Earth

Earth is formed of concentric layers rather like those of an onion. Because exploratory drilling has not yet penetrated the outer crust to a depth beyond about 7.5 miles, the nature of the deeper crust and the inner layers must be inferred from other data. Variations in Earth's magnetism as measured at different locations indicate distributions of mass within the planet. Compositions of meteorites, the kind of material that formed much of the original Earth, provide clues; so do measurements of the planet's radioactivity and heat-flow patterns in the crust. But the major source of information is recordings made by seismic (earthquake) waves as they pass through Earth's interior. Some waves travel through solids but not through liquids, other waves travel through both. Waves vary in strength according to distance from their ori-

New Land: The earliest lands were built of lavas from eruptions of volcanoes on the ocean floor. The process continues today, constructing islands such as (to name a few) the Hawaiian chain, Iceland, the West and East Indies, the Philippines, and the Galápagos Islands. The last of these, some of which are viewed here, from Bartolome Island, are only 2 to 3 million years old.

gin and the kinds of material they pass through. Thus wave behavior indicates locations, densities, movements, and other properties of Earth's layers.

The outermost layer, or lithosphere, consists of the rocky crust, which is rigid, and the upper mantle, consisting of somewhat less rigid rocklike material. Crust and upper mantle are separated by the "Moho," a border at which the transmissibility of seismic waves changes abruptly. Beneath the upper mantle are layers of rocklike composition in various states. These layers are the hot, more or less plastic asthenosphere; next is the transition zone, somewhat more rigid; then the nearly solid lower mantle; and finally the liquid outer core and the solid inner core, the cores being of silicon and iron. Seismic waves indicate that materials in all layers down to the lower mantle, at a depth of about 1,800 miles, are in slow motion, due to heating by radioactivity, density variations, forces associated with Earth's rotation, and the gravitational pull of Sun and Moon.

The crust, on and in which all landforms occur, is relative-ly thin, accounting for less than a thirtieth (at most) of the planet's radius. Its upper portions are mostly of granite-related rocks, relatively light, rich in silica and aluminum compounds, derived from magma (molten material) from the mantle and from crustal melts as well. The crust's lower portion consists mostly of relatively heavy basic rocks (those rich especially in iron and magnesium) formed from magma from the mantle.

Viewed more closely, the crust is an assemblage of more or less rigid, slowly drifting rock plates, 600,000 to 60,000,000 square miles in area, loosely fitting together. In all, there are 7 major plates, 8 of intermediate size, and about 20 small ones. Plates are 2 to 4 miles thick beneath the deep ocean trenches and as much as 120 miles thick under continents. In active zones along plate borders, continental crust is moving against other continental crust (as where the Indo-Australian Plate is meeting the Eurasian Plate, incidentally pushing up the Himalaya), ocean crust is colliding with continental crust (as in the contact of the Nazca Plate with the South Ameri-

The Crustal Plates: Plate boundaries are indicated by open and solid lines. Open lines indicate ocean-bottom ridges. Solid lines indicate ocean trenches, formed where plates are thrusting beneath other plates. Small triangles on solid lines, and the large arrows, show directions of thrusting.

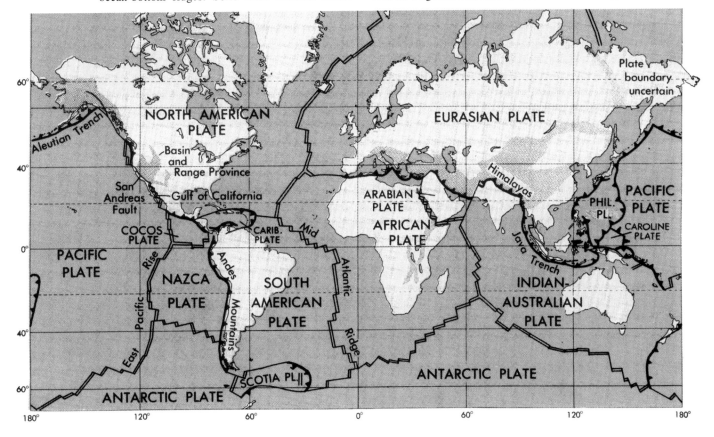

Evidence of Plate Movements: Much rock in the Canadian Rockies is reef limestone, consisting largely of remains of shellfish, corals, and other marine animals of types that grow only in subtropical waters. Several hundred million years ago, this limestone was forming on a sea bottom thousands of miles to the south; today, as a result of plate movements, it stands far to the north, with elevations of thousands of feet above the sea. This view, in Alberta, is north along the Trans-Canada Highway, with Cascade Mountain, a limestone mass, at center.

can plate, raising the Andes), and ocean crust is diving under other ocean crust, making trenches and creating volcanoes, as in the Aleutian Islands and West Indies. Plate borders are formed also along midocean ridges, where chains of islands are being built up by eruptions of volcanic materials from the mantle.

Some 30 per cent of Earth's surface is land – that is, the continents. These consist of slabs mostly of granite-related rocks, averaging about 2.6 grams per cubic centimeter in weight. These rocks, more or less fractured and crumpled, include sedimentary rocks (consolidated rock and mineral debris) and various metamorphic (altered) rocks, along with some basic rocks. The average elevation of the continents above sea level is 2,770 feet. Ocean bottoms, averaging 12,200 feet below sea level, are deepest at the trenches. (The bottom of the Mariana Trench, in the Mariana Islands area of the southwest Pacific, is 36,300 feet – 6.88 miles – below sea

level.) Ocean bottoms are on basic rocks, averaging 3.0 grams per centimeter in weight, except for lighter rocks fringing the continents. Bottoms show many features like those on land: volcanic cones and mountains, plateaus and plains, winding valleys cut by currents, and rolling hills.

Crustal Plates in Motion

As long ago as the sixteenth century, geographers and others studying maps of Earth noticed that sides of some continents, such as the west side of Africa and the east side of South America, could fit together rather neatly. They wondered whether the continents might be scattered fragments of a single land mass broken up by a cataclysm long ago. During recent decades geologists have found convincing evidence that during a period centered at about 180 million years ago a single great land mass, which they call Pangea

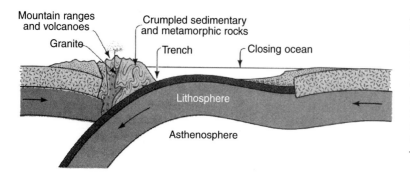

When Plates Collide: In this collision, ocean-bottom rock, which is relatively heavy, is subducting (diving beneath) continental rock, which is lighter. In the zone of convergence a trench forms where the edge of the continental mass is dragged down. Meanwhile, continental rock is crumpled, melted, and uplifted to form mountain ranges with cores of granite, and magma pushes up through the crushed zone to form volcanoes.

The Breakup of Pangea: Matching rock structures and fossils, and magnetically oriented rock materials, in present-day continents show that these were once a single great land mass. This supercontinent, known as Pangea, started breaking up about 200 million years ago. Drawings here show how the breakup probably progressed.

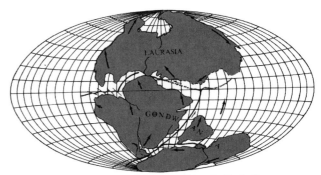

180 million years ago (Triassic Period)

(from Greek, "all of Earth"), did break up and the fragments moved hundreds and thousands of miles from their previous locations, becoming the present continents. More evidence indicates that similar breakups occurred earlier in Earth's history. The primary evidence is the existence of rock structures that match from continent to continent, such as those of northwest Africa and eastern North America, or northeastern North America and northern Europe. Next, the same genera of fossils have been found in land masses now widely separated; for example, fossils in South Carolina and southeastern New England that match those in parts of Spain, France, southern Great Britain, and northwest Africa. Evidence is plentiful that lands now under one kind of climate formerly were under another kind, as witness frigid Antarctica with its fossils of warm-climate plants and the hot Sahara Desert with bedrock grooved by glaciers. Finally, in many parts of the world there are ancient rocks containing magnetic iron-mineral crystals which show that the rocks have moved from their original locations. These crystals originally formed when the rocks were molten, and at that time the crystals lined up with Earth's magnetic poles; then they "froze" in position as the rocks cooled. Today these crystals do not line up with the poles; thus it is inferred that the rock masses containing them have moved.

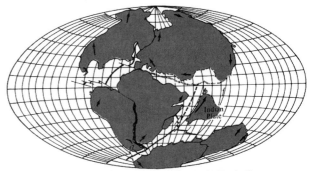

135 million years ago (Jurassic Period)

Plate movements continue today, occurring at rates up to 8 inches per year in the more active regions, notably the edges of the Pacific Basin. Plates appear to move mainly in response to magma pushing up through fissures in the mid-ocean ridges, spreading, and making new ocean floor, pushing the plates ahead or aside as they do so. Plates sideswipe one another, pull away, collide, and push over or under their neighbors. Continental and sea-bottom crust, riding on plates, subducts (dives under) other plates and down into the mantle and liquid core. There, in a recycling process, it is melted and consumed, becoming portions of magma that will rise again into the crust, cool, and form new crust.

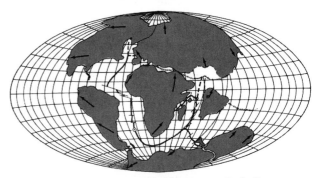

65 million years ago (Cretaceous Period)

Movements of plates and parts of plates are called tectonic (from Greek *tektonikos*, builder) or diastrophic (Greek *diastrophe*, distortion, and *strephein*, to turn). Causes of plate movements are not well understood, but it seems that movements could occur in response to a variety of forces, includ-

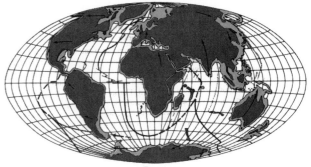

Present

ing (besides those already mentioned, as to Earth's inner layers) pressure from convection currents in the upper mantle, "hot spots" in the lower crust, spreading of ocean bottom as magma from the mantle wells up between plates and forms new rock, and upsets of balance in the crust that occur as rock material is altered and moved from place to place by erosion. Forces operating on or in the crust and in deeper structures may cause the fracturing that makes earthquakes.

Tectonics and Landforms

While plates move more or less as wholes, parts of plates also move, becoming more or less deformed. Rock masses of all magnitudes become warped, folded, or broken, pushed up or dragged down, or thrust sidewise. Mountains such as the Himalayas, Alps, and Rockies are created by crumpling and upheaving of crust as plates collide or slide over or under one another. Other mountains, such as Utah's Wasatch and California's Sierra Nevada, are crustal blocks that rise as neighboring blocks sink. Deep ocean trenches like those bordering the West Indies and Mariana Islands are created as one plate subducts another, dragging part of the ocean bottom down with it. As sectors of crust separate, portions break off the edges and sink deeper into the mantle, forming basins such as the Great Basin of the western United States, the Gobi Basin in China, and numerous smaller depressions.

Tectonic movements trigger other events. They open fissures between and within plates, allowing magma to surge up from the mantle or from zones of melting within the crust and then accumulate to make volcanic cones. Plates carry continental crust and ocean bottoms to different latitudes and longitudes, thus changing climates for those areas and creating new conditions for topographic change as well as for the existence of flora and fauna. Some highlands created by tectonic forces are high enough and cold enough to acquire glaciers, while portions of crust that have been depressed may become hot desert. As plates move sideward, rise or sink, bend or break, the ocean spreads over land in some sectors and withdraws in others. When land rises, its rivers flow faster, become more erosive, and create topography of relatively sharp relief; and as land sinks, its rivers become slower and less erosive, and relatively subdued relief results.

Among tectonic movements are continuous balancing actions among masses of material that form the planet. Some such actions are due to variations in gravitational attraction ("weight") among those masses. Thus the continents, which are mainly of relatively light rocks, are above sea level, while sea bottoms, being of relatively heavy rock, are below it. As already mentioned, portions of crust move up, down, or sidewise to maintain balance as erosional processes move rock material from place to place. Crust moves also in response to imbalances caused by plate collisions and, as noted earlier, by convection currents and other shifts of mass in the mantle and the transition zone. Thus Earth's major relief features – mountain ranges and plateaus, plains and basins and sea bottoms, and the rest – result from, or are strongly influ-

Scenery of Tectonism: The Stair Hole, in Lulworth Cove, Dorset, England, shows rock folds produced in the crust by opposing forces about 250 million years ago, and since uncovered by erosion. These formations are erosional remnants of a dome of limestone extending beneath the English channel into France. The cove, in weak rock, was hollowed out by the action of waves and currents which broke through stronger rock (*right*).

enced by, the balancing mechanism. This is known as isostasy (from Greek *isos*, equal, and *stasis*, a standing still).

Tectonic events deform and shape Earth's crust down to fine detail. The basic kinds of deformation include folding or bending, rupturing, fracturing, and faulting, and may affect any rock material, ranging from sand grains to crustal plates. Most landscapes as we see them have been more or less influenced by tectonic movements, and often the results of such movements are clearly visible.

For 4 billion years, then, the crust has been in the process of reworking and recycling by tectonic forces. Continental lands and ocean bottoms have been raised, sunk, and moved sidewise; they have been broken, folded, and covered or invaded by molten materials from the mantle or from the crust itself. The oldest portions of crust – those that have escaped recycling and remain little changed – are mostly in the interiors of continents; they are known as shields. But the continents generally are crazy-quilts of young rock masses jumbled with old ones, of diverse rock types and rock structures, of lowlands that became mountains and mountains that became lowlands, of ocean bottoms that became land and lands that became ocean bottoms, of lands that formerly were tropical or polar and now are polar or tropical, of crustal chunks that were hundreds or thousands of miles apart long ago and now are shoulder to shoulder. A piece of southern Alaska in Wyoming, a sliver of northwestern Africa in Georgia and Florida, a long strip of northeastern North America certainly broken off from northern Europe, Australia once part of Antarctica – such findings of accreted ("suspect") terranes are no longer surprises in geology.

Not only major landforms but minor features, too, owe much to tectonic events, directly or indirectly, whether they are flat-topped mesas or rolling hills, limestone caverns or sea cliffs, desert arroyos or mushroom rocks, volcanoes dormant or volcanoes erupting. Every landform is to be seen against the grand background of crustal movements.

The Continental Shields: Each continent (and Greenland and Australia as well) has a core of very ancient rocks called a "shield." The shield rocks, now partly exposed by erosion, have changed little in hundreds of millions of years. This drawing shows shields and their relationships to mountain ranges formed since the start of the Paleozoic. (From Flint and Skinner, *Physical Geology,* © 1974 John Wiley & Sons, Inc.; used by permission.)

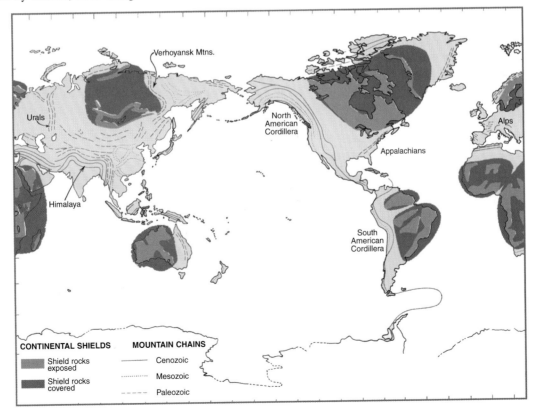

Regimes of Climate

As plates move, the climate at various locations changes accordingly, and landforms respond. Although the influences of climate on landforms tend to be subtle – much less dramatic than plate movements – they are ubiquitous and powerful in the length of geologic time.

Climate is the average weather at a particular location as observed or measured over a long period of time. It depends primarily on the rate at which heat is received at that location, directly or indirectly, from the Sun and from Earth's interior. Climate may involve any or all of many factors and aspects, including temperature, relative humidity, precipitation of rain or snow, elevation of the location above sea level,

Expressing Climate: Every landscape expresses, more or less, the local or regional climate. Scenery such as this, at Red Rocks Park, in California's Mohave Desert – bare and rugged rock masses, heaps of loose rock debris, lack of permanent streams, and sparse vegetation – is characteristic of many arid lands.

prevailing winds and wind velocities, nearness to the ocean or other large bodies of water, influence of ocean currents, amount of volcanic ash or other foreign materials in the atmosphere (these tend to block sunlight), presence of forest or other vegetation cover, and human activities such as agriculture and city or suburban development.

Existing landscapes reflect existing climate, and changes in climate change landscapes. Greater humidity and precipitation, for example, mean more rapid weathering and decomposition of rock, but less erosion by wind. More rainfall means more vigorous stream action and erosion, more transportation of erosion debris, more flooding, more landslides and soil creep. Colder temperatures may, by freezing water on Earth's surface, prevent formation of streams, but at the same time may cause the growth of highly erosive glaciers.

Climate can be said to produce its own landforms. In humid lands, highlands tend to become more or less rounded, with vegetation cover, but in arid lands highlands are likely to be angular. In humid lands limestone, if present, tends to be valley-forming, because of its solubility, but in arid lands it tends to form highlands, because it is frequently harder than other sedimentary rocks, such as sandstone or shale. On very cold and humid terrains water freezes, causing intense rock weathering and formation of glaciers, which enlarge valleys, grind down highlands, and when melting spread rock

Also Expressing Climate: Under very humid temperate and tropical climates, rock masses tend to be covered with thick vegetation and laced with permanent streams. Where exposed, rock masses often are seen to be somewhat rounded, showing results of strong weathering due to abundant moisture in air and soil. So it is in the Cascade Mountains of Washington state.

debris over the landscape. In the humid tropics warm temperatures speed decomposition of rock by chemical action, thus producing large amounts of rock waste which move down hillslopes by gravity and accumulate thickly on lowlands.

Climatic change often starts "chain reactions." A decrease of only about 11 degrees Fahrenheit in average temperatures in the northern hemisphere was sufficient to spawn the Pleistocene glaciers, incidentally lowering sea level by as much as 400 feet, changing the courses of major ocean currents, causing the reshaping of topography by erosion over millions of square miles, and increasing precipitation on lands bordering the ice. As temperatures rose again, these various processes went into reverse. Warming since the Pleistocene has converted the Sahara region, the Mideast, and the American Southwest from lands of forests, lakes, and streams into desert or near-desert. It was climatic change, caused by plate movements, that converted Antarctica from a subtropical region to a desert of ice. Climatic change occurring now is causing most of the world's glaciers to melt, so that sea level is rising and erosion patterns are changing along coasts all over the globe. Even on the smallest scale, as in the shadow cast on rock by a tree, or a breeze stirring sand on a beach, climate is at work on the crust, and all landforms reflect it.

From the Depths: New Rock

Much as in the early days of Earth, magma continues to rise from the mantle and from hot pockets in the crust, then cool and solidify within the crust or erupt at the surface to form new rock there. The generation and movements of magma are known as igneous activity (Latin *ignis*, fire). Magma erupted at the surface becomes what is called lava, and this cools to form what is known as volcanic rock (from Vulcan, Roman god of fire). Magma that solidifies within the crust becomes what is called plutonic rock (from Pluto, Greek god of the underworld). Multitudes of volcanic and plutonic rock forms are important elements of Earth scenery.

Most magma is from the mantle and emerges as lava from vents in ocean-bottom crust, along plate borders. Lava flowing out from fissures accumulates to form the network of mid-ocean ridges, such as those that wind north to south beneath the Atlantic Ocean and form plate boundaries beneath the Pacific. Lava from fissures also spreads to form new ocean floor, perhaps driving plates ahead of it. Lesser amounts of magma find their way up from the mantle and through the crust via pipelike passages. This magma becomes lava which accumulates around vents to form chains of volcanic cones, some of which, like those of Hawaii and Iceland, rise above sea level.

Lava erupts also from fissures and pipes in land masses, usually near present plate borders or near sutures (these are former borders along which land masses collided and united). Lava flowing from fissures may spread widely, forming plains and, over long periods, building up to make plateaus, such as the Columbia Plateau in the northwestern United States and the Deccan Plateau of India. Lava may issue from pipes as either flows or pyroclastics (materials shattered and blown out by volcanic explosions). Flows and airborne materials together may accumulate around vents to form volcanic cones ("volcanoes"). As lava flows and cools, it forms a wide array of interesting features, including ridges, pits, "pillows," and pinnacles – to name a few.

Large volumes of magma rising through the crust cool and

Igneous Activity: Volcanic activity has been modifying Earth's surface for more than 4 billion years; it occurs especially along plate borders. Cerro Negro Volcano, in Nicaragua, at a plate border, has been built up around a volcanic vent by eruptions of flowing lava and lava shattered by volcanic explosions. In this eruption the volcano is producing huge quantities of gas as well as pyroclastics, which form streaks on the cone.

Sculptured by Weathering and Gravity Movements: The Great Arch, in Zion National Park, Utah, has been formed by rockfalls from a cliff. The sandstone of the cliff has been weakened by leaching out of its natural limy cements by groundwater percolating down through it and emerging at the cliff base – a process called spring sapping. Weakened rock near the cliff base fractures under the pressure of overlying rock; then portions of it fall away, gradually enlarging the arch.

solidify before reaching the surface. These plutonic masses occur in almost all imaginable forms, including vertical or horizontal slabs, "balloons" and "eggs," strings, and pinnacles, depending on the composition of the magma and the nature of the rock structures invaded. Sizes of these forms range from mere threads to mountain masses. Such bodies are very common in the crust but are generally noticeable by ordinary observers only when uncovered by erosion or detected by drilling or seismic investigations. With time these processes reshape the original masses and may gradually decompose and disintegrate them. Although most localities on Earth have been covered by igneous rock during one period or another in the past, most of these materials have been removed by erosion, covered by other rock, inundated by advancing seawater, or recycled via the mantle. As older rock is removed, magma continues to rise from Earth's interior to form new rock on and in the crust.

Gradation: Leveling the Crust

The entire crust of Earth, including the surface and much of the rock beneath it, is subject to the continuous processes of weathering, gravity movements, erosion, and deposition. Acting more or less together, and influenced more or less by climatic, tectonic, and igneous activity, these processes tend to level, or "grade," the surface of the crust by moving rock material from higher to lower elevations. The building up of a surface by these processes is known as aggradation, and lowering of the surface is called degradation. The processes when considered together are termed gradational. They are described in detail in later chapters, but are outlined here, together, to show their close relationships with one another.

Weathering

Rock exposed to the atmosphere and associated conditions at or near Earth's surface is subject to weathering. In what is called chemical weathering, moisture, atmospheric gases, and organic compounds attack minerals in rock, alter them chemically, and thus weaken the rock, making it more prone to crumble. In mechanical weathering, rock formed at great depth under great pressure and later uncovered by erosion may break up because of expansion. Mechanical weathering

Desert Weathering: Grim Skull Rock, at Joshua Tree National Park, California, has been sculptured from quartz monzonite rock by rainwash, chemical weathering, temperature changes, and blowing sand.

also includes breaking up of rock by water freezing in rock crevices, by roots growing in crevices, and by temperature changes and wetting-drying episodes that break rock by causing it to expand or contract.

Weathering of rock is occurring everywhere on Earth's surface and to depths of hundreds of feet. Under humid climates, rock types that are relatively reactive with moisture – rocks containing iron or much calcium carbonate, for example – tend to weather more rapidly than similar rocks under semiarid or desert conditions. In cold lands where the thaw-freeze process is frequent, much rock is broken up as water freezes in rock cavities. In the humid tropics, large quantities of chemicals produced by the activities and decay of flora and fauna decompose rock relatively fast, and rock is disintegrated by growing roots. On terrains where the crust has been much fractured, weathering tends to be rapid because fractures allow relatively easy entry by weathering agents – gases, water, roots. Thus the nature and pace of weathering on any terrain depend on varying factors, notably rock type, climate, tectonic activity, and activities of living things.

Weathering tends to be slow, by human standards, but over geologic time its influences are powerful. It is the major cause of rock decomposition and also causes much rock break-up. It creates landforms by decomposing and disintegrating different kinds of rock at different rates. Rock debris produced by weathering constitutes much of the material that moves downslope by gravity and most of the sediments used by streams as tools to erode the land.

Gravity Movements

Everywhere on Earth, rock masses of all sizes – from bits of soil to mighty mountains – are subject to the pull of gravity. Rock broken up by weathering and other geologic processes is moved by gravity to lower elevations. Movements occur in many forms, such as rockfalls, rockslides, mudflows, subsidence (sinking of land), and creep. Mountain masses may move by gliding.

Gravity movements (also called mass movements, or mass wasting) are subject to a variety of conditions, which may reinforce or counteract one another. Movements naturally tend to be relatively frequent and voluminous where gradi-

A Slow March: Large rock fragments embedded in soil may take centuries to "creep" a few feet down a gentle slope. Small particles may move a few feet – or more – during a single rain. Freezing and thawing, wetting and drying, growth of plant roots, and activities of animals are among causes of gravity movements such as these.

ents (inclinations) are steep on loose materials. On cold lands, thaw-freeze usually produces relatively large amounts of rock for gravity to pull down, and thus one may see huge accumulations of rock debris – talus – at the foot of cold highlands. In humid tropics, chemical weathering may produce large amounts of rock debris, but movements of this material may be much retarded by vegetation growing thickly on slopes, so that little talus accumulates at slope bases. In deserts, weathering may be very slow because of the scarcity of water, but much of the debris that weathering does produce may remain on slopes, even if steep, or accumulate at the base of slopes, because running water is insufficient to carry it away. In regions subject to earthquakes, large amounts of rock material are periodically shaken loose and transported downslope until stopped by barriers, such as trees and walls, or by gentling of the inclination.

Erosion and Deposition

Most influential among the processes that degrade land is erosion, which can be defined for present purposes as the removal of earth material by agents acting on or near Earth's surface. The most powerful agent of erosion is water, in sheets and streams which by hydraulic forces (pressing and sucking) loosen chunks of rock from larger rock masses and transport these fragments to lower levels, meanwhile abrading rock surfaces with materials being transported. Valleys of all sizes are made mainly by these activities of water in league with weathering processes and gravity movements on

valley sides. Streams slice into broad highlands, and as valleys gradually widen, a landscape of scattered hills is created. Rock-shod glaciers grind mountain peaks into horn shapes, and ream out valleys to make glacial troughs and fiords. Sea cliffs are made and kept steep principally by the undermining action of waves and currents. Streams underground dissolve and erode limestone to make tunnels and caverns. Desert features such as "mushroom rocks" and yardangs are shaped by rainwash and blowing sand.

Erosion is influenced by many linked conditions. Most of these are aspects of climate, particularly precipitation, temperature, and wind action. Other influences include hardness and chemical compositions of rocks, arrangements of different rock types and structures in the crust, land elevations, and crustal fractures and flexures.

An unusual kind of erosion is done by impacts of asteroids and other objects from outer space. During Earth's past, impacts have made uncountable thousands of craters in the crust, some of them 100 miles or more in diameter. However, very large craters are made so infrequently (at intervals of millions of years), and erosion is relatively so rapid, that impact craters recognizable as such by ordinary observers are rare, and they do not represent a significant portion of present Earth scenery.

Because erosion is the most effective general agency in degradation and so often involves weathering and gravity movements in intricate ways, the term "erosion" is frequently used to include weathering and gravity movements also, and for simplicity it is often so used in this book.

Deposition, or laying down, of loose rock material by agents of erosion results in aggradation of Earth's surface. Examples include delta building by a river, deposition of rock debris by melting glaciers, beach building by waves and currents, and dune construction by desert winds.

Erosion by Waves and Currents: Near Big Sur, on California's coast, waves and currents aided by weathering agents undermine rock cliffs, cutting out coves, causing rockfalls and landslides, grinding up rock debris to make sand for beaches, and transporting sand along the shore.

A single landscape – that is, a collection or collage of landforms – usually expresses many gradational processes at the same time. On a mid-latitude humid terrain, for example, such processes will include perhaps a dozen kinds of chemical and mechanical weathering, plus soil creep, rock-falls and landslides, erosion by sheets and streams of water, and deposition of rock debris on valley bottoms. These various events tend to affect and be affected by one another.

Every landscape, whether in the tropics, mid-latitudes, or polar lands, has its own unique combination of gradational processes, conditioned by climate changes and by igneous and tectonic activity extending back into the remote past.

Landscapes as "Personalities"

Every landscape has what might be called a personality, reflecting the nature of conditions under which it originated and has since been shaped. A primary condition is the chemical composition of the rocks, because composition tends to determine vulnerability to erosion. Also fundamental as a condition are the rock structures – that is, arrangements or patterns assumed by minerals that make up rock and the arrangements or attitudes of rock masses that make up the landscape. Examples of structure include layering, fracture patterns, attitudes of rock masses relative to the vertical or horizontal, and locations of such masses relative to one another. Land elevation strongly influences stream velocities and, therefore, rates of erosion. Climate determines amounts and kinds of precipitation and fluctuations of temperature to which rocks are exposed. Vegetation cover more or less restricts both gravity movements and erosion by running wa-

Scenery of Deposition: In Denali National Park, Alaska, as elsewhere among cold mountains, weathering agents and glaciers break off much rock from mountainsides. Fragments descend by gravity, sheetwash, and stream action, and accumulate as glacial outwash on valley floors. Here a river braids (divides into interweaving channels) to negotiate its way through thick outwash.

ter and wind, but, on the other hand, may increase the pace of weathering. The length of time during which each process operates, and the timing of each process relative to others, are other determining factors. All such influences, in their various combinations, control the nature of a land's slopes – contours, stability, roughness, depth of soil cover, type and density of vegetation cover, and profiles.

Personalities of some landscapes are relatively simple, of others complex. The Great Plains of North America, for example, look quiet and geologically uneventful, suggesting little more than slow stream erosion and deposition. But Europe's Alps are rugged and diverse, expressing severe weathering and vigorous stream action, faulting and folding, a high rate of gravity movements, and intense glaciation. Well-drained terranes on limestone, as in parts of Kentucky

and Indiana, are bizarrely pocked with multitudes of sink-holes and caverns, resulting from solution of rock. Volcanic activity, a humid climate with torrential rains, and the surrounding sea have made much of Hawaii a scene of rugged ridges, precipitous slopes, and rushing streams.

Landscapes can be classed according to their origins – volcanic, tectonic, or glaciated, let us say – but every landscape, nonetheless, has a personality not exactly like any other in the world.

Landscape Histories

Given the diversity of geologic processes and conditions, the strong likelihood of interruptions by the unexpected, the great length of geologic time, and the fact that tectonic

Architecture of Weathering, Gravity Movements, and Erosion: Limestone and shale strata in Bryce Canyon, Utah, acquired multitudes of vertical fractures due to crustal stretching during the Colorado Plateau uplift. Rainwash and small streams following fractures dissolved and eroded rock masses into a "city" of spires. Irregularities are due to variations in resistance from layer to layer.

A Famous "Personality": Several million years ago, explosive volcanic eruptions in Cappadocia, Turkey, spread layers of volcanic materials over thousands of square miles. Here, in the Goreme sector, weathering and gravity movements, and removal of debris by streams, have shaped these materials into weird arrays of pillars, cones, and other remarkable features. In the past, people cut hollows in the soft volcanic rock to make dwellings, just as was done in similar rock at Bandelier National Monument, New Mexico.

movements and erosion tend to destroy evidences of the past, every terrain presents some degree of challenge to efforts at understanding. The geologic history of some landscapes can be relatively simple; for example, Cape Cod and Long Island were built mostly of morainic materials deposited on bedrocks by Pleistocene glaciers, and Florida is a sector of ocean floor uplifted in recent geologic time to become dry land. More commonly, landscape histories are complex. Sediments spread over the Great Plains by streams, glaciers, and wind cover the roots of very ancient mountains with complex histories hardly hinted at by present landscapes. Parts of the Rockies were originally raised some 65 to 55 million years ago, were cut low by perhaps 25 million years of erosion, then – as erosion continued – underwent two more cycles of uplift and erosion. The Folded Appalachians may have undergone many such cycles. Geologically speaking, the appearances of today's landscapes are largely the results of what has happened to them in only recent geologic time. What has happened recently – perhaps a lava flow, a glaciation, or a change of climate which has accelerated erosion – may have covered or removed the evidence of what shaped the land previously. So – a landscape that challenges our curiosity may look simple, but it may have an ancestry much more complex than appears.

How Long Does It Take?

"Geologic time" is a term popularly used to denote an almost unimaginably long duration. In fact, geologic processes produce landscape changes in time periods of almost any length. A terrain may rise 30 or 40 feet in minutes, as during the 1964 Alaska earthquake, or only a quarter of an inch per year, which is average for California's San Gabriel Mountains. A volcanic eruption can cover square miles with lava in a few days, as has occurred often in Iceland and Hawaii. Some 20,000 years could be required for weathering to remove Sierra Nevada granite to a depth of one inch, or for the Colorado River to cut the Grand Canyon 10 or 20 feet deeper. The thickness of rock being removed from the Appalachians by processes of degradation may average about an inch per 1,000 years. The rate of degradation tends to increase exponentially with increase in relief, so that in very high mountains, such as the Himalaya, degradation is proceeding at the rate of perhaps 6 to 10 feet per 1,000 years – depending, of course, on local rock types, climate, and other conditions that affect erosion rates.

Most major geologic changes take, by human standards, an awesomely long time. It's little wonder that thoughtful observers 200 years ago, not recognizing available evidence

After a Short History: Hawaii has relatively young landscapes. Built of lava and other materials erupted by oceanic volcanoes during the past few million years, the islands have been under steady assault by humid-climate weathering, gravity on the steep slopes, vigorous streams, and sea waves and currents. This view is northeastward over Honolulu, on Oahu.

transport of salt by streams from land to sea, did investigators begin to suspect Earth's true age. The figure of 4.5 billion years was calculated only a half-century ago, with the invention of radiometric dating, in which the age of a rock sample is calculated by noting what proportion of its radioactive content has decayed at its known rate of decay.

In geology, time figures in the millions become so familiar that our awareness of their magnitude becomes dulled; therefore some comparisons with historical time spans may be useful. The most recent continental glacier of the Pleistocene Epoch at its maximum covered the present New York metropolitan area about 17,500 Y.B.P. (years before the present), about three times as long ago as the invention of writing in ancient Sumer or the construction of Egypt's oldest pyramids. The Colorado River may have started cutting the Grand Canyon about 4 million years ago, when our early forebears were roaming East Africa. In a longer time frame the Mississippi Valley is thought to have originated as a crustal downwarp perhaps 25 million years ago. The rise of older parts of the Rockies and the Alps dates back 55 to 65 million years, and the earlier of the upheavals that produced the Hudson Highlands and Great Smokies may have occurred about a billion or more years ago. Layers of rock in the Pilbara region of western Australia show evidence of a landscape that existed there 3.46 billion years ago. The oldest rocks discovered so far, in northwestern Canada, have been radiometrically dated at slightly more than 4 billion years, or some 700,000 times the span of written history.

The maximum likely age for a major relief pattern (an arrangement of sizable highlands and lowlands) may be around 25 million years – that is, the time span probably required for high mountains to be eroded down almost to a

of Earth's age, speculated that mighty events such as the rise of lofty mountains and the opening of deep canyons must have been produced quickly and violently by catastrophes, such as Noah's Flood or perhaps earthquakes like the one which, reportedly with the aid of the horns of Joshua's army, brought down the walls of Jericho. Only through scientific study of fossils (which show evolutionary development of life forms from lower, older rock layers to higher, younger ones), of weathering and erosion, and of processes such as

After a Long History: Rocks in New York's Hudson Highlands originated as volcanic materials and miscellaneous sediments on an ocean bottom a billion years ago. During millions of years these materials sank, became buried under other materials to a depth of 25 miles or more, changed into rock, were invaded by molten materials and altered by heat and pressure, were folded and faulted, then were uplifted as high mountains. After several cycles of erosion and uplift, the original mountains have been eroded away; only the "roots" remain today. This view is north over the Hudson River toward Bear Mountain Bridge.

Rocks and Landforms of About the Same Age: In Kenya, the African Rift Valley exhibits volcanic rocks formed by relatively recent eruptions. Here landforms are almost as old as the rocks. This is Lake Elementeita, in the Rift Valley south of Nakuru. Beyond are young volcanic cones and, in the far background, an eastern scarp of the Rift Valley.

plain, assuming no substantial uplifts or other interruptions meanwhile. Some broad erosional surfaces, including areas in central Australia and on the great African Plain (areas where topographic change has been extremely slow), may be several hundred million years old. But certain it is that determining ages of landscapes in years is "iffy."

Although the age of rock in a landform can sometimes be quite accurately determined by radiometric dating, that age is not necessarily the age of the form. The form usually is younger – often very much younger. Hence, although much rock in the Hudson Highlands and Great Smokies may be a billion years old, the overall pattern of these mountains – that is, the pattern of major ridges and valleys – may date back only 15 million years or so. In fact, one may get a little confused when trying to decide which are the "real" Hudson Highlands or Smokies: the towering peaks of a billion years ago or the relatively modest hills of today? Those towering peaks are gone, victims of prolonged erosion. The present hills are but the roots of those peaks – roots of rock that formed at least 25 miles below Earth's surface and has since become exposed by several cycles of uplift, erosion, and renewed uplift.

Where Geologic Change Is Relatively Fast: Descending glaciers in mountains are highly destructive. They bulldoze loose rock material in their path, break off rock from valley sides, and, with rock fragments embedded in the ice, grind and wear down solid rock. Meanwhile frost action breaks up rock, and loose fragments gravitate to lower elevations. These glaciers are at work on a range north of the Matterhorn, in the Swiss Alps.

THE GEOLOGIC COLUMN

The geologic past is represented by a vertical chart showing divisions of the past in order, the oldest at the bottom and the youngest at the top. Each division is marked by certain major events. The approximate length of time spanned by each division is indicated in years as determined by radiometric dating of rocks corresponding to those divisions.

Era	Period	Epoch	Major Events with Special Reference to a Few Well-known Landforms	Millions of Years Before Present
CENOZOIC — Time of Recent and Modern Life and Landscapes	Quaternary — Age of Recent Glaciers and of Humankind	Recent	Glaciers partly melt; sea level rises; modern deserts appear. Humankind modifies Earth.	
		Pleistocene	Glaciers form extensively. Widespread erosion is controlled by fluctuations of sea level due to waxing and waning of glaciers. Rise and deformation of Alpine-Himalaya system continues.	2
	Tertiary — Intense Mountain Building and Volcanic Activity	Pliocene	Uplift and renewed valley cutting occur in older and younger mountain areas. Deformations along Pacific Coast begin. Detailed shaping of numerous modern landscapes begins.	12
		Miocene	Culmination of Alpine-Himalaya movements occurs; erosion does extensive leveling in Appalachians and Rockies. Shaping of modern landscapes continues.	25
		Oligocene	Uplifts and deformations are extensive in Alpine-Himalaya mountain system. Deposition continues in Rocky Mountain basins. Extensive faulting in Basin and Range province.	40
		Eocene	Sediments accumulate in Rocky Mountain basins. World-wide warm, uniform climate ends.	60
		Paleocene	Major structures of Rocky Mountains are completed. Broad framework of modern landscapes becomes established.	70
MESOZOIC — Time of Dinosaurs and Scenery Now Vanished	Cretaceous		South America and Africa separate; Australia and Antarctica also break apart. Rocky Mountain structures begin forming.	135
	Jurassic		Pangea breaks up; Atlantic Ocean basin forms. Volcanism becomes widespread in western U.S.	170
	Triassic		Rift-valley formation and volcanic activity become extensive in eastern U.S. and Canada.	225
PALEOZOIC — Early Life in Sea and on Land	Permian		Broad icecaps spread on southern continents.	270
	Carboniferous		Basic structures of Folded Appalachians and Alpine-Himalaya system take shape.	350
	Devonian		Vegetation begins to cover land. Earliest land animals develop.	400
	Silurian		Earliest known coral reefs form.	460
	Ordovician Cambrian		Oceans are widespread and marine life is abundant.	500
PRE-CAMBRIAN — Formation of Earth and Beginnings of Life	Late		Older Appalachians are forming; earliest known glaciation occurs.	600
	Early		Earth forms; crust develops with oldest rocks; life forms originate.	4000

The Geologic Time Frame

Spanning 4.5 billion years, Earth's history insofar as known has been divided into five grand eras, each consisting of shorter spans called periods, which in turn subdivide into epochs. A table of these divisions, with the youngest at the top and the oldest at the bottom, is known as the Geologic Column. It corresponds to the fact that in sedimentary rock (rock formed by consolidation of rock debris) the layers normally are older from top to bottom. Knowledge of the column is basic in general geology and, although not essential for an appreciation of scenery, does add the perspective of Earth history. A version of the column, listing some major geomorphic events of the past, appears on an adjacent page.

Pondering the geologic record, one may speculate about how Earth's crust may look a billion, 2 billion, or 5 billion years from now. Earth processes now occurring may continue with little change for hundreds of millions of years – perhaps billions. Today many geologists have arrived at the view that all forces acting collectively upon the crust are tending toward a state of equilibrium and the end of change. However, considering that Earth and Sun are both cooling and that cooling will mean changes in Earth's atmosphere, ocean, and crust, it seems that no state of equilibrium in the crust can be reached as long as Earth exists. In any case, there will be ample time for observers, whoever or whatever they may be, to view Earth's crust as a scene of fascinating change.

A Geologic Calendar in Rock: Rock layers in Arizona's Grand Canyon are pages in a book of the past. In the lower part of the canyon the river is cutting into billion-year-old rock, formed when only very simple marine organisms inhabited Earth. Upon this foundation are layer upon layer of younger rock strata, each formed from sediments deposited on lands and sea bottoms over long periods. The youngest layer, at the top, dates back to the Permian Period, 280 to 225 million years ago. With the aid of tributaries, the Colorado River has taken several million years to cut the canyon, as deep as a mile. This view is from Mohave Point, with a glimpse of the Colorado River (*low center*) cutting into the most ancient rocks.

3
Rocks: The Substance of Landforms

EARTH scenery – landforms collectively – is shaped from rock, either solid rock such as we see exposed in mountains and cliffs, or fragmented rock such as sand and cobbles in soils or on beaches. It is rock that forms Earth's crust. Drilling at any point on Earth's surface, whether on a continent or on the ocean bottom, is certain to encounter solid rock sooner or later – perhaps at the surface, maybe at a depth of 100 feet, possibly only at a depth of miles. Some rock as we come upon it may be relatively new and "pure," having recently formed from lava erupted by a volcano; other rock may be a hodge-podge of consolidated debris from erosion of earlier rocks; still other rock may consist of older masses that were compressed, heated, folded, invaded by other materials in the crust, transported, and later uncovered by erosion.

Rock masses of various types may lie over, under, or next to one another – or, indeed, mixed with one another. Some rock masses are located just where they originally formed; others have been transported by water, wind, glaciers, or plate movements from a distant place; still others may have formed at the surface but, during long ages, became buried under new rocks hundreds or thousands of feet thick.

Rocks as History

Most rocks at Earth's surface consist of materials with histories extending back hundreds of millions of years, even to the beginnings of Earth's crust. Perhaps all the original rocks of the crust have been recycled by tectonic activity, but some very old ones have been identified, in the cratons of the continents. The oldest known are certain gneisses in the Acasta River area of northwestern Canada, radiometrically dated as a little older than 4 billion years. These contain bits of the mineral zircon with an age of 4.3 billion years. Since the zircons are believed to be detritus (erosional fragments)

from still older rocks, it is inferred that there was rock on Earth's surface more than 4.3 billion years ago. The zircons may, indeed, be from some of Earth's original rocks.

Rock in any particular locality is evidence of geologic, environmental, or climatic conditions which existed there during the past. For example, certain limestone strata now miles high in Alberta, in the Canadian Rockies, are remains of reefs built up from clear, shallow sea bottoms by corals under a subtropical climate around 350 million years ago. Sandstone strata exposed at the base of the Palisades of New York's Hudson River include sands deposited some 200 million years ago on an arid terrain like that of New Mexico today. Digby Neck, a long peninsula in Nova Scotia, is part of a ridge of hardened lava produced by volcanic eruptions dating back about 190 million years, to about the time when northwest Africa was breaking away from northeastern North America. Thus in any particular locality the exposed rocks may represent environmental and climatic conditions very different from those prevailing today, as witness the fact that there are coal beds and coral reefs in Antarctica. In any one locality, furthermore, the sequence of rock strata, as we explore deeper and deeper, is a sort of book in which the story of the land changes from page to page.

The Makeup of Rocks

Rock comes in hundreds of varieties, according to the nature and proportions of the minerals they contain; thus rock identification in a precise sense is a task for experts. However, for ordinary observers rocks can be classified into a few relatively broad groups, each having a sort of personality in terms of its appearances and how its composition and structures influence forms of scenery. Thus granite often appears in rounded hills or towering "needles," basalt as

Massive Sandstone: The Great White Throne, in Utah's Zion National Park, is of quartz-rich Navajo sandstone, a type widespread in Utah, formed by natural cementation of dune sands deposited 175 to 200 million years ago. As the Colorado Plateau rose up, stretching and compression fractured the Navajo. Streams following fractures were speeded up by the uplift and carved the stone into the pillars and blocks characteristic of this region.

Mineral Makeup of a Rock Fragment: In this weathered specimen of banded gneiss, a metamorphic rock, crystals of the iron-rich mineral hornblende are interlocked with crystals of white feldspar and gray quartz. Crystallization was caused by heat and pressure miles underground. Like other rocks, gneiss varies somewhat in composition from specimen to specimen.

columns in cliffs or as pillowy or chunky lava flows, shale as thin, fragile rock layers exposed in roadcuts, and limestone as haystack-shaped hills or bizarre and beautiful features in limestone caverns. Recognizing major rock types, at least tentatively, with an awareness of their landforming characteristics, starts the ordinary observer on the way to an understanding appreciation of earth scenery. In the following pages these rock types are described with specific reference to the kinds of landforms they yield. If these descriptions seem too long, the remainder of this chapter can be skipped for now, and returned to when the need for more information about rock identification and characteristics is felt. For information beyond what is offered here, a field guide to rocks and minerals is recommended.

The solid portions of Earth's plates are called bedrock; the fragmented, loose rock covering bedrock in most places is known as regolith. Any bedrock exposed at the ground surface is called an outcrop. Massive landforms, such as mountains and rock cliffs, are mostly of bedrock; landforms such as beaches, sand dunes, and deltas are made of regolith. Regolith mixed with organic and other materials that allow it to support plants is known as soil.

Every rock mass consists of one mineral or, more commonly, two or more minerals. A mineral may be a chemical element, such as calcium or silicon, or a compound of elements combined in the same proportions in all specimens. Among the most common compounds in Earth's crust are calcium carbonate, with the formula $CaCO_3$ (calcite), made of calcium, carbon, and oxygen; and silica, SiO_2 (quartz), a compound of silicon and oxygen.

Each mineral species that has crystalline form has an arrangement of atoms which determines that form. The original form of a rock mass, on the other hand, depends on its mineral content and on the conditions under which it came into being; for example, under pressure deep in the crust, or under little pressure but much weathering near the surface.

Most rock becomes subject to one or more processes that tend to decompose, deform, disintegrate, or otherwise alter it. These processes are igneous activity, tectonic events, weathering, gravity movements, erosion, and deposition. Every rock mass has some degree of resistance to change, and this is strongly influenced by the physical and chemical properties of the mass, such as hardness, solubility, and degree of reactivity with other chemical substances; and such properties are in turn determined by mineral content. The structure of rock – that is, the arrangement of its minerals, the pattern of its fractures (if any), and its overall shape – also influences resistance. Resistance to change is a basic condition in the nature of Earth scenery everywhere.

Chemical and Mechanical Influences

Minerals are chemically active to varying extents. For example, iron-rich minerals, common in basic rocks, are prone to oxidize ("rust") under a humid climate. Silicon-rich minerals, common in granitic rocks, are more resistant to oxidation, but are vulnerable to chemical attack by water containing carbon dioxide. Water with carbon dioxide will dissolve calcium carbonate, the main ingredient in limestone. Thus each mineral has its characteristics of reactiveness with other chemicals in its environment, and every rock mass tends to be chemically reactive according to its mineral content.

Ordinarily, reactions of a rock's minerals with moisture, gases in the air, and other chemicals tend to weaken the rock physically. For instance, when iron in basic rock becomes oxidized by moisture, the oxide crystals as they form expand, causing the rock surface to fracture. Also, new compounds formed by chemical reactions in and on rock are likely to be softer than those replaced, and thus make rock more vulnerable to disintegrating action. As chemical attack proceeds, a rock mass weakens, decomposees, and disintegrates.

Rock of any kind may show different degrees of resistance in different types of environments, such as tropical or polar, dry or humid. Limestone and basalt, because of their vulnerability to solution and oxidation, respectively, tend to weather and erode faster under humid conditions than in the desert. Thus, in any given environment, landforms more or less express, or represent, not only the nature of the rocks but the nature of the environment.

For a convenient example of the chemical influence in land-shaping, take the Folded Appalachians, that long-eroded mountain chain which stretches from Newfoundland to Alabama. Its ridges are mostly on sandstone, conglomerate, and quartzite rocks, which originally formed from older, weathered rock debris and therefore are not very vulnerable to new weathering. Many valleys are in limestone, more soluble than the other rocks and thus relatively weak under the prevailing humid climate. Though mechanically strong, the limestone is relatively vulnerable to solution.

Mechanical strength determines the relative resistance of rock to wear and tear by blowing sand, glacier action, abrasion by rock fragments carried by running water, rockfalls

THE MOHS SCALE (in order of increasing hardness)		
Number	Mineral	Reference Material
1	Talc	
2	Gypsum	
3	Calcite	Copper cent
4	Fluorite	Brass pin
5	Apatite	
		Window glass
6	Orthoclase (feldspar)	Knife blade Tool steel
7	Quartz	
8	Topaz	
9	Corundum	
10	Diamond	

and landslides, and crustal movements. On the Mohs scale of hardness (ability to resist scratching) calcite, a crystalline form of calcium carbonate, is relatively soft, orthoclase feldspar and quartz relatively hard. These three minerals are common in rocks: calcite in limestone, quartz and feldspar in granite and a host of other rocks. Worth noting is that though a rock mass may have in it mineral crystals or grains that are hard, such as quartz pebbles in conglomerate, these may be bonded weakly, so that the mass as a whole is weak.

In some situations, landforms of chemically "weak" rock stand above those of stronger rock. Such an inversion can

---

Weathering by Oxidation: On the flanks of Mt. Lassen, a volcano in northern California's Cascade Mountains, lie many boulders of basalt. Moist air attacks the iron in basalt, forming a crust of iron oxide on the surface. Because this crust occupies more space than the iron did, it tends to flake off. Because the rock is essentially homogeneous and edges are more vulnerable than broad surfaces, prolonged weathering rounds the rock.

Weathering by Solution: Sandstone bedrock often becomes cemented by limy materials deposited between the sand grains by groundwater. Later, water working through the rock may dissolve some of this natural cement, causing the rock to disintegrate. Hollows in this wall of rock in Petra, Jordan, result probably from this process, which is known as cavernous weathering.

A Miniature Landscape: The quartz vein in this mass of granite is more resistant than rock around it; thus after hundreds of years of weathering the vein has become a ridge. It will continue to be a ridge as weathering proceeds, unless weathering reaches a zone in the rock where resistance increases. Such is the story of numerous landscapes on which variations in relief are determined by variations in rock resistance.

result where tectonic forces have raised weak rock masses and erosion has not had sufficient time to restore the topography to "normal." A mass of weak rock can stand for a while as a highland after a cap of strong rock, which has protected it, has eroded away. Another example might be a relatively young volcanic terrane where hard solidified lava flows lie on a valley floor and masses of cindery material cover higher elevations. Generally, however, highlands of consistently stronger rocks are to be expected where erosion has been long dominant. In other words, on "erosional landscapes" highlands generally mark areas where stronger rocks *have been*, and lowlands are seen where weaker rocks *have been*.

The Structural Influence

Every rock mass has structure, patterned or chaotic. Its crystals or grains may be clustered, aligned, or scattered. The rock may be porous or dense, fractured or unfractured. Any layers that exist in the rock mass will be horizontal or tilted. The mass will have a position or attitude relative to other rock masses, being beside, over, or under them. Such characteristics tend to make rock masses relatively strong or weak in given environments, and to influence the manner in which the masses are shaped by erosion or influenced by tectonic activity.

Fractures give access to air and water, so that any rock mass with fractures is likely to be less resistant than the same mass would be without fractures. A mass with rough or angular surfaces (these too are an aspect of structure) will, as a whole, probably weather and erode faster than similar rock with smooth and rounded surfaces, because the latter rock has a smaller proportion of its volume exposed. A mass of loose rock fragments is, as a whole, less resistant to weathering than a solid mass of the same kind of rock would

Differential Erosion: The wall-like mass here is a "dike," formed by solidification of magma injected into soft volcanic materials in cliffs around Crater Lake, Oregon. The dike is much more resistant to erosion than the surrounding materials; thus it remains standing high above them as erosion progresses.

Strong Layers on Weak: Here at Red Rocks Park, in California's Mohave Desert, the upper rock strata somewhat protect underlying, weaker strata from weathering and rainwash. Closely spaced vertical furrows here are signs of relative weakness.

be. If a highland consists of different rock layers with the strongest layer on top, the hill as a whole will probably resist erosion more than it would if a weak layer were on top. If the layers are tilted, erosion will be faster along upper edges of the layers than along lower ends, because water, with the help of gravity, will more easily penetrate between the upper ends.

Well-known landscapes demonstrate how structure can influence topography. Prolonged erosion of a highland of horizontal strata tends to produce a landscape with mesas and flat-topped buttes, as on the Colorado Plateau. Steeply tilted strata usually erode to a terrain of sharp-backed hills, such as the spectacular Dakota Sandstone hogbacks of Colorado's Front Range. A high, broad mass of granitic rocks with vertical joints spaced widely will probably erode to a landscape of rounded hills, such as New York's Hudson Highlands, while a mass of similar rocks with relatively close vertical joints will erode to tower-shaped or needlelike forms, such as "The Needles" in South Dakota's Black Hills. Erosion of a terrain of volcanic cones, jumbled lava flows, pit craters, and beds of pyroclastics (lava shattered in explosive eruptions) will produce a corresponding chaos of features – pillars, ridges, trenches, pits, platforms, knobs, and other forms, as in Hawaii or Iceland.

Influence of Tilted Strata: Mt. Rundle, in the Canadian Rockies, Alberta, is part of a rock mass of fossil-filled limestone, formed on a sea bottom several hundred million years ago and later raised to form mountains. It is in the Front Range, characterized by steep tilting of strata. Tilting largely determines the profiles. The scarp slope, left, on edges of strata, is rough because of their unequal resistance. The dip slope, right, parallels the downward inclination (dip) of the strata, and thus is smoother.

Influence of a Fold Structure: Mt. Kerkeslin, an ancient mass of quartzite rising to 9,970 feet in the Canadian Rockies, has been shaped from a huge syncline, or downfold. The limbs (sides) of the original fold have been mostly eroded away, to a depth of thousands of feet; the remaining rock, thanks to superior resistance, remains as a highland.

A Gallery of Rocks

Because rock type strongly influences the shaping of scenery, it is useful to know the general nature of each major type and to be able to recognize it in the field. The task is not too forbidding. Of the 2,500-plus varieties of minerals that exist, only 8 comprise the great majority of the world's rocks; and of the hundreds of varieties of rocks, relatively few are significant in forming terrain.

The major land-forming rocks occur in many variations, requiring technical knowledge for precise identification. For serious observers a field guide to rocks and minerals is recommended. However, exposures of the most common land-formers, such as shale and sandstone, granite and basalt, usually have aspects that help in recognition, especially by casual observers who have seen them in pictures or "in the flesh." With this understanding, the following descriptions and photographs of common land-forming rocks can be useful. Rock varieties that are not common in large masses but are of special interest in some environments – for example, pumice and scoria in volcanic areas, travertine in limestone caverns – are described on later pages.

The three broad categories of rocks are: igneous, formed from molten materials; sedimentary, consisting of consolidated sediments (loose mineral or rock material); and metamorphic, resulting from alteration of previously existing rock of any kind by heat, pressure, and invasions by other materials, without melting. Members of all categories are found on all continents. Some terrains show one type of rock only; others exhibit a mix, often in alternating layers or in masses more or less disrupted by tectonic activity.

Earth Pillars: Near Church Rocks, New Mexico, hard lava fragments produced by explosive volcanic eruptions act as caps protecting soft underlying materials from erosion. As erosion lowers the land, caps are undermined and fall off; then the pillars rapidly erode away. Pillars with such caps are sometimes called "demoiselles," or "hoodoos."

Highlands of Igneous Rock: The Bubbles, on Mt. Desert Island, Maine, are masses of granite. Being relatively homogeneous in composition and structure, these masses have eroded to rounded forms. Much of the smoothing was done by Pleistocene glacier ice advancing around and over them.

From the Zone of Heat: Igneous Rocks

The movement of molten materials on and in the crust is known as igneous activity, and the rocks formed by cooling and solidification of such materials are called igneous. Igneous activity created much of the crust originally, has occurred here and there throughout Earth's history, and continues in some areas to this day. Few localities have not been at some time the scene of igneous activity.

All continents have volcanic areas – some of them thousands of square miles in extent – where magma from Earth's interior has reached the surface. Such areas are along plate borders or along fractures within plates. Eruptions occur as lava flows, lava fountains, outbursts of lava fragments, and emissions of volcanic gases. Rock materials produced by eruptions may accumulate to form cones, plateaus, and mountain masses. Despite erosion, which is continuously destroying crust, igneous rocks still cover many terrains, testifying to a "fiery" past.

Within the crust, too, rock formed by igneous activity is plentiful. Magma from depth has penetrated fractures and cavities in the crust and has cooled there to make bodies of hard rock. Some of these have become exposed by erosion, tectonic activity, or excavation, and have been shaped into scenery of multitudinous forms.

Igneous rocks are of two categories: acidic (or felsic) and basic (or mafic).

Acidic rocks form from magma derived from crustal melts and, to a lesser degree, from the mantle. Relatively light in color and weight, these rocks consist mostly of quartz and alkali feldspar, often with small amounts of other minerals. They may cool and solidify within the crust or, reaching the surface, erupt from cones (not from rifts) within the continents or along continent fringes. Eruptions often are explosive, shattering the lava and scattering it over surrounding terrain.

Basic rocks, relatively dark and heavy, are rich in soda-lime feldspar and iron minerals such as olivine, pyroxene, and amphibole. The magma from which basic rock forms comes mainly from the mantle. It may cool and form rock within the crust; or, it may reach the surface and flow out or fountain out from cones or crustal rifts. Eruptions of basic lava occur mainly from oceanic volcanic cones and rifts, as in Hawaii, New Zealand, and Iceland, but may erupt also

from vents on continents, as in the northwestern coterminous United States.

Acidic and basic rocks grade into one another through many rock varieties. Some of these are difficult to identify, but the major varieties, such as gabbro and basalt (both basic) and granite and rhyolite (both acidic), are fairly recognizable. They are worth getting to know because each can occur in certain common kinds of landforms.

Originating Underground: The Intrusive Rocks

Rock formed within the crust by solidification of magma from the mantle, or from the crust itself, is called intrusive, or plutonic (from Pluto, Greek god of the underworld). The previously existing rock that is invaded is termed country rock, or host rock.

Intrusive rocks are relatively dense, having formed at depth under the confining pressure of overlying crust, perhaps miles thick. These rocks are relatively heavy when they contain iron or other heavy metals. Their mineral crystals, which interlock, are large enough to be visible without magnification, having formed at depth where cooling was slow, allowing ample time for crystals to grow. Intruded masses range from tiny particles to reservoirs miles in extent called batholiths. The commonest intrusive rock types are granite, gabbro, and diorite, each with variants.

Some people call any large, impressive mass of hard rock "granite." More precisely, granite is a compound mainly of feldspars and quartz, often with a minor amount of ferromag-

Once a Molten Mass Underground: Carrizzo Peak, near Carrizozo, New Mexico, is a body of magma that solidified at great depth, making a laccolith. Now uncovered by erosion, the laccolith (*in background here*) stands above a plain of relatively weak lava rock.

nesians. It is the intrusive equivalent of rhyolite. Granite commonly has formed where continental plate masses have collided, causing weathered rocks rich especially in silica and quartz to be subducted to depths of miles, there to be melted, perhaps invaded by molten materials from the mantle, and to become roots of mountains being thrust up by the collision. Thus granite is a rock formed at a depth of at least 25 miles, and it becomes visible only after erosion has removed covering material or tectonic activity has raised the granite above surrounding rocks. Granite and granite-related rocks constitute most of the materials in the cratons – that is, "cores" – of the continents.

Granite outcrops may be blocky or, more commonly, slabby or rounded, or pillarlike or needlelike where erosion has opened wide spaces between vertical joints. Granite with relatively large crystals is known (as are other igneous rocks with large crystals) as pegmatite. Common in scenery throughout the world, especially in highland scenery, granite appears in numerous forms, from small, loose, weathered grains to solid batholiths which are the roots, or cores, of some types of mountain ranges. More resistant than gabbro, diorite, and most other rocks as well, granite is usually a former of highlands.

A variant of granite is granodiorite, which is gray, with plagioclase and orthoclase about equal; it forms much of California's Sierra Nevada. Also related to granite is syenite, but this has no quartz and is relatively rare.

In the basic intrusive group the "type" rock is gabbro, consisting mainly of materials from the mantle that solidified within the crust. In fresh outcrops it is usually blocky or columned, dark gray, coarse-grained, very hard. Mostly of plagioclase feldspar, it often contains ferromagnesians – commonly pyroxene, sometimes olivine. It is the intrusive equivalent of basalt, having about the same chemical composition but larger crystals. Gabbro may weather to rusty brown or, in a dry climate, acquire a whitish crust. A relatively common and important variety of gabbro is diabase, which often forms dikes and sills, such as New York's Hudson River Palisades and England's Great Whin Sill.

Remnant of Another Cooled Underground Mass: The Palisades of the Hudson River, famed as scenery, are the edge of a great mass of magma that intruded between sandstone layers about 190 million years ago, then solidified and became exposed as covering sandstone eroded away.

Medium-grained Granite (at right): In this specimen of a common type, the grayish crystals are quartz; other crystals are orthoclase feldspar, which has been stained reddish by iron oxide, and hornblende, which is black.

Diorite (below): The dark minerals in this 5-inch chunk are plagioclase – more than is usual in diorite. Diorite forms at depth, where cooling is slow and crystals can grow large enough to be easily visible.

Pegmatite (below): Various kinds of igneous rocks that occur with large crystals in the groundmass are called pegmatite (from Greek *pegma*, "fastened together" – referring to the large crystals). This is a sample of granite pegmatite, with crystals of quartz (grayish) and feldspar (reddish).

Anorthosite (below): In this sample from New York's Adirondack Mountains, dark plagioclase feldspar has weathered to a whitish hue. Reddish specks are garnet. In some anorthosite specimens feldspar crystals have a lambent silky luster called "chatoyancy."

Welded Tuff: Frijoles Canyon, in New Mexico's volcanic Jemez Mountains, was the home of pre-Columbian Indian cliff-dwellers. The canyon has been cut in tuff whose particles were welded together by volcanic heat. Cavities later hollowed out by percolating groundwater were enlarged by the cliffdwellers. Erosional forms here resemble those in the Goreme Valley in Cappadocia, Turkey, which has a similar volcanic history.

Less common than granite and gabbro is anorthosite, made up mostly of plagioclase feldspar, originating mainly from the mantle. It constitutes much of the Adirondack Mountains in New York and, incidentally, is exposed widely on the Moon.

Diorite, intermediate between the acidic and basic extremes, including materials from both crust and mantle, is usually dark gray, blocky, coarse-grained, very hard. It contains mostly plagioclase feldspar and hornblende, often some pyroxene, sometimes a little quartz. Although less common than gabbro, it occurs in large masses, such as Vermont's Mt. Ascutney. It weathers to a dark gray and is less resistant than granite.

Originating Aboveground: The Extrusive Rocks

Extrusive rocks originate from magma that erupts onto the surface of the crust, whether on the ocean bottom or on land. Ocean-bottom vents usually produce lava of the basic type, most of which flows freely and widely, and cools to form solid rock. Some continental vents occasionally produce basic lava, but usually the magma reaching these vents contains more or less acidic material picked up during passage through the crust. The lava, being relatively viscous, often is

Tuff: Lava fragmented finely by volcanic explosions falls on surrounding land as "ash" and consolidates to form beds of "tuff." This may become welded by volcanic heat to form a more or less hard rock or may remain powdery, as here.

Rhyolite: Faint streaks in this specimen are flow lines, produced perhaps when the material, in a liquid or plastic state, was swept along in a rushing cloud of incandescent volcanic ash. A pink hue is common in rhyolites.

Prismatic Basalt: These typically fine-grained specimens from Paterson, New Jersey, formed from lava containing relatively little gas. The rock is dense and heavy, fracturing neatly into prismatic forms, on which 60-degree angles are common.

Amygdaloidal Basalt: This rock formed from lava near the surface of a flow, where escaping gases left cavities (amygdules). Cavities may fill with minerals, such as calcite and prehnite, deposited by subsurface water filtering through the rock.

erupted explosively and is shattered and scattered over surrounding terrain. The shattered material, ranging from fine powder ("ash") to fragments weighing tons, may become welded by volcanic heat into massive rock.

Some extrusive rocks are relatively new, little eroded, and at the surface. Others are old, much eroded, more or less covered with erosion debris or vegetation. Some became buried long ago under layers of other rock, such as sandstone or limestone, and then uncovered by erosion.

When massive (very compact, without fractures), most extrusive rock consists essentially of crystals. These are usually too small to be seen without a magnifier, because cooling at Earth's surface is rapid and allows relatively little time for crystal growth. The extrusive rock obsidian is a glass, with no crystals at all (it is amorphous). However, some extrusive rocks form (as do the intrusive group) under conditions that allow phenocrysts (large crystals) to grow; porphyries (igneous rocks with closely scattered phenocrysts) may form.

Rhyolite, the "type" rock in the acidic extrusive group, erupts mostly from volcanoes on continents or along their fringes. It can be massive or fragmented, reddish to light gray or white, consisting mainly of quartz and orthoclase feldspar, occasionally a little mica and ferromagnesians. Crystals in the groundmass are mostly microscopic, but scattered large crystals may be present.

Massive rhyolite can be very tough. It is the extrusive

equivalent of granite. Often it shows a flow pattern, acquired probably as magma pulverized by an explosive eruption swept along as a fiery cloud (*nuée ardente*), accumulated on the ground, and was welded solid by volcanic heat. Like granite, rhyolite is relatively resistant when massive.

Among basic extrusive rocks the commonest is basalt, which erupts mostly from vents in ocean bottoms, occasionally from continental volcanic cones and fissures. It flows freely over the ground surface or is blown out from the vent

Columnar Basalt: Columns with prismatic cross sections may form as basalt lava cools, contracts, and fractures. Cross sections are polygonal, often hexagonal (six-sided) – a form frequently chosen by nature (as in honeycombs and mudcracks) because it allows the closest fit among many parts. These columns are at Devils Postpile National Monument, California.

Ropy Pahoehoe: Currents in a basalt flow often push the lava into coils. These occur on flows that are quite fluid, not too gassy, slow in cooling and thus relatively slow to harden and become covered with crust. This example is from Idaho.

Andesite Porphyry: In this Mt. Vesuvius specimen, plagioclase feldspar crystals are prominent among darker minerals, probably pyroxene and hornblende. Andesite was named after the Andes Mountains, where it is especially abundant.

as pyroclastics. Outcrops of massive basalt often are blocky or in scenic prismatic columns. Pillow shapes are seen where the lava poured into the sea, as in Hawaii, or emerged from a vent underwater. Surfaces of flows may include forms such as thick ropelike coils, broad black "pies," or, if the lava was very gassy, clinkers and cinders with cavities that may later fill with other minerals, such as calcite or quartz deposited by groundwater.

Massive basalt is usually gray to dark greenish gray or black, often with conchoidal (curved) fracture surfaces. As the extrusive equivalent of gabbro, basalt is very fine-grained, chiefly of plagioclase feldspar and pyroxene or other ferromagnesians such as olivine, and with little or no quartz. It weathers to a rusty brown or, under a dry climate, develops a whitish limy crust. With its iron content, massive basalt tends to weather faster than granite, but may form highlands on terrains where weaker rocks, such as shale, underlie it.

Andesite, intermediate in composition between rhyolite and basalt, is the extrusive equivalent of diorite. It is erupted mostly from volcanoes along continental fringes, such as Mt. Vesuvius in Italy and volcanoes of the Andes Mountains in South America. It is massive or fragmented, reddish to green or gray. Visible and microscopic crystals are present, often producing a speckled appearance; porphyries are common. The minerals are mainly plagioclase feldspar (andesine) with some hornblende, pyroxene, or biotite, often a very small proportion of quartz. Andesite is very hard when massive, and moderately resistant. It has been called felsite when difficult to distinguish from rhyolite; but felsite is a term now out of favor. Closely related to andesite is dacite, which includes less quartz.

Highlands of Extrusive Rocks: The Davis Mountains of western Texas are eroded remnants of piled-up lava flows and ash deposits. In this view from the south, the mountains appear rugged and chaotic, thanks to variations in resistance and structure, typical in masses of volcanics. In the background note the columnar jointing.

Diverse Sedimentary Strata: This outcrop, in a roadcut on the Appalachian Plateau in Pennsylvania, is typical of stratified sedimentary rocks. Layers are horizontal or nearly so, as usual with plateau topography. They grade from coal to limestone to shale, then back again, according to environmental changes during the period of deposition.

From Debris: The Sedimentary Rocks

Sedimentary rocks cover more than 75 per cent of Earth's land surface, and also portions of the ocean bottom fringing the continents. Occurring mostly as upper layers of the crust, sedimentary rocks form by accumulation, consolidation, and cementation of sediments. These consist of rock debris from weathering and erosion, minerals precipitated from water (especially seawater), and remains of decomposed plants and animals. The sediments may have been deposited on land surfaces, lake and stream beds, or sea bottoms. Much sedimentary rock has originated where tectonic activity depresses the crust and allows the ocean to invade land. It is in sedimentary rocks that most fossils (plant or animal imprints, casts, or remains) are found.

Much sediment occurs as particles, called clay (less than $1/256$ millimeters in diameter), silt ($1/256$ to $1/16$), sand (over $1/16$ to 2), granules (over 2 to 4), pebbles (over 4 to 64), cobbles (over 64 to 256), and boulders (over 256).

The process, called lithification, that converts sediments into rock occurs by compaction of sediments under pressure and by natural cementation by substances such as silica, lime, or iron oxide deposited by percolating subsurface water. Most sedimentary rocks are in layers, usually called strata (plural of Latin *stratum*) or beds, separated by bedding planes. Layers when first formed are mostly flat, but may later be warped, folded, or broken by tectonic activity. A group of layers with distinctive characteristics is called a formation, such as the Morrison Formation in Wyoming.

Shale, the most common rock at Earth's surface, makes up about half of all sedimentary rocks. It is made of clay or silt particles deposited in quiet water on river, lake, and sea bot-

A Display of Sandstone: Various kinds of sandstone make up much of the Colorado Plateau. Layered sandstone masses in the Church Rocks area, east of Gallup, New Mexico, have been shaped mostly by erosion into blocks, pinnacles, and rounded forms, all rather typical of this versatile common rock.

Assorted Sediments: A roadcut in the White River Valley, Texas, exposes layers of sand, gravel, and cobbles. Finer materials here are of a caliber characteristic of sandstone; coarser materials suggest conglomerate (which here is noticeably less susceptible to gullying). If at depth in the crust, subject to great pressure and natural cementation, these varied sediments would become rock.

toms. Shales are compacted but not cemented; they include "claystone" and "mudstone." The rock siltstone is somewhat similar but is coarser-grained, has been naturally cemented, has thicker layers, and is harder.

Particles in these rocks are smaller than $1/16$ inch in diameter, the smallest size distinguishable by the unaided eye. Many kinds of minerals may be present; usually quartz and clay minerals predominate. The rock occurs generally in thin beds, smooth on the bedding planes and relatively soft, scratching easily and fracturing into slabs, flakes, or splin-

ters. Colors include gray (most common), black, red, blue, brown, and green. Fossils may be present; also, ripple marks made by waves, currents, or wind. Being relatively soft, shales tend to form lowlands.

Sandstone consists of sand particles deposited by running water near highlands, by waves and currents near a shore, by wind, or at the foot of cliffs by gravity movements. The particles, larger than in shale, are $1/16$ to 2 millimeters in diameter and distinguishable without magnification. They are bonded more or less strongly by natural cements. Sandstone grades into siltstone, which is finer-grained, or conglomerate, which is very coarse.

Constituting a quarter to a third of all sedimentary rocks, sandstones come in many varieties, including such diverse minerals as quartz, feldspar, basalt (as in black beaches), gypsum, and coral. Red and brown hues are common; also white, gray, and yellow. Some surfaces show ripple marks; some exhibit fossils. Masses of sandstone with strong natural cementation can become uplands.

Orthoquartzite, a very strong kind of sandstone, derives commonly from quartz-rich sandstone in which grains are cemented with silica deposited by groundwater. The cement is so strong that the rock breaks through grains rather than around them. Fracture surfaces may have a grainy look, but are smooth to the touch, unlike the gritty fracture surfaces of most sandstones. Typically pinkish to white, orthoquartzite is sometimes gray. Outcrops usually are blocky. This rock is a ridge-former, extremely resistant, as in New York state's Shawangunk Mountains. Orthoquartzite closely resembles metaquartzite, a metamorphic rock described below.

Conglomerate, known by laymen as puddingstone, is formed of water-worn sand and cobbles more than 2 millimeters in diameter (even boulders may be included). The material was originally deposited by swift streams at the foot of highlands or by waves and currents on beaches. The larger-caliber particles, often of quartz, consist of relatively resistant portions of the disintegrated rock from which the conglomerate formed. The name "fanglomerate" may be used

Shale: The black color of these strata, which were tilted by mountain-building forces, is due probably to the presence of swamp sediments. Thin layering is common among shales.

Ripple Marks: A sandstone slab on Bearfort Mountain, northern New Jersey, shows ridges made by currents on a Devonian beach. Ridges were covered by sand and turned to stone along with the rest of the material. The rock mass was later tilted to the vertical and finally uncovered by erosion.

Thin-bedded Sandstone: The fine grains of this specimen were deposited probably in a slow stream The layers are thinner than in most sandstones. Iron oxides provide the red hue.

Crossbedded Sandstone *(at right):* Zion National Park features crossbedded sandstone formed from ancient sand dunes. Crossbedding was caused by changes in velocity and direction of winds that transported and deposited the sands about 200 million years ago, when the region was desert.

Conglomerate *(below):* This Triassic rock mass from southeastern New York State consists of stream-worn cobbles in a mass of sandstone. Coarseness of the sediments indicates deposition by a swift stream (swift enough to transport cobbles) descending a highland or by strong waves on a steeply inclined beach.

Massive Dolostone: This fine-grained stone from a Devonian sea bottom in Wisconsin contains relatively few shells. It is limestone that was altered to dolostone shortly after deposition. At extreme right is the fossil of a trilobite, one of the commonest life forms of the Paleozoic Era.

Shell Limestone: Some limestones are rich in fossils. Predominating in this specimen from an Ordovician sea bottom near Cincinnati, Ohio, are the broad shells of brachiopods and fragments of various bottom-dwelling organisms. The specimen is about 6 inches wide.

if the rock formed from assorted debris deposited as an alluvial fan. Conglomerate with small-caliber particles grades into sandstone.

The resistance of conglomerate hinges on the strength of its cement. As the rock weathers and erodes, pebbles and cobbles may stand out on the surface and later, when they fall out, leave cavities. Some conglomerate masses, varying in resistance because of their miscellaneous makeup, erode into bizarre bulbous and cavernous forms.

Sedimentary breccia is, like conglomerate, a naturally cemented mass of rock debris, but it consists mainly of fragments that are angular, not water-worn and rounded. Fragments may have been talus or rock broken up by crustal movements, landslides, or volcanic eruptions.

Limestone, most common of carbonate rocks, consists mostly of calcium carbonate ($CaCO_3$). It forms by cementation of chemical precipitates, mostly from seawater, or from limy remains of coral, algae, shellfish, or other organisms. Colors are white to gray or bluish, even black. Limestone can be hard, but even when hard is scratched by steel. It will fizz when touched with hydrochloric acid (HCl) – a test that helps with identification.

The term "limestone" may include the similar rock dolostone. Very common, this consists entirely or partly of dolomite, a mineral like calcite but in which magnesium partly replaces calcium. Dolostone is often bluish gray, sometimes white with a tinge of yellow, brown, or pink. It tends to be less soluble than limestone.

Most limestone forms from limy mud, which may or may not include visible shells. Coquina is a variety made up of shelly material. Chalk consists of limy protozoan remains. Oolitic limestone has small, spherical concretions that grew layer by layer, onionlike, when rolled by waves in shallow water. Crystalline limestone is a variety that crystallized after it consolidated, but was not marbleized by pressure. Travertine, noncrystalline and massive, is deposited by surface water and groundwater in springs and caverns.

Crystalline Limestone: This fragment of nearly pure limestone was crystallized by heat and pressure to form calcite, a crystalline form of calcium carbonate. This crystal is imperfect, but the sides intersect at angles typical for calcite.

Mudstone: This rock, finer-grained than shale, formed probably on the bottom of a pond or a lake from layers of mud deposited in quiet water. Mudstone characteristically is very soft and crumbly. A reddish color is common.

Rock Gypsum: Gypsum may occur in massive form, as rock, as in this specimen, which includes crystals known as selenite. Gypsum occurs also as sand, as on the ancient lake bed (now dry) at White Sands National Monument, New Mexico.

Limestones are more or less soluble in water that contains carbon dioxide. Thus these rocks under a humid climate usually are valley-formers. Masses that are elevated and thus have swift drainage, with rapid removal of water containing carbonate in solution, may erode to form solution features such as sinkholes and caverns.

Another soluble sedimentary rock, less common than limestone, is gypsum. It consists of calcium sulfate that dissolved in bodies of water and later was deposited by evaporation. Often interbedded with limestone and shale, it may be of almost any color. Solution in gypsum beds may create caves, sinkholes, and other such forms.

A Behemoth of Conglomerate: Huge monoliths at Meteora, Greece, have been carved by erosion from limy conglomerate masses heaved up from an ancient sea bottom during the current collision of the African and Eurasian plates. Surfaces have been smoothed and pitted by solution. Bulbous forms like these, with large cavities, are common in much-weathered conglomerate masses, and in some sandstones as well.

A Slate Slab: The black band in this specimen from southern New York State is the edge of a layer of the shale from which the slate formed. This slab broke across that layer, demonstrating "slaty cleavage."

Phyllite with Secondary Folds: In this 5-inches-wide specimen from Vermont, ripples were made in a large fold by opposing forces during metamorphism. The shine comes from numerous flakes of mica.

Changed Rocks:
The Metamorphic Family

Rock is called metamorphic if it has been substantially altered in composition, texture, or internal structure by heat, pressure, or infiltration and enrichment by invading materials, without melting. Rock at depth, especially, is subject to such changes because of the weight of overlying crust and because of heat, pressure, and intrusions of magma associated with plate collisions and mountain-building. Metamorphic rocks form a large portion of the "core rocks" of the continents – that is, the cratons.

Metamorphism can include alterations in color, texture, structure, and hardness, along with formation of new minerals. Metamorphism is called thermal when it occurs locally by magmatic intrusions, and regional when it occurs over broad areas in association with mountain-building. Some metamorphic rocks, notably gneiss and metaquartzite, are relatively strong, often forming highlands.

Slate, the most common metamorphic rock, derives from shale. Somewhat lustrous, with microscopic interlocking crystals, it consists mostly of clay minerals with some mica and quartz. It has slaty cleavage, tending to fracture along layers of platy or oval mineral grains aligned by pressure during metamorphism. Colors are bluish gray, or perhaps green, red, or brown. Surfaces of slate may be banded, the bands being edges of bedding planes (of the original shale); these may not parallel cleavage planes. Folds may be visible. Usually this rock is stronger than shale, less hard than steel, and may form ridges in regions of weaker rocks.

Slate Outcrop: This metamorphic rock in northern New Jersey formed when layers of shale were compressed, crumpled, and tilted by mountain-building forces. White bands and streaks are of quartz that precipitated from hot fluids injected during metamorphism.

Folds in Mica Schist: Visible here are cross sections of folds made by opposing forces during metamorphism, when the rock became so hot as to be somewhat plastic. S- and Z-folds are common in schist.

"Tortured" Rock *(upper right):* Perhaps more often than any other rock, schist has been squeezed, stretched, twisted, broken, dislocated, and invaded by other materials. An example is seen in this roadcut in California's San Gabriel Mountains, a zone of intense rock deformation near the San Andreas Fault.

Metaquartzite "Steps" *(right):* Quartzite usually weathers to blocks, often with conchoidal (curved) fracture surfaces. Edges tend to remain sharp, despite weathering, because of high resistance.

Orthoquartzite *(below):* In this specimen (here to compare with metaquarzite) layering and colors of the original sandstone were preserved as groundwater deposited quartz between rock grains and cemented them. The grainy or sugary look is seen on many quartzite fracture surfaces.

Phyllite, derived from slate, is more metamorphosed than slate and flakier, with a silvery luster due usually to the presence of tiny crystals of mica and chlorite. It splits into sheets, often wavy, with the surfaces sometimes paralleling bedding planes. Phyllite is relatively resistant.

Schist (German, "glittering") can be derived from igneous rocks, shale, or other metamorphic rocks, notably slate and phyllite. More metamorphosed than phyllite, schist has visible interlocking crystals, mostly of mica, hornblende, and quartz, arranged as sheets, usually wavy and contorted. These

Banded Gneiss: Bands are edges of folded alternating layers of different minerals. In gneisses derived from sedimentary rock such as shale or sandstone, layering may correspond to original bedding. As in schist, S- and Z-folds were produced by forces operating on the rock mass while plastic. The coin is a cent.

may parallel original bedding planes. Schist commonly is fractured along the sheet surfaces.

Any color or combination of colors, occasionally with stripes, may be seen in schist. Its many varieties are named after the predominant minerals – for example, mica schist or hornblende schist. Schist may be weak or strong, its hardness usually increasing with quartz content.

Metaquartzite is a rock consisting mainly of quartz. It has been recrystallized at depth by heat and pressure due to regional crustal movements, as in mountain-building, or to nearby igneous activity. Fracture surfaces are smooth to the touch because the fractures cut uniformly through both the grains and the cement. In strength and appearance metaquartzite may resemble the sedimentary rock ortho-quartzite.

Gneiss (German, "banded"), one of the commoner metamorphic rocks, comes in many varieties. Deriving from metamorphic rocks of clayey or sandy origin, or from igneous rocks, it often acquires new minerals by injections of magma. Interlocking crystalline grains are visible. Often bands of granular feldspar and perhaps quartz alternate with darker schistose bands, frequently of hornblende or mica. Sometimes resembling schist, gneiss is relatively heavy and may have almost any color or color combination. Varieties of this rock, such as granite gneiss, hornblende gneiss, and muscovite gneiss, are named for minerals that predominate in them. Gneisses are usually blocky or slabby in outcrops, and relatively strong, commonly forming highlands among diverse rocks.

Marble is metamorphosed limestone or dolomite, composed mostly of recrystallized calcite or dolomite mineral, or both, with visible fine to coarse grains that interlock. Grains may give fracture faces a sugary look. Marble when pure is white; impurities may make it gray, blue, yellow, pink, black, or green. Marble is sometimes mottled, sometimes with wavy streaks or bands. It is mechanically strong if not deeply weathered. Like limestone it will fizz when touched with hydrochloric acid. Marble may contain fossil fragments, usu-

Marble Specimen: The crystalline texture in this blocky fragment from Mt. Timpanogos, Utah, was acquired during metamorphism of the original limestone. Bands colored by impurities often make marble, like this specimen, very decorative.

Billows in Marble: Ridges and furrows on this outcrop in New York City are on edges of stronger and weaker layers. Smoothing has resulted from solutional weathering and rain-wash, and from abrasion by Pleistocene glacier ice.

ally distorted by metamorphism. Being soluble, it tends to be a valley-former in humid areas.

The Rock Cycle

Thanks to the broad sweep of geologic processes, driven by motions of the crustal plates, plate collisions, and continuing climatic change, rock masses are undergoing some phase of creation, deformation, destruction, metamorphism, and deposition just about everywhere on and in the crust.

Igneous rocks when subjected to heat, infiltration, and pressure may be converted into metamorphic rock. When exposed to weathering, gravity movements, and erosion, they decompose and disintegrate, and fragments may become consolidated to make sedimentary rock.

Sedimentary rocks may melt and then solidify to become igneous rock. Or – they may be converted by metamorphism into metamorphic rocks, or may disintegrate and reconsolidate to become sedimentary rocks once more.

Metamorphic rocks may be converted by melting and solidification into igneous rocks, changed by weathering and erosion into the makings of sedimentary rocks, or changed by further metamorphism to become metamorphic rocks of a different kind.

This set of recycling processes, known as the rock cycle, can be visualized in a diagram. With this in mind, we can observe the world of scenery with awareness of continuing changes in the substances of which scenery is made.

The Rock Cycle: Rocks commonly are involved in a maze of processes – creation, alteration, and destruction – with circumlocutions and shortcuts in repeating cycles. The diagram here summarizes highly complex relationships. Igneous rocks (*left*) are made by solidification of magma. Uplift above sea level exposes rock to erosion. Resulting sediments are transported and deposited, and may become sedimentary rock. This when lowered may become buried, converted into metamorphic rock, and melted to form new magma. Sedimentary and metamorphic rocks may, like igneous rock, become subject to uplift and erosion.

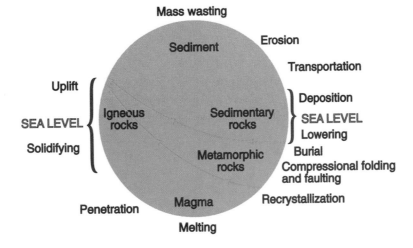

Weathering of Granitic Rock: Widely spaced joints and chemical action by atmospheric gases and moisture (often present as dew), aided by working of surface grains by sharp temperature changes, account largely for the shaping of quartz monzonite at Joshua Tree National Park, in California's Mohave Desert. The rock's homogeneity favors weathering to rounded forms. Rainwash and sandblasting by desert winds erode the rock. A hundred thousand years ago this group of monoliths probably was a solid mass with a few widely spaced joints.

4

The Works
of
Weather

CLIMATE, which can be defined as the average state of the atmosphere in a region over a long period, is a partner of tectonics in the working and reworking of Earth's crust. The state of the climate at a particular place and time is known as weather. The crust everywhere is subject to actions of weathering agents, directly or indirectly, and the results of these actions are called weathering. Ordinarily the term "weathering" doesn't raise thoughts of momentous happenings. Unpainted walls of seacoast cottages weather to gray in the salt air, marble statues become pitted and stained in the acid-laden atmosphere of cities, walls break up from "frost," and iron fences rust – but such events seem relatively minor. They hardly suggest what a fundamental, powerful factor weathering is in the world environment.

As a geological term, weathering covers all those natural processes that alter, decompose, and disintegrate rock as a result of its exposure at or near Earth's surface. These processes, which are to be carefully distinguished from gravity movements, erosion, and deposition, are often subtle, usually slow, hardly noticeable. However, occurring everywhere most of the time, on and in the crust, during geologic ages they work great changes in rock, produce huge amounts of debris, and strongly modify topography. Rugged mountain peaks, "goblin" rocks in the desert, potholes in city streets –

all testify to weathering as a rock sculptor. Talus at the foot of cliffs, dunes in the desert, deltas and beaches along shores – these show that weathering is the provider of huge quantities of fragmented rock for transportation by gravity, running water, wind, and glaciers, and for the building up of new landforms while old ones are being torn down.

"Changeable as the weather" is a familiar saying. Weather does change day to day, even hour to hour; also, in any particular place, it changes over thousands or millions of years, altering the conditions to which rock is exposed. As tectonic activity moves a land to a different latitude, elevates or lowers it, shifts ocean currents, moves the land toward or away from the ocean – or, as Earth for a time receives more or less heat from the Sun – corresponding changes in weathering result. Most rock at or near the surface of the crust has undergone changes in weathering during the past, and may at present exhibit features suggesting past climates rather than the present one. Coral reefs in the Canadian Rockies, glacier-sculptured mountains in New York state, numerous dry lake basins in New Mexico – such phenomena are evidence of different climates in the past. Recognizing them adds much to the appreciation of Earth scenery.

Weathering is basically of two kinds: chemical and mechanical (physical). In chemical weathering, rock is

Differential Weathering: In New York's Hudson Highlands, weathering of edges of layers of differing resistance in this outcrop of granite gneiss produces a surface with ridges and furrows. The ridges are edges of relatively resistant layers. Generally the ridges are on quartz or feldspar and the furrows are in a ferromagnesian (iron-rich) mineral, such as hornblende.

es it to swell; but the distinction between the two basic kinds of weathering is important for the understanding of what actually happens.

The Chemical Attack

The action of chemicals accounts for most of rock weathering. Such action may involve hydrolysis, hydration, oxidation, carbonation, ion exchange, solution, or any combination of these. Details concerning chemical reactions belong in a chemistry text, but an outline of the major reactions involved in weathering, with their effects on landforms, is in order here.

Water, the universal solvent, is constantly at work on the crust and within the crust to depths of hundreds of feet. When water is combined with carbon dioxide (of which the atmosphere has huge quantities), its reactions with rock greatly accelerate. Water with carbon dioxide alters feldspars, which are abundant in granitic rocks, to form silica and clay minerals. Ferromagnesian minerals, common in basic rocks, are attacked by water to produce iron oxides, especially

altered and more or less decomposed by reactions with other chemical substances, especially water. In mechanical weathering, rock is disintegrated by forces associated with events such as expansion of water as it freezes in rock crevices, temperature changes, removal of pressure on rock, and wetting and drying. Often chemical and mechanical weathering occur in the same place at the same time, as when water saturating a mass of rock dissolves portions of it and also caus-

Grain by Grain: Weathering, though relatively slow, is an important ally of streams and gravity in the wearing down of mountains. Here in Kings Canyon, in California's Sierra Nevada, weathering is disintegrating the massive granite grain by grain. In this photograph and the adjacent one, note the tendency of a homogeneous rock to weather to rounded forms.

Balanced for a Purpose? Balanced rocks, common in scenery, often seem to have been positioned by human hands, possibly for religious or ceremonial purposes, or just by whim. Actually, most balanced rocks result from weathering or erosion. These granite specimens, in an Arizona desert, are remnants of rock masses that weathered spheroidally in place.

Weathering of Stratified Rock: Rock masses with layers horizontal or nearly so tend to preserve flat tops as they become weathered and eroded. But different layers weather and erode at different rates, according to their resistance. Camel Rock, near Santa Fe, New Mexico, is capped by a resistant layer, which weathers less rapidly than softer rock beneath and therefore hangs over it.

limonite and hematite, and react with water charged with carbon dioxide to yield iron carbonate. Water thus charged is relatively swift in dissolving carbonates in limestone, dolomite, and marble.

As crystals of new minerals form crusts on rock, they usually expand. Expansion tends to detach the crystals from rock surfaces, and they may drop off as flakes or grains. Further, the new minerals are likely to be softer and lighter (less dense) than the originals and thus more susceptible to disintegration and to removal by further weathering, mass movements, and erosion.

Under certain conditions, such as where joints are very close and extend deep into bedrock, or where highly reactive materials are involved, chemical weathering has been known to penetrate bedrock to depths as great as 900 feet. Weathering is most active in warm, humid regions, producing the red soil called laterite, rich in aluminum and iron oxides. Even in deserts, where water is scarce, chemical weathering can be substantial, and it is believed to be more destructive there than mechanical weathering.

Chemical weathering is represented dramatically by large solution features, such as sinkholes and caverns in limestone terranes. More common, on terranes of almost every description, is chemical decomposition of rock surfaces, with consequent crumbling and exfoliation of weathered materials. Features such as pits and furrows separated by ridges are produced according to unevennesses in the resistance of the rock mass exposed. As air and water penetrate through joints and between rock grains, the interior of rock comes under attack, joints widen, and the rock crumbles.

Familiar on many surfaces, most noticeably on highland summits and slopes, are heaps and scatterings of large rock fragments called boulders of disintegration. These usually are remains of bedrock broken up by weathering (especially chemical weathering) over thousands of years. Often smaller-caliber material in such locations has been removed by rainwash, wind, and gravity, leaving large boulders in place. Such boulders, being products of weathering, differ from the boulders called erratics, shaped and deposited by glaciers.

On some surfaces, boulders of disintegration stand as tors. These are heaps, stacks, or pillars of boulders, usually granite, often shaped like biscuits or muffins. Especially well known in Cornwall and Devon, in England, they are fairly common also on granite terranes in localities as varied as California's Joshua Tree National Park, the central plain of Spain, and the desert near Aswan, Egypt.

Apparently a tor originates as an underground rock mass that has nearly rectangular jointing and is close enough to the

Weather Pits: In this outcrop of granite gneiss, the groundmass of relatively resistant quartz and feldspar contains scattered clusters of hornblende and other minerals of less resistance. Weathering of these minerals is relatively fast. Hollows occur where clusters have been removed.

surface for substantial chemical weathering to occur, especially in joints. As erosion removes overlying material, the mass gradually becomes exposed. Weathering in joints accelerates, dividing the mass into blocks. These weather fastest along edges and at corners, tending to become rounded. Small-caliber weathered material is swept away by surface processes; larger remnants remain in stacks.

Because tors are fragile and uncovered, they are relatively short-lived compared to solid and more massive structures. On some terrains, such as portions of the Rockies and the northern Older Appalachians, most of the tors that existed before the Pleistocene were scraped off by glacier action.

The Mechanical Attack

Frost Action

Under such names as ice prying, ice wedging, ice cracking, and frost action, the work done by water freezing in rock and soil is dominant in very cold lands and is second only to chemical weathering elsewhere. Its effectiveness depends on factors such as porosity of the rock or soil, closeness of fractures, ease of entrance by water into fractures, amounts of water and air present, weakening of rock by earlier weathering, and frequency or intensity of freezing. Generally, frost action is greatest on slopes facing away from the sun.

Rock Shattering

Rock that is porous can be shattered as water filters into it and minerals precipitate from the water, forming crystals that expand as they grow. Likewise, as water confined in crevices and between grains in rock freezes, the crystals expand with astonishing force – more than a ton per square inch. Even very strong rock can hardly resist such stresses and strains. Grains and flakes loosen and disaggregate from the rock surface. Hairline fractures widen, allowing more water and air to enter. On a larger scale, water freezing in crevices forces apart rock masses that weigh many tons.

The intensity of rock shattering by frost action in a given locality depends on various conditions, such as porosity of the rock, its jointing, exposure, levels of humidity, and frequency of thaw-freeze. Shattering may be at maximum on humid subpolar or temperate terranes rather than polar lands, where thaw-freeze is less frequent. The higher European Alps, Rockies, and other very cold highlands display jagged

"Frost Flowers": Water freezing in soil forms spearlike crystals, which heave the soil and break it up. Fragile, easily crushed underfoot, soon melted in the sun, "frost flowers" hardly suggest the full magnitude of the force exerted by the freezing process in rock crevices.

peaks, huge accumulations of broken rock (talus) at the foot of cliffs, and broad expanses of frost-broken rock called felsenmeer (German, "sea of rock"). Some highland regions that are temperate today display such topography produced during the Pleistocene Epoch, especially toward its close, when thaw-freeze was increasingly frequent. Today frost action continues to be important in land-sculpturing wherever thaw-freeze episodes are frequent and snowfall is abundant.

Frost Heaving and Thrusting

As water in the ground freezes and expands, it pushes surrounding material in directions of less resistance or, perhaps more frequently, toward the freezing front (the advancing edge of the frozen zone). The ground surface heaves up or thrusts sidewise. A common example of heaving is the raising of a hump on a blacktop road. Thrusting is seen where a retaining wall becomes tilted and cracked as water accumulates and freezes behind it. Incidentally, soil subjected to thaw-freeze is loosened and made more subject to erosion.

Frost heaving and thrusting produce remarkable landforms in arctic and subarctic regions, especially on the swampy terranes called tundra. The ground to a depth of many feet remains frozen all year (hence the term "permafrost") except for a little summer melting in the top layer. Various features are produced at the surface as, from season to season, the ground is worked by thaw-freeze cycles.

A common feature in tundra areas is so-called patterned ground: soil and stones arranged in polygons, circles, and other forms. Frost heaving and, apparently, movement of soil upward toward the ground surface by convection (upward movement due to warming) sort loose surface material to make rings, with larger-caliber stones at the edges, and with diameters up to 10 feet or so. Similar rounded forms, without sorting and with a vegetation cover, also occur; so do small polygons, up to 30 feet in diameter. On a slope, rings may become elongated by creep – that is, a slow downslope movement of rock material by gravity – to make "garlands." Sorting may also produce stone "nets," "steps," and "stripes," the last of these extending as much as 300 feet. During a thaw, the ground may subside, leaving the larger stones at the surface probably because the fine-grained materials are more cohesive with one another.

Weathering in Vertical Strata: Bedding planes separating sedimentary rock layers are planes of weakness. Where layers are tilted, water penetrates down between them more easily, decomposing and disintegrating them. Weathering in Virginia's Appalachians has shaped these tough vertical sandstone strata into spectacular "fins."

Frost Heaving: In winter, wet soil may freeze, expand, and rise up around heavy stones, causing them to appear sunk. With a thaw, the soil sinks again. Such working of soil opens it to air and moisture for use by living organisms and also facilitates movement of rock materials downslope.

During freezing, tundra terranes frequently crack into blocks with polygonal forms, up to 300 feet in width. These differ in nature from polygons produced by sorting. Cracking results from stresses produced during freezing. During thaws, water accumulates in the cracks; then, as the temperature drops again, the water refreezes, forming ice wedges in the cracks. Wedges, tapering downward, are often 9 to 10 feet wide, as much as 30 feet deep. As wedges melt, the cracks may fill with sediment.

Watery tundra often features the mounds or hillocks known as pingoes. Pingoes on floors of lake basins are produced as a rising permafrost table traps water in a soil mass, the mass freezes, and the mass then rises hydrostatically – that is, because with its high proportion of ice it is lighter than the water in which it is immersed. Other pingoes form on gentle slopes over artesian springs (springs of water that has traveled underground from a higher elevation and thus emerges at the surface under pressure). Emerging water freezes and forms the core of the pingo. Pingoes usually occur in clusters, are circular or oval, and can reach over 200 feet in height and more than 3,000 feet in diameter.

Melting in the top layer of tundra causes subsidence of the soil, producing basins in which water collects to form thermokarst lakes (known also as thaw lakes). These are common in the northern half of Alaska. Some are temporary; some have become permanent because of climatic warming since the Pleistocene Epoch..

In some large lakes in temperate as well as cold regions, "ice push" is a familiar phenomenon. Ice accumulates thickly and expansion pushes it shoreward, carrying along rock

Felsenmeer: On New Hampshire's Mt. Monroe, as on other summits of the White Mountains, thick felsenmeer – broad expanses of frost-broken rock – is a familiar sight. It was produced mostly toward the close of the Pleistocene, when thaw-freeze must have occurred most days of the year.

Not a Volcanic Cone: A pingo, resembling a little volcano, rises from a flat tundra landscape in Canada's Northwest Territories. In the top is the typical crater, resulting from cracking and melting of ice. Watery surroundings, including thermokarst ponds, characterize pingo country.

Stone Rings: At Thule, Greenland, large numbers of stone rings have formed in the freeze-thaw layer. Some are tens of feet in diameter; some are more nearly polygonal than circular.

Polygonal Ground: In an old U.S. Geological Survey photograph, polygonal ground flanks the winding course of the Meade River, about 35 miles southeast of Barrow, Alaska. Thermokarst lakes appear in the upper part of this picture as black, roughly oval areas.

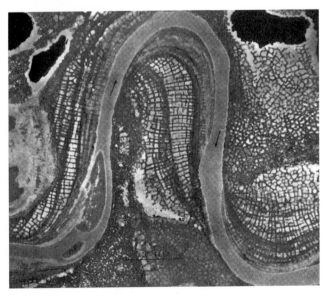

Decomposed by Groundwater: In Palo Duro Canyon, Texas, groundwater moving along bedding planes between sandstone layers dissolves natural cements that hold the sand grains together. Cavities form as cement dissolves out. Sizes and numbers of cavities vary from layer to layer.

Mudcracks in Clay: Shrinking as it dries, mud often cracks into polygonal patterns, such as hexagons and pentagons, which allow closer fit than circular forms would. In contrast, drying sand has more space between grains; thus when wet it swells less than clay, and when drying it cracks little if at all.

"Dog Mesa": Somewhere over the Colorado Plateau an air traveler photographed a mesa with sides in which coves have been hollowed out by groundwater sapping.

Spatter Marks? A sandstone slab in Arizona has depressions that may seem to result from impacts of raindrops. Of course, raindrops cannot pit hard rock. These depressions result from occasional wetting by desert rains, solution of portions of the rock, and removal of the dissolved material by wind and rainwash.

fragments picked up from the lake bottom. These can become heaped up into ridges, called ice ramparts, along edges of the ice.

On terrains that were tundra during the Pleistocene and have since been relatively undisturbed, stone rings and other relics of tundra weathering may survive. In the United States such features are seen on New England's northern mountains as well as mountains and plains in northern parts of the Midwest and West.

Wetting and Drying

Soft, porous rocks, such as some shales and sandstones, swell slightly as they absorb water. Structural strength is lost as rock grains are pushed apart by pore pressure – that is, pressure in groundwater due to the weight of surrounding rock and water. Minerals dissolved in the water may precipitate and, expanding as they crystallize, exert further pressure within the rock. These pressures tend to produce fractures or lines of weakness, and cause outer portions of the rock to fall away.

Rockfalls due to this sapping action leave recesses – often called alcoves or niches – in cliffs, sides of roadcuts, and steep riverbanks, especially on shale and sandstone terranes. Recesses are most likely to develop near the base of the cliff or slope, because more water accumulates behind the surface there and pore pressure is greater than in the rock above.

Recesses are commonly arched at the top, because the arch is the structural form that best supports overlying rock. The kind of recess usually called a niche has a pointed arch, as in Gothic cathedrals and in niches that hold religious figures in churches. Niches are frequently and impressively developed in the sandstones of the Colorado Plateau, notably at Zion Canyon and Mesa Verde.

A "Cathedral" Feature: In Zion Canyon, Utah, groundwater seeping through the sandstone of the canyon wall weakened the rock and allowed portions of it to fall, leaving a niche. This process is speeded by exfoliation due to pressure release. Such features are very common in sandstone cliffs of the Colorado Plateau.

Temperature Changes

Rock at or near the ground surface undergoes heating and cooling as the ambient temperature varies. Heating causes rock to expand; cooling causes contraction. Alternating expansion and contraction result in strains and stresses which cause a rock surface to spall (flake off) or undergo granular disintegration. This is most likely if moisture is present, weakening the surface by chemical action. Much spalling is caused by forest fires.

The greater the temperature change and the more rapid it is, the greater can be its destructive effects on rock. These effects naturally tend to be strongest on barren terrains, such as polar lands and deserts, where there is little vegetation to moderate temperature changes and where rock may bake in the sun all day and chill quickly after sundown.

Desert travelers have reported "popping" of rocks during the cool of evening after a hot day. The noise is said to result from spalling as rapidly cooling rock surfaces contract over a still-hot interior. Such reports have been challenged, because rock conducts heat slowly and thus a high temperature gradient between surface and interior would be unlikely. However, if the surface is cooled suddenly by rain or by condensation (which does occur in shaded places in deserts), and if these cooling effects are added to the usually rapid evening cooling, spalling could indeed occur. A like event may be seen when water is thrown on hot bricks in a fireplace.

Pressure Release and Sheeting

Bodies of rock that form thousands of feet, perhaps miles, below the surface have high pressure "built in." As erosion gradually removes overlying material, confining pressure on the rock bodies is reduced, they expand, and plates or slabs a few inches to a few feet thick break off along the surfaces. Rock fracturing due to pressure release, or "unloading," is known as sheeting and is common in granite and some other massive rocks formed at depth. Breaking off of sheets and slabs is called exfoliation. It differs from occasional dangerous "explosions" of rock in deep quarries, which occur when rock bulges out because of the weight of overlying rock.

Plates, or sheets, that break off sheeted rock masses are much thicker than flakes produced by chemical action. Usually, sheets grow thinner toward the ground surface. Some follow the original rock "fabric," or structure; some do not. Sheets near the surface of a slope become loosened by weathering and fall or slide down.

Some sheets parallel the ground surface, and thus may curve over highland summits. Thus are produced domes such as Georgia's Stone Mountain, Virginia's Lookingglass Rock, and world-famous Sugarloaf, which looms over Rio de Janeiro, Brazil. When rounded, a landform presents minimum surface to weathering and erosion. Rounding is favored by homogeneity of the rock and grain-by-grain weathering.

An Exfoliation Dome: North Dome, in California's Yosemite Valley, is a classic example of a highland shaped mainly by exfoliation due to pressure release. Because the rock is homogeneous, weathering creates a spheroidal form. Half Dome, also in Yosemite, is the remaining half of a dome shaved by a glacier moving through the valley.

Exfoliation vs. Granular Disintegration: In exfoliation due to pressure release, platy or shell-like masses usually a few inches to a few feet thick break off bedrock. Grains and small flakes also break off, as a result of other kinds of weathering. A mix of plates, flakes, and grains appears on this surface of granite gneiss in New Jersey's Ramapo Mountains.

Sheeting can be important in the recession, or retreat, of cliffs on massive rock, as on the Colorado Plateau or the High Plains. Towerlike forms such as those of sandstone in Monument Valley, Arizona, owe their verticality mainly to vertical sheeting, but also somewhat to sapping by ground-water at the base of the monoliths.

Organic Agents

Animals and plants may not appear to be weathering agents, affecting rock and landforms, but they are.

Boulders "split" by trees are a common sight in rocky, forested areas. Actually trees cannot split massive rock; rather, roots penetrate existing crevices, expand as they grow, and force rock masses apart. Root growth also works soil and thus may cause rock fragments to move downslope.

Animals such as prairie dogs, woodchucks, moles, earthworms, and ants, constructing their underground dwellings, dig holes and tunnels in soil, opening it to air and water, and loosening it for easier removal by gravity movements and erosion. Elephants in the wild dig pits in their quest for salt, and wild hogs dig for tubers. Elephants, buffalo, and other large animals dig out waterholes, and some, rolling on wet or dry ground, make wallows for mud or dust baths. Various

Constructive Action: In tropical lands, earth towers and mounds are built up by termites. In Kenya, termite mounds like this are often dug into and hollowed out by mongooses and aardvarks seeking a favorite food. Later the mound may be taken over by a hyena mother for raising a family.

Small Excavators: Prairie dogs are not powerful but are many. With time they loosen much soil and make it more susceptible to erosion. Animals that dig, from ants and worms to aardvarks and elephants, are significant in land-shaping.

Two Climates on One Hill: The climate of any small area, such as a hillside, a swamp, or a meadow, or any portion of these, is a microclimate. It may differ radically from surrounding microclimates. On this hill, for example, the south side, exposed to hot sun, is dry and bare; the north side, exposed to little or no sun, is relatively humid and covered with vegetation. Such differences strongly affect kinds and rates of weathering.

Root Prying: Tree roots cannot penetrate massive rock, but some species are amazing in their ability to penetrate crevices and, given time, push heavy rock masses apart.

animals build mounds which become notable scenery, such as termite hills in East Africa. Certain organisms, notably lichens, live on rock, obtaining sustenance from it and gradually decomposing it. Grazing animals help to determine the nature and amount of vegetation cover.

Presumably most of the millions of kinds of organisms that exist come near or into contact with rock at least occasionally, depositing chemical materials on or near it. After death, organisms decay, producing more such materials that come into contact with rock. The combined effects of the activities of all organisms (even ignoring human works) are vast. Carbon dioxide produced by decay of flora and fauna (not to mention the output of factories, automobiles, and jet planes) is in itself a prime factor in weathering and the shaping of scenery over the ages.

Weathering vs. Time

An expert has commented that the creation of an inch of soil by weathering can take 10 years – or 10 million years. Certainly the variability of rates of weathering is very wide. Rock type and rock structure, climate and microclimate, chemical composition of the local air, amount of vegetation present, activities of organisms, and rate of removal of weathered material are involved, as well as gravity movements, erosion, and deposition. Modern air, with its loadings of carbon dioxide, sulfur dioxide, and other highly reactive gases, has accelerated the pace of weathering almost everywhere.

Botanical Weathering: The lichen here is finding nutrients on granite gneiss and incidentally is depositing chemicals that tend to decompose the rock.

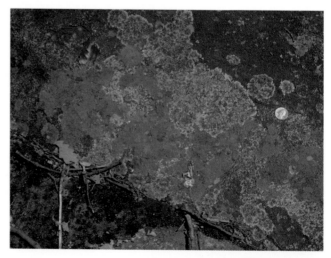

Where landforms rather than just soil are involved, the roles of weathering and erosion are so intimately associated with each other that they can hardly be considered independently. Thus, deciding just how long it would take for weathering to produce a certain variety of landform is all but impossible.

A few examples of weathering rates will show how they can vary: In Egypt, surfaces on granite in temples at Giza have weathered to a depth of about 0.2 inch – a very noticeable change – in 5,400 years, while granite at Aswan, in a less industrialized area, still looks fresh after 4,000 years. Glacier-scoured, relatively fresh granite at Narvik, Norway, has lost about 0.2 inch in 10,000 years. Limy shale at Giza has been reduced to rubble in 1,000 years. Coastal limestones in salty air in Puerto Rico and Australia are weathering away to a depth of 40 inches per 1,000 years. Lettering cut into typical old sandstone and slate gravestones in New York City has been obliterated in a century or less. Marble in temples at Paestum, Italy, and similar marbles in other Mediterranean countries are deeply pitted after 2,000 years – but most of this weathering has occurred during the industrial era. The cherished stone walls of some readers of these lines may last 10 years, or 100, "depending."

A few general principles about weathering rates under varying conditions can be risked: Chemical weathering in most localities is dominant over physical weathering – that is, over ice and root wedging, sheeting due to pressure release, disintegration due to temperature changes, and so forth. Because water is relatively reactive, a humid climate favors chemical weathering; so do warm temperatures, to the

Granite "Remembers": A century and a half after these letters – perhaps the initials of lovers – were carved into this mountain granite, they are still sharp and clear, despite rain, frost, and sun.

extent that a rise of only 18 degrees F. can double a region's weathering rates. Weathered material lying on a rock mass tends to protect it; therefore the faster this material is removed by gravity movements and erosion, the faster the underlying rock will weather, other factors being equal. Generally, weathering is most rapid on humid, warm terranes and slowest on cold lands, mountains (except where frost action is intense), and deserts.

After Nearly 2,000 Years: Limestone was used by the Romans when they constructed their theater in what is now Merida, Spain. In this photograph the original limestone tiers of seats are in the background; restored tiers are in front. Weathering has made the difference.

Scenery of Gravity Movements: In alpine mountains where thaw-freeze is frequent, ice-prying creates huge amounts of rock debris. Much of this accumulates in ravines, moves downward, and forms talus cones as it comes to rest on valley floors. The rate of debris production is high relative to the rate of debris removal by running water, because water is frozen much of the time; hence cones can become very large. This scene is at Bow Lake, Alberta, in the Canadian Rockies.

5

Scenery of Gravity Movements

ALL objects on Earth – grains of sand, fields of boulders, glaciers, mountain masses, the entire crust – are subject to the constant pull of gravity. When support is removed, or when gravity's pull exceeds resistance to that pull, the object moves downward – falling, sinking, sliding, rolling, flowing, tumbling, slumping, spreading, creeping, or toppling. Such movements differ from locality to locality, according to prevailing conditions, and are usually interrelated and more or less overlapping. Collectively they are known as gravity movements or mass movements, and the overall process of land degradation by such movements is often called mass wasting. Gravity movements rank with weathering and erosion as powerful land sculptors.

A Universal Maker of Landforms

Gravity movements result essentially from elevation of portions of Earth's crust. If the crust had been stable since its formation more than 4 billion years ago, by now it would have been just about leveled by erosion, and gravity movements, if any, would be minimal. On the contrary, continuous activity in Earth's deep interior has kept crustal plates and portions of plates moving sidewise, up, and down, over and over, maintaining the basic condition for gravity movements – that is, differences in elevation. Associated with plate movements there have been igneous activity, erosion and deposition, and isostatic balancing, all of which can produce changes in elevation.

Not surprisingly, gravity movements tend to be swifter, more voluminous, and more widespread along plate borders and along major fractures within plates. These are zones in which sections of crust held under tension or compression suddenly break free and move along faults, producing earthquakes. In these events land elevations change, crustal rock is shattered, and loose material on slopes gravitates swiftly downward as rockfalls, landslides, debris flows, avalanches,

From a Mix of Gravity Movements: In Cascade Canyon, in Wyoming's Teton Range, rock fragments loosened by weathering fall, topple, and slide down a ravine cut by streamwork and avalanches, forming a talus cone below. Note large-caliber fragments near the bottom – typical of talus cones.

and other such movements. Lands ringing the Pacific Ocean, regions of the Mediterranean and Mideast, and parts of southern Asia, notably India and Tibet – these are along plate borders and most subject to the extremes of earthquakes and gravity movements. Daily the media report new earthquakes, landslides, and avalanches in and near the Andes, Alps, Himalayas, and other tectonically active mountain ranges. In such regions gravity can be so active as to outperform running water as the major sculptor of landforms.

Gravity Movements: Noticed or Not

Probably every landscape shows some evidence of gravity movements. These may be as obvious as rock debris at the foot of cliffs, scars left on cliffs by landsliding, heaps of rock at the foot of gullies, mud spreading from mouths of canyons, masses of sediments miles wide creeping down gentle slopes toward valley floors, and depressions in land due to solution and removal of underlying soluble rock by groundwater. Many features produced by gravity movements, such as minor slumps and ripples in soil, are hardly

recognizable when formed or, through time, they are rendered unrecognizable by erosion or by vegetation cover. Large movements, such as rockslides or mudflows involving thousands of tons of material, leave evidence impressive in volume and form, may be recognized easily, and are not soon eroded away. Less obvious is scientific evidence that even in recent time blocks of crust the size of mountains have broken loose from larger masses and plunged down onto valley floors, as in Switzerland, or to the sea bottom, as in the Hawaiian sector. Other mountain masses, during millions of years, have slowly slid miles from their places of origin.

A standard textbook lists no less than 35 conditions that facilitate gravity movements. Most important is plenty of easily moved regolith, especially clayey materials, which tend to be slippery in part because of the platy shape of the grains. Also important are steepness of slope, inclination of rock strata parallel or nearly parallel to the slope, frequent or torrential rains, intense weathering (especially ice prying and chemical action), sapping by groundwater, undermining by erosional agents (running water, waves, glaciers, wind), earthquakes, and destabilizing of rock by human or animal activities and by plant growth. Such conditions translate into high rates of gravity movements on various terrain types, such as cold mountains where frost action breaks up much rock and slopes are steep; or volcanic mountains covered thickly with volcanic ash and dust, ready to be turned by rain or melting into mud; or arid lands, where rain may be torrential (though infrequent) and vegetation is insufficient to anchor soil; or hilly tropical regions where rain is abundant, chemical action intense, and rock decomposition rapid.

Gravity movements are intimately involved with erosion and weathering. Streams and weathering agents loosen and break up rock, remove support, and thus prepare material for descent. Interrelationships among the agencies in a given situation may be so subtle that their respective roles are difficult to distinguish from each other. For example, detachment of a fragment from bedrock may involve simultaneous weathering and gravity action. Descent of the fragment may then occur directly by gravity. During descent, weathering continues. Meanwhile erosion by running water, waves and currents, advancing glacier ice, or wind action may further wear the fragment and help to move it.

Restraining Gravity Movements: All over the world, farmers know that soil on steep slopes can be lost and crops ruined by rushes of stormwater and by the constant pull of gravity. Here, near the Nepal-Tibet border, farmers terrace the slopes, slowing both runoff and soil creep.

Obviously, gravity movements are involved in events of environmental importance beyond the creation of landforms. Landslides from cliffs undermined by marine or stream action or blasting; mudflows that bury villages; sinking of land due to removal of groundwater or oil by wells; debris slides on deforested slopes; snow avalanches on mountainsides; floods caused when slides dam rivers; and tilted or broken retaining walls along highways – these are examples. Clearly, an understanding of gravity movements has many applications in environmental management.

As a footnote to all the above, we can note that in discussions of gravity movements often the same term is used to describe both the movement and the landform it creates; for example, "landslide" can mean either the sliding of earth material down a slope or the pile of material thus formed at the bottom. In these pages the question of whether the action or the landform is being referred to can generally be understood from the context.

Landforms from Vertical and Nearly Vertical Movements

Possibly the most familiar of gravity movements are falls, seen commonly as rock fragments falling from a cliff and as the accumulation of such fragments below.

Rockfalls, Earthfalls, and Debris Falls

The fall of a rock fragment or fragments is called, simply, a rockfall. The fall of mixed rock fragments and soil is an earthfall. The term debris fall is applied to vertical movements involving rock fragments, soil, vegetation, and just about anything else that's loose. The term may be used also for any fall of materials larger in caliber than sand.

A fall of rock involving rotational motion is called a topple. A typical topple (to coin a phrase) occurs when a rock mass poised at the edge of a cliff is pried by frost action and somersaults down. Another example might be the fall of a

Control of Gravity Movements by Steepness of Slope: Along Echo Cliffs, in northern Arizona, rock fragments fall where the cliff is vertical, then slide and tumble where the slope moderates. Large boulders, with greater momentum, tumble far out onto nearly level ground, then move by creep.

Ready to Topple: At Chiricahua National Monument, Arizona, erosion of layered rhyolite bedrock masses with close vertical jointing has shaped a forest of pinnacles, many with "heads" and "hats" ready to topple. Irregularities in profiles result from variations in resistance of the layers.

Rockfalls: New York's Hudson Palisades are of closely jointed diabase, which offers many crevices for frost action. With its iron content, diabase is subject to relatively rapid chemical weathering. With high rates of both mechanical and chemical weathering, rockfalls are frequent along the cliffs.

balanced rock from the top of a pillar. Many rockfalls start as topples.

Falls of rock and other loose material are common where cliffs are being undermined by stream or wave action, movement of a glacier, and human activities such as quarrying and excavating. Sea cliffs, notably those along the California coast, are notorious for rockfalls as well as other gravity movements. These may result from undermining by wave action, sapping by groundwater, and flaking off of cliff rock as salt spray penetrates it and salt crystals grow within it. Where rock strata or joints incline down toward the cliff face, loose slabs may slide forward and fall.

Talus Sheets, Aprons, and Cones

Rock fragments that have descended from a cliff or other steep slope and have accumulated at the bottom are called talus. Talus is likely to consist mainly of sharp-edged rock fragments, little worn by weathering or erosion. Mixed with them may be fragments transported by running water. Some amount of talus is likely to be seen at the foot of cliffs or other steep slopes about anywhere, but the greatest quantities occur in mountains where frost action has been intense. Talus accumulations in mountains such as the Canadian Rockies and Alps are spectacular.

Talus commonly has the form of a broad apron extending out from the lower part of the slope; hence "talus apron." Talus spreading out from the foot of a gully commonly has the form of one half of a cone sliced vertically, top to bottom; hence "talus cone."

On talus slopes the caliber (overall size) of the rock fragments usually increases toward the edges. Large fragments have more momentum from the fall than small ones have; thus large ones can slide or tumble farther from the cliff

Talus Aprons: Almost any broad spread of talus might be called an apron. In this scene, along the Lost River south of Challis, Idaho, the term "talus aprons" seems well deserved. Accumulations are thick because weathering and gravity movements produce talus at a high rate while running water, in this dry region, removes it at a low rate.

base. Also, large fragments move over the other material with less friction. On the contrary, in alluvial fans and cones, which consist of sediments deposited by running water, the caliber of fragments tends to decrease toward the edges, because the ability of running water to transport sediments diminishes as steepness of slope diminishes.

An extensive mass of rock falling or sliding down from a cliff usually leaves a scar. This is a rock surface showing less weathering than adjacent surfaces and differing somewhat in color; it may be in a recess. Distinct scars are likely to be relatively new, indicating recent falls.

Subsided Ground

Sinking of earth due to removal or weakening of support is called subsidence. One kind is settlement, resulting from more or less gradual compacting of underlying material; for example, when wet soil at the surface dries and shrinks, creating a depression, or when frozen ground melts. The settlement that has caused the famous belltower in Pisa, Italy, to lean about 5.5 degrees or more from the vertical began during construction, in the 12th century; it has been due to failure of a clay layer beneath it. (In 1990, because of the increasing lean, the tower was finally closed to the public.) In Mexico City, subsidence due to removal of groundwater by wells has caused collapses of streets and buildings. For a like reason, during the period 1925-1977 part of the floor of California's San Joaquin Valley sank 30 feet. Again because of groundwater removal, the basin floor beneath Tucson, Arizona, was reported in 1991 to have been sinking ½ to 2 inches annually. The islands of Venice, Italy, have been sinking for a similar reason.

Collapse, a more or less sudden movement, is another kind of subsidence. An example is the falling of the roof of a cave or tunnel made in soluble or structurally weak rock by subsurface streams. Collapse causes sinking of the surface above the cavity. Remarkable assemblages of collapse sinks pock the Karst Plateau of Bosnia, the "cockpit" country of Puerto Rico, and the vicinity of Mammoth Cave, Kentucky. Sinks form also in lava flows when roofs of lava caves and tubes collapse. In mining areas, sinks result from failures of shaft and tunnel roofs.

Subsidence Effect: The Leaning Tower in Pisa, Italy, was built on weak clay. Gradual failure of the clay causes the tower to lean increasingly. Engineers have been unable to to stop the process.

Mountains That Glided: The Mythen Peaks, overlooking Lake Luzerne, at the northern edge of the Swiss Alps, are fragments of sedimentary rock formations that originated far to the southeast. Separated from the main mass by erosion, these fragments glided gravitationally to their present location.

Features from Lateral Mass Movements

Spreading and Gliding

When a mass of loose material is placed on a level or nearly level surface, the mass because of its own weight may spread outward from the center, which usually is the thickest part of the mass. An example is the behavior of a pail of mud poured onto a pavement: the individual particles, moving more or less independently, behave as a flow. Spreading of this kind occurs where mudflows from mountain valleys emerge onto a plain, and where relatively fluid lava pours from a volcanic fissure.

A different kind of lateral movement happens when the material consists of separate blocks that move on a definite slip surface. Blocks in a pile do not spread as a flow; they glide over a surface lubricated by basal material in a liquid or plastic state. As in spreading, the force causing movement can be the weight of overlying material.

Talus blocks accumulated on an icy or wet surface may glide. On a much larger scene, gliding has happened where huge mountain masses have become detached from larger units and, because of the overlying weight, have moved miles or tens of miles. Gliding of very large bodies is likely to be extremely slow, occurring in tiny increments during thousands or millions of years, but there appear to be exceptions: in 1972, in Daghestan, in the Caucasus, a mountain whose base became saturated by heavy rains moved 12 miles in 8 days. Large blocks that glided have been identified in many highlands, including the Alps and the Rockies – for example, in the Antler Mountains of central Nevada. West of New York's Hudson Highlands are several hills, among them Sugarloaf, consisting of detached Highland blocks of granite that glided miles westward over younger sedimentary rocks. Gliding movements are inferred because there is no sign that the blocks were moved by thrust faulting. Some geologists say gliding may involve a degree of flow, resulting from deformation of the mountain mass by its own weight.

Movements on Slopes

Most downward movements of earth materials occur on slopes between the vertical and the horizontal. Generally, the steeper the slope, the more rapid is the movement, though

Slope Movements: Along a small watercourse in Belize, a clayey bank has become saturated, lost structural strength, and collapsed. A low scarp has been left by slumping. Slumped material (*foreground*) shows signs of further slumping, flowing, and spreading.

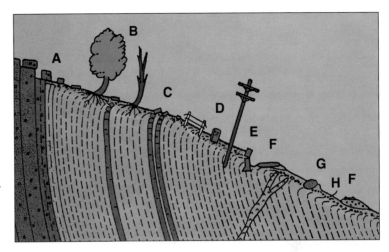

Evidences of Creep: **A**, moved joint blocks. **B**, trees with curved trunks concave upslope. **C**, downslope bending of rock strata; **D**, displaced posts, poles, and monuments. **E**, broken or displaced retaining walls and foundations. **F**, roads and railroads moved out of alignment. **G**, turf built up against downside of creeping boulders. **H**, stone line at approximate base of creeping soil. **A** and **C** represent rock creep; all other features shown are due to soil creep.

other factors may cause exceptions, as when plant roots or roughness of a slope impedes sliding or other kinds of gravity movements.

A basic condition of most slope movements is the angle at which loose material moving down will come to rest on similar material; this is called the angle of repose. Generally, the angle of repose for loose materials is 33 to 37 degrees on natural slopes. After a period of adjustment the inclination will be somewhat lower; then it is called the angle of rest. The angle of repose for dry dune sand is 33 degrees (somewhat less if wet); the angle of rest is about 31 degrees. The exact angle in any given situation depends on numerous factors, such as internal friction, cohesiveness of the loose material, caliber and angularity of the particles, degree to which particles are sorted as to size, and moisture content.

For sliding solids, the critical inclination is the angle of slide, usually about 40 degrees – higher if the surface is relatively rough.

The three major categories of slope movements are creeping, sliding, and flowing.

Features Resulting from Creep

Imperceptibly slow movement of a rock fragment down a sloping rock surface or in talus, as a result of temperature changes, freeze-thaw, chemical decomposition, or other destabilizing events, is called rock creep. The movement may include a little sliding but mainly involves slight lifting of the rock in a direction perpendicular to the underlying surface, followed by lowering of the rock vertically, so that

there is a net downward gain. Fragments of different sizes creep at about the same rate, except that very large ones may creep faster because they have so few points of contact with others and friction is therefore minimal.

Rock creep usually occurs in tiny increments, but movements may occur so frequently that considerable distance is covered in a short time as measured on the geologic scale.

Rock Strata That Were Bent: Where a country road in Vermont's Green Mountains cuts through marble strata, creep has caused the strata to bend into the roadcut (*at right*). Edges of strata appear in this view.

Animal Paths? Small soil terraces on hillsides often are thought to be paths made by animals such as sheep and cattle. (One wit has called them "cowtours.") In some instances these terracettes, or lynchettes, may actually be small slumps. These specimens are in a pasture on glacial drift in Vermont.

Often, boulders poised on rock surfaces above roadcuts, well away from the edge, look stable but within a few weeks or months creep to the edge of the cut and tumble down to the roadway.

Soil creep is a very slow movement in which individual particles of rock or soil move together as plastic or as viscous fluid, with mixing and deformation within the mass. The main destabilizing factor is likely to be wetting and drying, freezing and thawing, water pressure in the soil pores, or lubrication by moisture. Soil creep decreases with depth, be-

cause resistance to creep increases as the weight of overlying soil increases. Soil creeping downslope may have embedded within it large rocks with tops that project above the surface like ships on a slanted ocean. The entire assemblage may be moving down, an inch per year – or per century.

Tilted telephone poles and fences on slopes betray soil creep; so do trees with curved trunks, the curve being a sort of compromise between the tendency of the tree to grow vertically and the tendency of creep to tilt it downslope. Note: Some trees, of course, curve not because of creep but because they are reaching out for sunlight.

Another example of soil creep is seen where rock strata intersecting a slope have been bent downward. More familiar, and somewhat similar, is the tilting of a retaining wall by creep. Creep pressure is maximum at the top, minimum at the bottom. As the wall ages, wetting and drying compact the soil behind the wall, more soil is added on top, pressure on the wall increases, and wetting reduces the soil's structural strength and its ability to stand without collapsing and spreading. Freezing of wet soil behind a wall further increases pressure. Some retaining walls have "weep holes," to allow trapped water to escape.

Creep on a Slope of Nearly Bare Rock: The slope here is on sedimentary strata tilted to about the angle of repose. Fragments of weathered rock move down intermittently when disturbed by events such as frost-prying, rainwash, and temperature changes.

Creep is important in the shaping of many hillslope profiles. On relatively gentle slopes mantled with much loose material, creep is likely to play a larger role than running water in moving material downward. Running water needs substantial gradient in order to transport much load.

Slide Features

Slope movements which are relatively rapid, and in which a fairly coherent mass descends over a smooth surface, such as a bedding plane or a joint plane, are known as slides. These tend to occur on wet surfaces, not so much because of lubrication by water as because of the pressure of water that has penetrated between the sliding mass and the surface beneath it. This pressure, due to overlying water, tends to buoy up the sliding material. Slides, and flows of earth material as well, tend to be commoner on slopes that face away from the sun, because these are likely to be damp.

Rockslides

Movements consisting mainly of rock fragments of cobble size or larger are termed rockslides. They can be started by destabilizing events such as earthquakes, undercutting by a stream or a glacier, thawing of ice that has held rock masses

A Small Rockslide: In parts of Colorado's San Juan Mountains, accumulations of clayey volcanic materials become slippery when wet and descend as rockslides and mudflows. Here a mass of boulders and clay has slid down the side of a roadcut.

Creep vs. Tree Growth: On this slope on loose earth at a 30-degree angle, trees grow curved because soil creep tends to tilt them as they try, by nature, to grow vertically. Curvature can result also from a tree's tendency to grow toward light.

together, wetting of the surface beneath loose rock, the fall of a tree. Slides often are touched off by blasting. Among mountains slides involving millions of tons of material can exceed 100 miles per hour, sometimes crossing a broad valley and thrusting up the other side.

A rockslide slows as the slope moderates, and stops when resistance to sliding becomes equal to the gravitational pull. Rock at the front stops first, and material still coming behind it may push it up to form a heap like an upturned toe.

Where the slope steepens beneath a big, fast-moving slide in a narrow valley, the downrushing mass may leave the ground like a skier at a jump. Some observers say the mass compresses air beneath it, making a cushion on which it rides down. But similar slides have been identified on the Moon, where air is lacking. The rise of material above the surface could be due to turbulence. In any case, some materials do flow as a high-velocity avalanche, partly above the surface. Where the valley widens or ends at a lowland, the material spreads, slows, and crashes down.

One such slide occurred in 1903 near Frank, Alberta, on Turtle Mountain's east slope. Slabs of limestone there are perpendicular to the slope but there are joints parallel to it,

Scars Left by Debris Slides: Steep slopes on bedrock with widely spaced joints allow few rootholds. Vegetation there is vulnerable to soil creep and sliding, which leave slopes bare. So it is on Giant Mountain, in New York's Adirondacks.

creating a serious hazard. One day when miners were blasting a coal vein in the mountain, a half-mile-wide block of limestone on the slope broke loose and started sliding. Gaining speed and breaking up, the mass flowed down as an avalanche, razing most of the town and killing 70 people.

The Frank disaster demonstrated dramatically the hazard for people, roads, and buildings near the base of steep inclines with rock strata or joints paralleling the slope. If the rock is clayey, wetting causes it to lose cohesiveness and become slippery. The hazard is less if the surface is rough, because roughness impedes sliding.

Rockslides are common on volcanoes of the explosive type, where slopes often are steep and covered with loose materials such as volcanic ash and other pyroclastics, which become slippery when wet. One notable example is Mt. St. Augustine, which rises at Cook Inlet, Alaska. Studies of debris layers around this volcano indicate the occurrence of at least 111 major slides during the past 1,800 to 2,000 years. The most recent slide, triggered by an eruption in 1883, carried the north side of the mountain – perhaps 10 per cent of its bulk – into the sea.

Some rockslides begin as the slipping of a huge block or plate of bedrock which soon breaks up to form a rockslide. On the islands of Hawaii a number of such slides occurred in prehistoric time, including at least one from the Koolau Range on Oahu; the evidence is in the aprons of slide debris on the sea bottom around the islands. A famous bedrock slide in historic time occurred in Switzerland in 1806 when, according to an observer, Rossberg Peak "split," and a huge mass of the mountain fell away, broke up, and swept down into the Goldau Valley, burying four villages and taking hundreds of lives. Slides of such magnitude undoubtedly have occurred widely in the past and account for many of the mountain profiles we view today.

Rockslides commonly leave hollows or scars on the surfaces where they originate. A scar usually consists of a shallow depression in an expanse of relatively fresh, little-weathered rock.

A Debris Slide with a Toe: On this slope facing a lake in the Canadian Rockies a minor slide of soil and rock fragments up to a foot in diameter ends in a toe. The light band across the toe is a hikers' trail.

A Roadside Slump: Slumps often occur on slopes with wet, unconsolidated soils and little covering vegetation. Here, clay at a country road's edge was undermined by a rainwater stream and collapsed, with the backward rotation typical of slumps.

Debris Slides

Rockslides that are wide and involve debris, such as trees, wrecked buildings and automobiles, and other miscellany, are called debris slides. They occur often on slopes deforested by clear-cutting or by fires. In New York's Adirondacks, fires over the years have destroyed trees and other vegetation, reducing soil stability. Debris slides start easily here because the bedrock joints are widely spaced and offer few holds for roots. The bare rock expanses are themselves called "slides."

Debris movements that follow narrow tracks, as in mountain valleys, are better called debris avalanches.

Mudslides

A mass of wet, small-caliber, clayey material on a slope can start sliding because of further wetting, an earthquake, or some other destabilizing event. A mass containing much water can move at a velocity up to 55 miles per hour. Mudslides are most common in humid regions and on intermediate slopes, from 27 to 45 degrees, such slopes being gentle enough for much sediment to accumulate and steep enough for sliding. Often a mass starting to move as a slow soil slip or slump loses structural strength and becomes a mudslide. In turn, the mudslide may pick up enough rocks and vegetation to become a debris slide. Thus many kinds of gravity movements change into other kinds as they develop.

Mudslides are common on muddy hillsides in volcanic regions, such as Colorado's San Juan Mountains, and in coastal areas of southern California, where soils are clayey and earthquakes – even those of small magnitude – frequently destabilize slopes. In May 1998, in a typical slide, a mass of mud on the deforested side of Italy's Mt. Sarno, near Naples, rushed down and smothered five villages, killing at least 121 people and leaving 1,000 homeless.

Soil Slips and Slumps

In soil slips, descending material consists largely of soil rather than rocks of cobble size. The movement is essentially parallel to the slope. Soil slips occur often on sea cliffs of weak rock that are being undermined by wave action, and along riverbanks of sand and clay undermined by stream erosion. Often the soil slip carries much debris.

In a slump, a mass of weak material slides down over a curved surface with more or less backward rotation. This phenomenon is demonstrated when a weary person slumps down in a soft sofa, with back curved and knees up, or when cement that is too wet is pushed into a large crevice and collapses. Usually there is a rounded hollow above a slump. A block of rock on a slippery surface can perform the same maneuver, becoming what is called a slump block.

Slumping is seen often in weak bedrock and loose material along cliffs and sides of roadcuts. It is common also in walls of volcanic craters, resulting often from removal of underlying, supporting magma during eruptions.

Soil Slips in a Canyon Wall: In the very steep walls of Hanapepe Canyon, on Kauai Island, Hawaii, soil slips in weathered, weak volcanic materials are frequent.

Flow Features

Gravity movements in which particles move more or less independently, as in a stream of water, are flows. Most are made possible by wetting of the material, which may consist of rocks, snow and ice, mud, or various kinds of debris. Velocity of flow can range from 200-miles-per-hour avalanches to movements on frozen ground that take weeks or months to cover a few hundred feet.

Down and Up: In this slump at Bottomless Lakes State Park, New Mexico, the slump block and the curved surface on which it has slipped are noticeable. The native rock, gypsum, is riddled with solution cavities, and thus is structurally weak and susceptible to gravity movements.

In recent years much has been heard about liquefaction in soils, especially granular ones. In granular soils there is considerable space between grains and cohesion is weak; further, these soils can contain considerable water. When the soil is saturated and subjected to a sudden shock, as during an earthquake, blasting activity, or a landslide, pore pressure abruptly increases and the soil mass loses all structural strength, becoming essentially a liquid and flowing as such.

The flowing of water-saturated regolith over frozen ground is known as solifluction or, better, gelifluction. The term solifluction is now sometimes used for a flow of saturated earth over any kind of terrain.

Snow Avalanches

Snow avalanches are movements that start as slips or slides and become swift flows. They are common in mountains where snow accumulates deeply on inclines of about 30 degrees or more. Loose, fluffy snow with little cohesion may start downslope at the slightest disturbance, sweeping up more snow as it goes, producing a large cloud, and on reaching the base of the slope may cross a valley and move partly up the opposite side.

Loose-snow avalanches tend to be relatively small and harmless, but snow consolidated into a slab can be dangerous. At an inclination between 30 and 45 degrees a slab can break loose along a fracture line and rush down as a single mass preceded by a strong windblast, causing deaths and property damage before coming to rest. All too familiar are reports from the Alps and other high, snow-covered terrains that skiers and climbers have lost their lives in an avalanche.

A common condition leading to snow avalanching is weak bonding between the underlying snow layer and the layer over it. Weakness can result from causes such as melting or the presence of fragile crystal forms at the interface of the layers. Possible triggers for the start of snow avalanches are

Slumping and Sliding: In the west wall of Kilauea Iki crater, Hawaii, gravity movements of poorly consolidated volcanic materials have been caused by heavy rains and by earthquakes associated with volcanic activity. High on the slope at left is a massive slump, in which backward rotary motion is noticeable. At center is a debris slide, started perhaps as a soil slip.

Avalanche Chutes: On a mountainside above Lake Louise, in the Canadian Rockies, avalanching of snow, ice, and rock has widened and straightened a row of gullies, and has built a large talus cone (*lower left*) at the lake's edge.

numerous, including additions to snow already in place, chunks of snow falling from snow cornices or from trees, wind gusts, and disturbance of the snow by animal or human activities.

Snow avalanching occurs often in mountain ravines. Rock fragments carried by downrushing snow erode the sides and bottoms of the ravines, enlarging and straightening them to form the deep, parallel ruts called avalanche chutes.

Rock and Debris Avalanches

Very large and rapid movements of rock fragments following a narrow track and behaving more or less as a flow are known as rock avalanches. These events, occurring in steep mountain valleys, usually start as some other kind of gravity movement, such as a small rockslide or mudflow, then grow in size and velocity into massive movements. Usually a rock avalanche picks up enough debris en route so that the name debris avalanche applies. As it levels off and slows at the foot of the valley, it may become a debris flow.

Lahars

Probably the swiftest and most destructive kind of gravity movement is the lahar. This is a mix of rocks, mud, and melt-

ing snow and ice rushing down a valley in the side of a large, snow-capped volcano of the explosive type. It commonly starts as the result of a volcanic explosion which shatters the side of the cone, melts much snow and ice, and destabilizes accumulations of ash and other volcanic materials on the side of the cone. On the steep slopes characteristic of explosive volcanoes, the velocity of a lahar may exceed 200 miles per hour, and the blast of air created along the sides may

Classifying Landslides: The term landslide covers many different kinds of movements. A chart is helpful toward distinguishing them, understanding their relationship to one another, and indicating the kinds of material in or on which they occur.

TYPE OF MATERIAL	TYPE OF MOVEMENT (Increasing Speed) ---------►			
	SLIDE		FLOW	FALL
	ROTATIONAL	PLANAR		
BEDROCK	ROCK SLUMP	ROCK SLIDE / BLOCK SLIDE	ROCK AVALANCHE	ROCK FALL
REGOLITH	EARTH SLUMP	DEBRIS SLIDE — DEBRIS AVALANCHE	DEBRIS FLOW	SOIL FALL
SEDIMENTS	SEDIMENT SLUMP	SLAB SLIDE — EARTH FLOW (Increasing Sediment Size)	LIQUEFACTION FLOW / LOESS FLOW / SAND FLOW	SEDIMENT FALL

A Mt. St. Helens Lahar: After its catastrophic eruption May 18, 1980, Mt. St. Helens continued its activity intermittently. During the eruption of March 1982, masses of mud, ice, and melting snow were dislodged below the gap in the crater wall and poured down the north flank of the mountain, reaching the valley of the North Fork of the Toutle River.

One of the Finest: Long ago a 6-mile-long earthflow in the San Juan Mountains, in western Colorado, dammed the Lake Fork of the Gunnison River, creating Lake Cristobal. Radiometric dating of wood in the flow indicates that the flow occurred in the 13th century.

After 300,000 Years: About 25 miles north of Mt. Shasta, California, the main highway crosses a broad, hummocky plain, the nature of which long puzzled geologists. Recently the plain, covering 250 square miles, was identified as a debris flow from the volcano about 300,000 years ago. Humps mark locations of blocks and slabs of volcanic rock among materials of smaller caliber.

sweep away trees like blades of grass. A lahar may start as a column of debris and gases rises thousands of feet above the volcano, then collapses to form a white-hot torrent rushing down the mountainside.

Lahars often make grim news. In Peru, in 1970, an earthquake near Chimbote shook loose a half-mile-wide slab on the side of 22,205-foot Mt. Huascarán, a volcano. The slab broke up and mixed with snow and ice as it catapulted down, sweeping up more and more material. Concentrating in a valley, 35 million cubic feet of debris achieved a velocity estimated at more than 275 miles per hour and leveled every-

thing in its path. Where the valley broadens at the low end, the mass spread and became a debris flow. The final toll was scores of villages destroyed and a loss of life estimated at 20,000 people.

A larger but less catastrophic lahar occurred in 1980 on Mt. St. Helens, in the Cascade Mountains of Washington state. An explosion blew out the side of the cone, starting rockslides. These were joined by mudflows of volcanic ash mixed with water from ice and snow melted by volcanic heat. Soon thousands of tons of material were racing down the mountainside at more than 200 miles per hour. Finally

Aspects of a Slope: Steepness of slope determines (1) whether erosion, transportation, or deposition is dominant, (2) which kinds of slope wash and mass (gravity) movements are operating, and (3) the nature of the slope resulting from these conditions and processes.

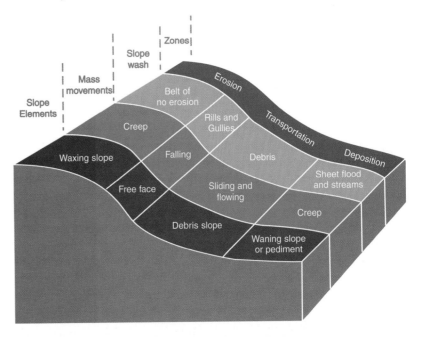

Mudflow Closeup: Rivulets of rainwater in gullies in the wall of a sand quarry saturate the clayey sand, causing it to flow. Note the characteristic lobes and "toes" in this flow.

spreading out as a debris flow, the 3 billion cubic yards of material covered about 3 square miles. Because the lahar occurred in a wilderness area, few lives were lost and damage to property other than the forest was minimal.

Earthflows and Debris Flows

A flow of moderate to slow velocity in which most of the particles though wet remain in contact with one another, providing some degree of internal friction and cohesion, is an earthflow. With its content of 27 to 60 per cent water, it is more fluid than a slump, less fluid than a debris flow or a mudflow. Its velocity of flow may range from inches to several feet per day. Earthflows are among the most common slope movements in highlands. Usually an earthflow leaves a dish-shaped scar in the slope and spreads out below into a shape like that of a tongue, toe, or teardrop.

Earthflows grade into debris flows, which contain more water and move more easily. The higher fluidity of debris flows compared to earthflows allows higher velocity and, therefore, the transport of larger rock slabs, boulders, trees, and other miscellany.

Debris flows are common on steep slopes of clayey material, especially those with little or no vegetation. They can cause relatively rapid cliff retreat and much damage to houses, roads, and other property at the edges of cliffs or below them. A debris flow was the final phase of the Mt. Huascarán lahar. Debris flows have been generated by hurricane winds in the southern Appalachians; for example, in Nelson County, Virginia, in 1969.

About 300,000 years ago a debris flow involving some 15 cubic miles of material occurred on the north slope of Mt. Shasta, a volcano in the Cascades. Reaching the lowland, it spread out to form an odd-looking, hummocky plain 270 square miles in extent. The true nature of this topography, on which the humps are huge chunks of transported volcanic rock, was not discerned by geologists until a few years ago.

Mudflows

Flows of mud can occur on any kind of slope, and have been very notable on volcanic cones. However, the term mudflow has been reserved by many geologists for the swift flow of mud down a narrow, steeply inclined desert valley.

In deserts there is relatively little vegetation to anchor soil, which is commonly mantled with dust. Rain is infrequent, and when it does come the droplets tend to become dust-covered balls, which merge to make a flow. From this mass little moisture is absorbed by the ground. Thus a flow once begun can race down a mountain valley, achieving a velocity of 30 to 55 miles per hour, gathering up more and more soil and debris. Finally reaching the lowland, the flow spreads out and stops.

Heavy rains on desert mountains, as in the American Southwest, often start mudflows that completely cover roads, destroy livestock, and inundate towns. The high velocity and density of mudflows enable them to transport large boulders, which make them the more dangerous.

In some desert areas, such as California's Mohave, mudflows may be easy to identify, because vegetation on them is sparse and their outlines are clearer. When spread out on lowlands, mudflows are broad and flat-topped, with narrow, lobed edges. On arid terrains, 50 per cent or more of material in alluvial fans may be from mudflows.

Rock Glaciers and Gelifluction Forms

On cold mountain terranes, broken rock accumulating in valleys or ravines may include considerable snow and ice. When some of this melts, water fills crevices between the rock fragments, reducing the effectiveness of rough edges in maintaining the stability of the mass as a whole. At the same time the water somewhat buoys up the immersed rock. Thus a rock glacier may form and begin to move.

Rock glaciers are active today in alpine areas throughout the world. On some mountain terranes that are temperate today, such as the Middle Rockies (particularly the San Juan Mountains) and the highlands of northern New York and New England, rock glaciers that formed during the Pleistocene can be seen in higher valleys as long tongues of densely clustered boulders many feet deep.

Cold terrains underlain by permafrost are subject to the kind of solifluction called gelifluction. This is the flowing or sliding of saturated surface soil on ice. Occurring on slopes as gentle as one degree, gelifluction spreads regolith as a sheet over areas ranging from a few square feet to square miles in extent. On steep slopes, regolith may be stepped; its borders usually are marked by large stones and sometimes by vegetation.

Gelifluction can occur on any terrain underlain by permafrost. Evidence of gelifluction, especially concentrations of boulders in slight depressions, can be seen today on middle-latitude highlands that were tundra toward the end of the Pleistocene. Examples are common in the Northeast as well as in mountainous regions of the West.

Scenery of Gelifluction: This broad field of boulders, 5 to 8 feet deep, at Hickory Run State Park, Pennsylvania, is a relic of periglacial conditions that prevailed in the late Pleistocene. Sandstone blocks broken off edges of nearby highlands by frost action moved over slightly inclined, wet, frozen ground, collected in the broad, shallow depression here, and came to rest.

Gravity Movements in the Canadian Rockies: On a mountainside in Jasper National Park, rock debris gravitating from steep slopes mixes with melting ice and snow and gradually moves down. The descending tongue of debris (*low right*) seems to be a rock glacier.

A Potpourri of Slopes: Switzerland's Valais Alps, like many other parts of the European Alps, are rock masses severely bent, warped, folded, broken, and displaced by crustal movements. Slopes resulting from these deformations have been shaped by frost weathering, gravity movements of all sorts, rainwash and stream action, and glaciation. In this region one may find almost any kind of slope profile that could exist.

6
Hillslopes

MOST natural landscapes consist entirely or almost entirely of slopes, ranging from the slightest inclination to the vertical, and including overhangs. The overall appearance of each landscape tends to be determined by the nature of its slopes.

All natural slopes are created by the operation of one or more of the agents involved in tectonic and igneous activity and in gradation. More specifically, some slopes are created as portions of crust become contorted, broken, displaced, jumbled, or otherwise altered, so that cliffs, ridges, domes, basins, and their many variants are created. Slopes originate when volcanoes erupt lava and pyroclastics, which may cover surrounding land to depths of hundreds or thousands of feet. Streams, with the cooperation of weathering and gravity movements, gradually slice up the land, making slopes of every description, and they make additional slopes by depositing erosional debris. Glaciers grinding and plucking the land make still more slopes. Wind makes slopes by blowing dust and sand to and fro. Slopes are made by the uplift of sea bottoms to become dry land, and by wave and current action along shores.

Once created, slopes tend to be changed either by the agencies that created them or by other agencies, or by both. Thus a cliff created by faulting may be reshaped by further faulting. It may be partly buried by a lava flow. Almost certainly, from the moment of its creation it will be altered more or less by weathering, gravity movements, and erosion by running water, windblown sand and dust, or perhaps glacial action.

Slopes are subject to relatively sudden change during earthquakes, volcanic eruptions, floods, landslides, and glaciation. Slopes on unconsolidated material, such as sand and gravel, can be altered quickly by running water during a single storm. Most slopes, however, change slowly. A hillside on solid rock may look the same year after year, decade after decade, century after century. A slope on unconsolidated material anchored by vegetation may show little change over decades.

The agencies that create crustal features are in league to change them and perhaps eventually destroy them. However swift or slow the changes, all slopes are, in the perspective of geologic time, temporary – even those that endure for a million years.

Slopes, especially larger ones, may not show clearly the nature of their origins and subsequent modification. The

Vertical Slopes: The buttress-like forms of Cathedral Rocks, in Capitol Reef National Park, Utah, exhibit vertical slopes. Such slopes, characteristic of much Colorado Plateau scenery, are associated with horizontal strata, vertical jointing, and the dry climate.

Slopes Made by Fault Movements: Some slopes result primarily from tectonic activity along faults. Such is true in the Gulf of Kotor, on the Adriatic Coast. The Gulf formed as a large crustal block or blocks sank relative to other blocks. These slopes are subject to reshaping by weathering and erosion and to breakup by more tectonic activity.

geologist attempting to trace out the history of a slope will look for signs of tectonic, igneous, and erosional events. He or she will note, to the extent practicable, the types, relative resistance, and arrangements of visible rock masses; the texture of loose materials on the slope; the local climatic conditions; the kinds, amounts, distribution, and movements of flora and fauna, and the slope's exposure as to sunlight and precipitation. While taking account of all such possibilities, the geologist must consider also how long the slope is likely to have existed, and must keep in mind that land-shaping processes have probably removed much of the evidence as to how the slope originated. (Many a slope is but a relic of its

former self.) Finally, the geologist is wary enough to remember the principle of equifinality: similar features can be created by different factors.

Slopes are of interest not only as fundamental parts of nature, varying infinitely in form, but as features that strongly influence man and his activities. What kinds of slopes are unsafe for construction, how to control drainage and erosion on bare or forested slopes, how slopes figure in water supply and flood control, slope conditions that favor landslides, whether or not to try contour plowing, what measures are needed to prevent slumping in a roadcut – these are some of the matters addressed by slope science. An engineer must calculate with care and skill the forces tending to destabilize the slope and those tending to stabilize it (that is, those that resist movement). On the soundness of the engineer's calculations much may depend, such as safety of dams, highways, and clusters of houses, not to mention human lives.

Slopes are best understood when classified according to how they originated and what processes have dominated their recent history. Slopes created by eruptions of volcanic materials are seen on volcanic cones and lava flows. Slopes made by faulting are noted on the sides of tilt blocks and grabens. Most slopes are associated with the works of running water; they are sides of channels, valleys, and divides. It is these that are called, strictly speaking, hillslopes. An understanding of them helps toward an understanding of streamwork and other land-shaping agencies as well.

Shaping Hillslopes: Surface Processes

On a landscape shaped essentially by prolonged erosion it is hillslopes that we see between us and the horizon. These range from broad mountainsides to little stream-banks, from sides of badland gullies to cliffs on buttes and mesas, from

Influence of Rock Types: In Fremont Canyon, Capitol Reef National Park, streams have cut through sandstone strata and beds of volcanic ash. On the sandstone, slopes are relatively steep, sometimes vertical, with rock benches formed on more resistant strata; runoff and rockfalls are dominant processes. Slopes on ash, which is softer, are gentler and convex, creep being dominant.

Slopes on a Granite Mountain: On one sector *(left)* of Champlain Mountain, Mt. Desert Island, Maine, joints are widely spaced, limiting streamwork and frost action. Most weathering is by pressure release (unloading), with resulting exfoliation of large slabs. These creep down a relatively smooth, slightly convex slope. In a different sector *(right)* of this mountain the granite has closely spaced joints, weathering divides it into numerous blocks, and the slope surface is rough.

profiles of mushroom rocks to the bizarre forms of hoodoos. The processes that shape these slopes, as streams continue their work below, are creep, rainsplash, runoff, interflow, throughflow, and groundwater flow.

Slow But Sure: Creep

Creep is familiar to us already as a gravity movement, but it must be further scrutinized here in more detail. It is, as we recall, the gradual, very slow downslope movement of loose or poorly consolidated earth materials. The two major kinds are rock creep and soil creep. Rock creep is the movement of rock on rock, such as movement of boulders on a slightly inclined surface. Soil creep is the movement of a soil mass with non-recoverable deformation; that is, the original form of the mass is not recovered after creeping. Usually creep is noticeable only after a period of days, weeks, even years.

Despite its slowness, soil creep is very influential, often dominant, on most hillslopes. Its variability is remarkable, depending on factors as different as chemical and physical properties of soil particles, amount of water in the soil, steepness of slope, type and density of vegetation, temperature conditions, location of soil particles with respect to vegetation, amount of moisture in the soil, and activities of animals on the slope. A few examples will suggest the great variability in rates of soil creep, with respect to other surface processes, in various environments.

On hillsides with a cover of trees or other vegetation, rainsplash and sheetwash are only minimally erosive, because branches and leaves absorb impacts of raindrops and roots hold the soil. On such slopes creep tends to dominate. Thus in humid temperate regions, on vegetated slopes, soil creep is likely to be 5 to 10 times as erosive as rainsplash and sheetwash. Even so, in these regions where slope inclination, or steepness, is as much as 20 to 30 degrees, downward movement of the upper few inches of soil (the most movable portion) by creep is likely to average only small fractions of an inch per year, barring events such as heavy rains or earthquakes that cause landslides.

In dry regions, where anchoring vegetation is absent or sparse, creep may be of slight effect. There sediments are likely to be moved by rainsplash and runoff (runoff includes both sheetwash and surface streams) 5 to 10 times faster than on well-vegetated humid slopes.

In Norway, in a periglacial area (an area near glaciers), soil on a 19-degree slope was observed to have moved about 10 inches in a few weeks; but this movement was due more to gelifluction than to creep. In Sweden, under somewhat more temperate conditions, gelifluction movements averaged some 2 inches per year at the surface, becoming only half that at a depth of about 8 inches and ⅛ inch at 2 inches. In contrast, in the humid tropics, where many soil-destabilizing conditions exist – torrential rains, movements of animals, rapid root growth, and so on – creep may exceed the usually high rates in periglacial areas.

Thus creep rates vary widely. And creep is slow. Even at

A Mix of Slopes: In the Samburu area of Kenya, scattered highlands (examples of monadnocks or inselbergs) are remnants of lofty mountains almost entirely eroded away. Metamorphic rocks in these highlands weather and erode irregularly, because of their deformations, and thus present diverse profiles. The Samburu terrane could be called a peneplain, if peneplains do exist.

the relatively high rate of an inch or two per year, creep alone would require centuries or millennia to work significant changes on a wide, long slope. But creep has allies, and geologic time to work in.

With Every Rain: Rainsplash

In popular literature, rainsplash has sometimes been put at or near the top of the list of land sculptors. Actually, raindrop impacts have little sculpturing effect on most lands. They cannot break up bedrock, and their impacts tend to be absorbed by vegetation. Only on slopes of bare, loose earth, as on cleared fields and on deserts, can raindrops be effective earth-movers.

Under close observation, clay and sand particles hit by raindrops are seen to leap outward from a slope, some falling back higher on the slope but most falling lower down, thus making a downslope gain with each leap. The proportion of particles making this gain may be as high as 90 per cent on a 25-degree incline, and greater on steeper inclines. Water already on or in soil tends to weaken effects of impacts but not eliminate them.

Runoff: Sheets and Streams

Rainwater and meltwater that do not sink into the ground, but move on the surface or just beneath it, are called runoff. This includes sheetflow, the rivulets into which sheetflow soon divides, and the larger streams into which the rivulets pour. Runoff does not include water that sinks into the ground and travels through the zone of aeration but without reaching the saturated zone – that is, the undergound zone in which all available spaces are occupied by groundwater.

The capacity of runoff to erode and transport rock on a slope increases with the depth of the water. It increases also with inclination, but at 25 to 30 degrees of inclination the

Slopes Made by Vigorous Streams: In Hawaii, basaltic lavas are relatively vulnerable to weathering and stream action, especially where lava was left porous by escaping gases. Because of heavy rains and high gradients, stream erosion is very rapid relative to gravity movements, and valleys tend to be cut deep and narrow. These erosional forms are in Iao Valley State Park, on Maui.

Solving a Slope Problem on a Farm: On a South Dakota plain, elevated ground near an oxbow has been plowed and planted along contour lines (lines following points of equal elevation). Furrows are barriers to runoff and gravity movements.

rate of increase begins diminishing, and at about 45 degrees the capacity to erode and transport often begins to decrease. One likely reason is that as the inclination angle increases, the length of slope receiving a given amount of rain decreases and gravity movements intensify. But many factors are present and any one of them can be influential.

On most slopes, descending sheetwash finds chains of depressions and cuts into them, thus forming rills (small channels). These tend to be parallel or nearly so near the top of the slope, and they grow in numbers downward. Channels lengthening by headward erosion (that is, erosion in the upstream direction) intercept other channels, and the channels with the greater gradient at interceptions capture the water. Thus an efficient drainage system, usually dendritic in pattern, is cut. Often the pattern is neatly demonstrated in gently inclined sides of a new roadcut in sand or clay.

Of special interest to engineers and others concerned with slope erosion is the fact that runoff increases in volume and erosive power toward the bottom of a slope; thus, if damage from erosion is to be minimized, slope steepness should decrease toward the bottom.

Shaping Hillslopes: Processes Underground

Rainwater and meltwater that penetrate the ground but do not reach the water table become either interflow or throughflow. Interflow is water that travels through the ground above the water table for a time and then joins another body of water, such as a stream or a lake. Throughflow is water that travels underground without reaching the saturated zone, and then emerges at the surface. Water that penetrates down to the saturated zone becomes groundwater.

Interflow, moving just below the surface, tends to destabilize slopes by lubricating soil and causing it to swell. Swelling involves loss of structural strength and thus may result in soil slips or flows. The water may penetrate between down-dipping rock strata, pushing them slightly apart

Slopes on Soft Earth: In Golden Canyon, Death Valley National Park, swift temporary streams slice easily through clay and sand. In this box canyon, the wall near the top is vertical, with parallel furrows cut by streams. Slopes, being on sticky earth, maintain inclinations of 45 degrees and steeper.

Throughflow: Meltwater making its way down through joints in granite gneiss emerges from the wall of a roadcut and freezes. The water did not penetrate down far enough to reach the water table; thus it did not become groundwater.

A Slope on Cavernously Weathered and Eroded Bedrock: In Utah's Fishlake Mountains, a hillslope on volcanic rock displays numerous cavities produced by solution weathering and interflow or throughflow. As erosion lowers the land, cavities made underground become exposed and a very rough slope surface results.

by pore pressure (pressure in interconnecting rock cavities due to the weight of overlying water) and causing strata to start sliding. Slopes on tilted strata, fractured rock masses, or clay are particularly susceptible to such failures; for example, the clayey sides of volcanic portions of the San Juan Mountains, in Colorado, and the cliffs on tilted and crushed strata at Pacific Palisades, on California's coast.

Originating near a hilltop, throughflow may work downward within rock and soil, turn outward, and, under the pressure of water and earth around it, destabilize or break off por-

tions of the slope surface, perhaps creating niches or arches. Where the slope is on relatively soluble material, such as limestone, the flow can make systems of tunnels, or "pipes," just beneath the surface. During heavy rains, pipes may carry much water, some of which may emerge at lower levels as springs. Roofs of pipes may collapse during a rainy period, leaving a field laced with gullies.

Groundwater flow may affect slopes much as throughflow does. Being usually at greater depth, groundwater is less likely to affect slope surfaces directly. However, where bedrock is limestone, groundwater may dissolve out caverns and tunnels, and as erosion removes overlying rock, cavities become exposed and the hillslope becomes pocked and rugged.

Niche Architecture in Sandstone: Near Lupton, Arizona, a cliff exhibits niches created by the dissolving and weakening action of subsurface water and rockfalls.

Slopes on Horizontal Strata: Sandstone strata here yield not only broad, flat-topped landforms but tower features as well, depending on the jointing. In this view in Arches National Park, Utah, erosion produces broad features where joints are mainly horizontal, tower forms where joints are predominantly vertical.

Hillslope Profiles

How hillslope profiles take form and evolve through time has been a favorite subject of study, and of disagreement, among geomorphologists. So many factors are involved in hillslope events during thousands or millions of years, and often so much of the evidence of past events has been erased, that hillslopes can be easily misunderstood, and generalizations about hillslope history and slopes in particular can be chancy.

A textbook provides an example of how relatively minor differences in the properties of rocks can strongly influence slope profiles. The eastern section of Badlands National Park, South Dakota, displays sharp-backed ridges and mazes of small valleys, with slopes ranging from about 39 degrees to the vertical. Rainwash is dominant, steepness of slope tends to increase with valley size, and valley walls are retreating with unchanging steepness as erosion proceeds. This terrane is on the Brulé Formation, which involves layers of soft clay

Slopes on Rock of Varying Types: In Badlands National Park, South Dakota, the Chadron Formation (*left*) is mostly of loose sand and gravel. Creep dominates; steep slopes cannot be maintained. The Brulé Formation (*right*) has much soft clay but enough relatively resistant shale to support steeper slopes. Where other factors govern, e.g., widely spaced or narrowly spaced joints, slopes can be gentle or steep on masses of the same rock type, such as granite domes and "needles."

Symmetry of Slopes: In a Great Smokies vista, north- and south-facing slopes show a remarkable similarity, suggesting homogeneity of the mountain rocks. The slopes are at an inclination indicating that creep rather than erosion by running water is dominant.

and silt but enough relatively resistant shale strata to give structural strength and make possible the sharp erosional forms. Farther west in this park the terrane is likewise deeply incised, but valley sides are generally convex, from about 24 to 40 degrees; ridges are few, and round-backed. Here is the Chadron Formation, mostly of rather loose sand and gravel. Creep dominates, and slope steepness diminishes as the slope retreats under erosion. Thus in the Badlands a seemingly minor difference in the nature of the rocks has made the difference between steep, rugged slopes and smooth, undulating ones.

The influence on slopes of the attitudes of rock structures is demonstrated in canyons of the Colorado Plateau. Canyon walls, usually on edges of shale and sandstone strata, retreat mostly by granular disintegration and by exfoliation of slabs

along joints parallel to the cliff face. These joints are due to progressive pressure release (unloading) as deeper and deeper rock becomes exposed by stream downcutting. Given these conditions, along with the frequent presence of resistant strata acting as caprock, valley walls maintain verticality as they retreat, and probably will do so until towering, extensive rock masses have been reduced to small, scattered mesas and buttes.

Analyzing Hillslopes

Hillslopes have three basic components: length, steepness, and overall geometry.

A hillslope's length – the distance over which the slope extends without substantial interruption – is determined largely by the density of drainage on it. The narrower the spaces between streams, the more rapid the erosion and the faster the reduction of the slope, other factors remaining equal. These other factors include rock type and resistance, attitudes of rock layers (if any), types and density of vegetation, activities of animals, climatic conditions such as exposure to sunlight and frequency of thaw-freeze, and – as a matter of importance – the length of time the slope has been under attack by the elements.

Steepness of valley sides is strongly influenced by the rate of downcutting by valley streams relative to the rate of weathering and gravity movements. Downcutting tends to steepen slopes; weathering and gravity movements tend to reduce steepness. Where downcutting is dominant, slopes tend to be steep. Rates of downcutting, weathering, and gravity movements depend more or less on those factors, mentioned above, that tend to control a slope's length.

The geometry of hillslopes is sketched less easily. To begin with, it is clear that where creep is dominant, the curve of the slope is likely to be slightly to moderately convex, starting

In Classic Style: Grasmere, Westmoreland, in the English Lake Country, shows slopes far along in development. The very steep, probably straighter slopes of earlier times, when streams were cutting the original topographic pattern, have given way to gentler, convex slopes on which creep now dominates.

Exfoliation from Canyon Walls: In Utah's Zion Canyon, cut deep by the Virgin River, sandstone walls rise nearly vertically. Slabs peeling off the canyon walls fall to the canyon floor. Slab formation, favored by vertical jointing, can occur because of pressure release (unloading) or spring sapping. The falling away of slabs helps to keep the canyon walls steep.

nearly level at the top, steepening to about 34 degrees (the angle of repose) halfway down, and then extending almost as a straight line to the bottom. With time, the convex slope loses steepness but maintains its convexity.

Where sheetflow and rainwash together are dominant, a slope profile may be almost a straight line. During erosion, slope steepness is maintained, or nearly so. Where, on the contrary, stream action is dominant, the profile tends to be more or less concave, starting at the vertical, moderating half to two thirds of the way down, then diminishing markedly near the bottom.

Of much importance to slope profiles is whether debris accumulates at the bottom of the slope or is removed by erosion. If debris does accumulate, it gentles the slope near the bottom and protects underlying rock from erosion. If debris does not accumulate, rock at or near the slope's base remains exposed for erosion and the entire slope tends to maintain its inclination as it retreats.

Many hillslope profiles are compound, often being convex at the top, then straight, and finally concave leading to straight. Differences result from variations in the efficiency of each erosional agent according to variations in slope conditions, such as rock types and structures, moisture content of rock and soil, covering vegetation, and animal activities.

To repeat: a valley being cut by a stream in relatively re-sistant rock, such as granite or massive basalt, is likely to be steep-sided. Rock resistance keeps the stream channel narrow; thus the stream has sufficient velocity to cut effectively even in strong rock. On such rock, rates of weathering and mass wasting are relatively low. Where there is an approximate equilibrium between stream cutting and valley-wall erosion, the average slope probably will be 35 degrees or more. Where a valley is being cut in weak material, such as poorly consolidated sand and gravel, the valley walls may be

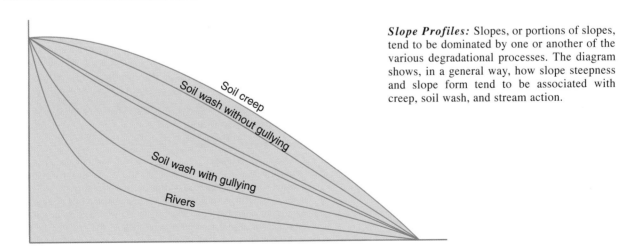

Slope Profiles: Slopes, or portions of slopes, tend to be dominated by one or another of the various degradational processes. The diagram shows, in a general way, how slope steepness and slope form tend to be associated with creep, soil wash, and stream action.

steep but are unlikely to be vertical. Wall inclinations may become gentler as regional elevations generally are lowered by erosion and as valley walls retreat.

A hillslope on homogeneous rock, with all surfaces about equally subject to erosion, is likely to have a "standard" form, corresponding to theoretical models. In nature, some hillslope profiles fit models imperfectly, and often there is a visible reason. A rugged profile is likely on edges of strata of uneven resistance, or where jointing is irregular. If facing valley walls differ in steepness, the reason may be that the stream in the valley is cutting down into an inclined rock

stratum and migrating sidewise down the incline, undercutting the valley wall on the lower side of the incline and making the wall on that side steeper than the facing wall.

A relatively resistant horizontal stratum, or cap, at the top of a cliff will protect the covered portions of weak underlying strata but not the exposed edges. Weathering, runoff, and mass wasting will cause undercutting of the cap, overhangs will fall now and then, and the cliff will be kept vertical. As already explained, verticality is favored also by exfoliation of slabs along vertical joints parallel to the cliff face, especially near the bottom. Such is the story of many steep-walled valleys being cut in the horizontal strata of the Colorado Plateau. Here the caprock is often a hardened flow of basalt, which is generally stronger than the shales and sandstones that predominate in the region.

As a cliff or other very steep slope retreats, it may leave behind a platform of bare bedrock extending outward from the cliff's base. This erosional feature, with an inclination up to 7 degrees (usually less), is called a pediment. Sheet-wash may be the dominant erosional agent. Pediments are most noticeable at the base of steep mountainsides, mesas, and buttes on arid terranes. They can occur on humid lands also, but there they are usually more or less concealed by vegetation. Pediments may extend and eventually coalesce to form "pediplains."

Hillslopes Through Time

Many formulas have been proposed for predicting the evolution of hillslope profiles over long periods. Factors involved may be so numerous, so variable, or so difficult to measure, estimate, or foresee that prediction is tentative at best; so also for efforts to understand precisely how existing slopes have come to look the way they do. Still, some generalizations can be risked.

A Desert Pediment: In the Basin and Range province of the Southwest, pediments – those broad, long, gently sloping rock platforms extending from the base of mountains – are common. This pediment, in the Mohave Desert of southeast California, is mostly covered (as is usual with pediments) by sediments.

Basic is the fact that steepness of slopes tends to be associated with strong relief – that is, with relatively high-standing divides and deep valleys. Slopes tend to become gentler as erosion lowers regional elevations, reducing stream gradients and rates of downcutting. Beyond that the plot thickens, and the understanding of a slope's history can become very difficult. As already indicated (and worth repeating), slope development is affected not only by age but by rock composition and structures, climate, water content of slope materials, vegetation, and conditions relating to transport and deposition of slope materials. Any or all of these conditions can change with time, and as erosion proceeds. Further, slope development often is interrupted by some major event such as regional uplift, volcanic activity, or glaciation. In many instances environmental changes leave a slope as but a relic of what it was in the past; it has been shaped by processes no longer operating, and much – perhaps most – of the evidence of its past has been removed by erosion.

Changes in slope profiles and in topography due to climatic change are dramatically illustrated in northern Africa and the Mideast. Areas now known as Egypt and Israel, for example, were humid, forested lands during the Pleistocene Epoch, until around 6,000 years ago. As the climate warmed and the glacier front in Europe retreated northward, regions to the south became dryer. In classical times the land was still relatively well-watered and fruitful, but since then streams and forests have waned and vanished, and desert conditions, with their own special influences on landforms – influences that tend to create rugged, steep slopes – now prevail.

The Colorado Plateau offers further examples of slopes whose "normal" development has been interrupted by an extraordinary event. Regional uplift beginning a few million

Effects of a Lava Flow on a Slope: South of Shiprock, New Mexico, layers of basalt capping a mesa protect the underlying sandstone from erosion and thus are at least partly responsible for the steepness of this tableland's bordering slopes. Lava flows in this vicinity came from volcanoes of which famed Ship Rock and neighboring volcanic necks are remnants.

Slopes on an Erosional Terrane:
After millions of years of erosion, creep is more effective than runoff in shaping most slopes on New York's Hudson Highlands and New Jersey's Ramapo Mountains. Valleys cutting across the northeast-southwest "grain" of the highlands have walls with inclinations mostly of 20 to 30 degrees, with creep dominant. Typically the hillsides parallel to the grain have the steeper and straighter slopes, with runoff dominant. Irregularities in slopes are due to numerous factors, among them glacial erosion and variations in rock type and structure. The view is over the Wanaque Reservoir in Ringwood, N.J., in the Ramapos.

A Protective Cap Crumbles: On the side of a low hill in Petrified Forest National Park, Arizona, a relatively thin resistant crust on weak clays is undercut by creep and rainwash. Overhangs break off, and fragments creep downward. Rock debris accumulates because in this desert environment there is little running water to move it away.

After Prolonged Erosion: In desert mountains west of Tucson, Arizona, many millions of years of erosion have produced a wide variety of slopes, corresponding to variations in the resistance of the diverse volcanic rocks of which the mountains are made. As often occurs in desert highlands, these slopes tend to be steep because, even under the very dry climate, running water is the dominant erosional agent.

Slopes Modified by Glaciation: On Mt. Desert Island, Maine, north-south valleys cutting through east-west ridges were scoured and much deepened by Pleistocene glaciers. The result was a topography of north-south ridges with very steep east and west sides and gently sloping ends. On the steep sides frost weathering and rockslides are dominant, producing thick talus. The ridge ends are dominated by weathering and creep, and tend to have convex profiles. This view, northwest from Pemetic Mountain, includes portions of three major ridges and two major valleys.

years ago gave streams increased gradient and, therefore, greater downcutting power, so that new, narrower, steeper-sided valleys were cut in the floors of old, wide ones. Drying of the climate subdued weathering and gravity movements, thus in effect increasing the dominance of stream downcutting. This accounts substantially for the sharpness of relief – high divides and deep, steep-sided valleys – that exists today.

Another demonstration of radical environmental change is seen in the Hudson River's gorge in the Hudson Highlands. Pleistocene glacier ice pushing and grinding through the original winding valley cut off projecting ridges, straightened the valley, smoothed and steepened the valley sides, and deepened the valley by hundreds of feet.

Changes in Hudson Highlands slopes due to Pleistocene conditions are but a detail in the history of the Appalachians generally. The Older Appalachians are roots of mountains that existed about a billion years ago; the Younger, or Folded, Appalachians were folded and raised about 225 to 275 million years ago. Both ranges have been worn low, then restored by uplift, possibly half a dozen times. In each cycle, uplift has accelerated the streams, increased their downcutting power, and thus helped to produce strong relief; but with time the land has been lowered again, the slopes gentled, and rugged topography gradually replaced by a relatively low, rolling terrain. In like manner around the world today, lands are being raised, streams are cutting down, and new slopes are being created – slopes that, however strong the rock they are on, are likely some day to vanish.

Slope Development Through Time: Arizona's Grand Canyon demonstrates (among multitudes of other phenomena) the evolution of slopes during several millions of years. As the Colorado River has cut down to depths as great as a mile, rock strata of varying resistance have become exposed in the valley walls. Thus the walls have stepped profiles, with nearly vertical slopes on edges of stronger layers and gentler slopes on edges of weaker layers. The view is from Pima Point.

A Mountain Stream: Among the most scenic of rivers and their valleys are those in glaciated mountains. Here is the Maligne River, in Jasper National Park, Alberta, in the Canadian Rockies, with the sawtooth ridges of the Queen Elizabeth Range beyond. In July the river is swollen with water from melting mountain snows.

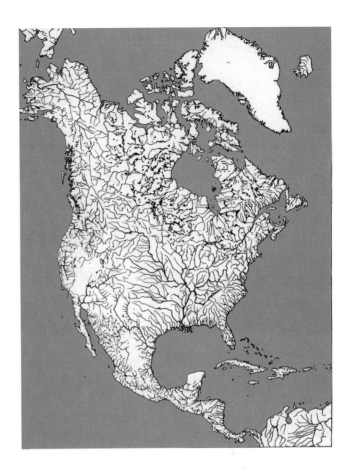

Drainage Picture of a Continent: The extent to which sculpturing of the crust is dominated by running water appears in this sketch of the drainage system of North America. Every line represents a valley in which water is cutting into the crust or depositing sediments. Only large watercourses are depicted here; millions of smaller ones also are in the system.

7

Scenery Shaped by Running Water

IF THE mythical man from Mars, that red desert planet, did indeed visit Earth, he probably would be amazed at – above all – our wealth of water. Not only do oceans cover some 70 per cent of our globe, but the lands are laced with sparkling streams and lakes, high mountains and polar regions are crusted with ice, and the ground to depths of thousands of feet holds a veritable ocean of water in storage. Further, Earth has an ample atmosphere, which, thanks to the planet's rotation and changes of seasons, makes possible the hydrologic cycle. In this series of events, water is evaporated from the oceans into the air, moved by wind currents over the land, precipitated as rain or snow, and then returned to the ocean by surface and underground routes. In the course of all this, Earth's water supports life in multitudinous forms and, in its travels on and within the planet's crust, sculptures rock masses to make scenery.

Among the phenomena of nature, none has impressed man more than streams. Although inanimate they are easily imagined as alive. The larger streams (commonly referred to as "rivers") appear to have powers far exceeding man's, with a remarkable ingenuity in dealing with obstacles, and moods ranging from peaceful to violent. As providers of lifegiving water, rivers have been deified since earliest antiquity, notably by the ancient Egyptians, Greeks, and Romans, all of whom personified the Nile; by the Hindus, who have long worshiped the Ganges; and by the American Indians, who revered the Mississippi River as the Father of Waters. Today, among us moderns, rivers are still accorded high respect as providers of water, as highways for travel and commerce, as carriers of floodwaters, as borders between states and nations, as sites for water sports, and as one of the most attractive and fascinating expressions of nature. On a lesser scale, but hardly of lesser importance, are the countless brooks and the millions of temporary trickles that lace every landscape at times of rain and snowmelt. These minor streams not only feed rivers and replenish groundwater but, working in multitudes, are significant agents of topographic change. Little streams are of major concern in land use – for the highway engineer designing roadcuts and culverts, the farmer trying to prevent washouts in a cornfield, the contractor planning a parking lot, and the homeowner vexed by a gullied lawn and water in the cellar. The activities of running water touch human interests in countless ways – as, indeed, they touch just about everything in this world.

Valleys in a Mountain Terrane: Streams have cut deep in the Great Smoky Mountains. Downcutting and gravity movements are about in balance. The drainage system is thoroughly developed; the land consists mostly of valleys. These, being in rocks of similar resistance, look similar here except for the V-shaped depression (*center*) – a nivation hollow, formed by snow accumulation and frost weathering. The view is from Big Witch Overlook on Skyline Parkway, North Carolina.

Draining the Continents

Of Earth's estimated 326,000,000 cubic miles of water, about 97.2 per cent is in the oceans and nearly all of the remainder is in lakes and inland seas, in ice caps and glaciers, and in the ground. At any given time, in the hydrologic cycle, about 3,100 cubic miles of water is in the atmosphere and 300 cubic miles in stream channels. These appear to be very small allotments for the processes of weathering and streamwork. However, in a year about 102,000 cubic miles of water passes through the atmosphere and some 9,000 cubic miles of the water that falls on land as rain and snow returns to oceans. That amount of water in stream channels, traveling hundreds or thousands of miles, is sufficient for an impressive amount of land sculpturing, given geologic time.

Streams in the higher portions of the continental drainage networks are relatively small, but as they extend downward they join other streams to make larger ones, which in turn descend to join still larger ones, so that most of the water finally reaches the ocean as relatively few large, powerful rivers. Thus the Mississippi River, North America's largest, 2,348 miles in length, receives the waters of streams from 1,243,000 square miles, about 40 per cent of the area of the coterminous United States, and it discharges into the Gulf of Mexico about 133 cubic miles of water and 517 million tons of sediments per year. The Amazon, 3,912 miles long, drains some 2,300,000 square miles, and the Yangtse, 3,602 miles long, drains 756,000 square miles. Such numbers not only indicate the might of great rivers but suggest what a large number of streams, including billions of temporary trickles, are involved in continental drainage and land-shaping.

The work of running water on and in the crust depends ultimately on the triad of basic conditions that influence topographic change: climate, tectonic activity, and – in some instances – igneous activity. Climate determines whether a terrain will be wet or dry, cold or warm, with strong influences on stream volume, on weathering and gravity movements, and, therefore, on streamwork. Tectonic activity influences streamwork by deforming crustal blocks and moving them up, down, or sidewise. Volcanic eruptions build up some terrains with lava, thus making new slopes and often blocking valleys, and upsetting existing drainage systems.

A Modest-sized Mississippi: The Father of Waters, rising in Lake Itasca, northwestern Minnesota, extends to the Gulf of Mexico. Along the way it increases in volume and width as other streams join it. This view of the river, of modest size at this location, is southeast of Minneapolis-St. Paul.

Erosion by Sheetwash: In the Colorado River Valley north of the Grand Canyon, sheetwash on gently sloping terrain removes loose material from around boulders fallen from the cliffs. Boulders partly protect material beneath them from erosion; thus are formed pedestals on which the boulders rest.

Intrusive igneous activity causes the crust to bulge upward or sink, again creating or changing slopes. Any valley as we find it, whether it be a majestic lowland like the Mississippi Valley or a little gully somewhere in the mountains, has a history more or less determined by climate, tectonic activity, and, perhaps, igneous activity.

Rainsplash, Sheetwash, and Streams

Water from rains or the melting of snow and ice exists everywhere on the earth, in some places at the surface and in all places beneath the surface. Whatever its location, it is usually moving and in some fashion shaping the crust.

As already observed on hillslopes, impacts of raindrops falling on hard rock have negligible erosive effect, but raindrops falling on loose earth – silt, sand, gravel – make small depressions. These are insignificant if on a level surface, but on a slope raindrop impacts make soil particles leap, slide, and roll, moving them downward. This process during long periods can cause substantial amounts of loose material to move down slopes and shape topography significantly.

Again as observed on hillslopes, water that sinks into the earth percolates down through rock fractures and between rock grains. That portion of it which does not become interflow or throughflow sooner or later in most localities reaches either an impervious (watertight) rock layer or the top of the groundwater zone, also called the saturated zone, where all cavities are already filled. The top surface of this zone usually is the local water table, though in some rock structures of limited extent water may be perched at a higher level by special conditions, such as an impervious rock layer with

pervious layers above and below it. Reaching a broad impervious layer or the water table, descending water spreads sideward, moving through fractures and pores in the rock. Continuing to spread, some water finds outlets at the ground surface, emerging there as springs that feed wetlands, lakes, or streams; the rest continues to move underground toward the ocean, however distant.

Rain falling on a hillslope or meltwater flowing on it may initially form sheetwash. This moves loose rock debris downward, perhaps incidentally exposing and eroding bedrock surfaces en route. Sheetwash soon finds chains of depressions, in which it gathers to form streams. Streams join other streams, and these join still others, forming a drainage system or systems on the slope.

Most streams are temporary. Originating during rainstorms or where ice and snow are melting, they flow for a time but

Groundwater Circulation: Water from precipitation percolates through the zone of aeration, A, where cavities in the rock contain some air, to the zone of saturation, S, where cavities are full of water. The top surface of the saturation zone is the water table, which rises or falls as precipitation varies. Groundwater follows curved paths in seeking a level.

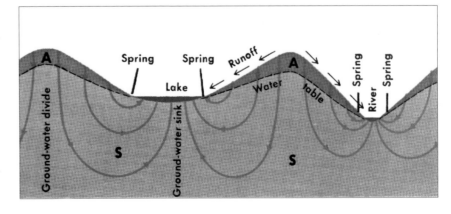

The Course of a Now-and-Then Stream: In arid southwest Texas, almost all streams are temporary. This arroyo south of Alpine was dry even in April. When flowing, the stream is "influent" – that is, part of it sinks down to the saturated zone below. The vertical banks, common in arroyos, indicate that the stream, though only intermittent, tends to be torrential, keeping the banks steep by undercutting.

most of our attention and, in most localities, are the chief sculptors of scenery. They usually start at perennial springs or, in some cases, where glaciers are melting. Spring water is likely to be either throughflow or groundwater that has been traveling through regolith or bedrock and is now emerging at the surface.

Streamwork: A Closer Look

Swelling and shrinking, flowing smoothly or turbulently, rushing around obstacles, cutting into bedrock, transporting and depositing erosion debris, permanent streams follow regimens which are very complex and no less fascinating.

The chain of depressions or elongated trench followed by a stream on a slope is the stream's valley. The part of the valley occupied by the stream during normal, or usual, flow is the channel. The channels of some streams, typically those

soon are likely to be absorbed by the ground or evaporated into the atmosphere. As they flow, they more or less erode their channels and transport rock debris. Before drying up they may reach another stream, a lake, or the ocean. Numbering in the millions per square mile (if we count all the trickles), temporary streams account for much if not most of the erosion and deposition occurring on Earth.

Surface streams that are permanent are the ones that catch

Desert Gullying: At Zabriski Point, Death Valley, California, temporary streams in masses of sediments shape steep-sided gullies and sharp-backed ridges. Rainfall is insufficient to support vegetation that would limit erosion. The gullies are small examples of "V-valleys," common in steep slopes on soft rock.

Classic Streamwork: When swollen by heavy rains, this stream erodes its bank on the outside of a curve, undermining a tree and causing boulders to slide down into the channel. Such streamwork is most noticeable where materials are loose, but it occurs also, more slowly, in bedrock.

having steep gradients (inclinations) in highlands, occupy the entire floor of the valley. The channels of other streams, notably those with low gradients on lowlands, wind across a broad valley floor, occupying little of it.

Streamwork is controlled by the gradient (degree of slope) of the stream's channel and by the volume of streamflow. The greater the gradient or the volume of flow, the greater in direct proportion is the ability of the stream to erode its channel and transport sediments (loose rock material) on the channel floor. These sediments are the stream's "load." The finest, lightest rock fragments are transported in suspension. Rock fragments too heavy for the stream to lift at times of normal flow are the "bedload"; they bounce, roll, or slide along the channel bottom when the current is strong enough. Portions of the load that are immersed are made lighter by the weight of the water displaced, and thus they are easier to move than materials not immersed. At times of high water and rapid flow, a large proportion of the stream's load may be in motion; at low water, little if any. If the stream is swift enough to keep most of its load moving downward, more or less bedrock will be exposed on the channel bottom – a typical situation for swift highland streams. If the stream is relatively slow, as is usual on lowlands, sediments will probably cover the channel bottom.

Where a watercourse is on bedrock, its valley is deepened and widened as the stream abrades the rock, wearing it away by pushing or dragging sediments over it, and as hydraulic pressure is exerted along fractures and bedding planes. Meanwhile some bedrock may be decomposed by its chemical reactions with water and air. As the stream cuts down, channel and valley walls are worn back by weathering, rainwash, and undercutting by the stream, and debris from these processes gravitates into the channel, adding to the load. Deepening and widening of channel and valley tend to be accompanied by straightening and by smoothing of the channel bottom, because projections are more vulnerable to erosion. The more cohesive the material into which the stream is cutting (e.g., clay is more cohesive than sand or gravel), the deeper and narrower the channel is likely to be.

In time, there may be reached a sort of equilibrium, or balance, involving a stream's gradient, channel characteristics, and the rate at which erosion debris is shed into the channel.

Bedload: Rock fragments on this streambed are relatively large because the gradient is steep and streamflow has been swift enough to carry smaller-caliber material away. At times of heavy rain or melting, flow will be strong enough to roll or tumble large fragments downward, further rounding them by abrasion.

Cutting in Bedrock: The Pecos River north of Del Rio, Texas, has cut a gorge in hard limestone. The fact that rivers can cut into rock, making their own valleys instead of simply occupying previously made depressions, seems well demonstrated here.

Then the stream is just able to transport its load, and it is said to be "graded."

The state of a stream with respect to controlling conditions may be expressed in a valley's cross-sectional form. Generally, where a stream is cutting into resistant rock, such as massive granite, the channel is relatively narrow and deep. The flow is relatively swift, because the stream must con-

A V-Valley in the Swiss Alps: The valley cut by this stream has the relatively steep sides and V-shaped cross section characteristic of many valleys that have been cut into relatively resistant rock and in which gradient is moderately steep. Downcutting and gravity movements are in approximate equilibrium.

centrate its energy and flow rapidly to cut into the rock, and because not much debris is shed from valley sides that also are of resistant rock. In such channels the load is light and bedrock is likely to be exposed on the channel bottom. But where a stream is cutting into weak materials, such as clay or a soft sandstone, the channel tends to be relatively wide and shallow, because the valley sides are shedding erosion debris rather rapidly into the stream, preventing it from cutting down and causing it to spread and move over sediments instead of cutting into them.

While a valley is being deepened, widened, and graded by stream action, it may also be lengthening upward by the process of headward erosion, in which, again, weathering, gravity movements, and slopewash all are involved. Upward lengthening may continue until the valley intersects a higher valley or reaches nearly to the top of the slope – rarely to the top itself.

A valley's longitudinal profile (the profile as seen from the side) tends to be steep and nearly straight in the highest segment, less steep and also concave upward in the middle segment, and nearly horizontal toward the lower end. In the highest segment, downcutting by the stream is dominant; in the middle and lower segments, downcutting wanes and gravity movements progressively become dominant. This longitudinal form is likely if conditions relative to climate and rock type are about the same throughout the stream's course. If these and other conditions do change, the profile is likely to differ from the norm.

Valley to Valley

As valleys are followed to lower and lower elevations, toward the ocean, the ultimate destination, valley forms may change in ways reflecting a gradually decreasing gradient. Consider a sequence of valleys with a gradually decreasing gradient, under conditions – especially climate and rock type – that remain nearly constant.

Start with a narrow, zigzagging highland valley on a steep, rocky slope. Moving down from its place of origin, the stream works to straighten, deepen, and widen the valley.

Headward Erosion: With each rain, this small valley in the unconsolidated earth of a farmer's field is lengthened upward, as well as widened, by stream downcutting and gravity movements.

Panorama of an Ancient Valley: In the vicinity of Rich Creek, Virginia, near the border between the Folded Appalachians and the Appalachian Plateau, the New River flows in a broad valley many millions of years old. The view here is east toward the Folded Appalachians. Despite its name, the New River is one of the oldest in the eastern United States, having originated (on a different landscape) as long ago as the early Tertiary Period.

Boulders gravitating from the valley sides onto the channel bottom tend to block or divert flow, producing rapids. Where the stream passes from a strong rock layer onto a weak one, the more rapid erosion of the weak layer may cause a waterfall or rapid to develop. Where the valley bottom is nearly level for a short distance, the stream may spread to form a pond or wetland.

The highland valley eventually joins another one which is much larger. Its stream, fed by more springs, also is larger. In this valley the gradient is gentler. Instead of zigzagging, the valley changes directions in sweeping curves. Its floor is occupied only partly by the channel. Flanking the channel on both sides are slopes of talus leading up to the base of the valley walls. Also along the channel sides at intervals are small floodplains, which may get a new cover of sediments whenever the stream floods over the channel banks.

In this larger valley, erosion has removed or reduced projections from the valley sides and bottom, eliminating rapids and waterfalls. Although streamflow may be fairly rapid, it is not very turbulent, because the channel is relatively smooth and straight. Where bedload is visible, it is seen to consist mostly of rock fragments smaller than those on the mountain streambed. Altogether the scene is less rugged than in the upper valley. A long period of erosion, covering perhaps hundreds of thousands – even millions – of years, may have brought this lower valley and its river to their present state. Barring substantial interruptions, the prospect is that as the river cuts down further, continuing to lose gradient and downcutting powers, and as the valley continues to be widened by slope erosion and gravity movements, the river will eventually lose all or nearly all of its power to cut down and will be spending all or nearly all of its energy transporting sediments.

This wider valley may lead into a still wider one, perhaps miles wide, extending in sweeping, majestic curves, seen best on a clear day from a highland summit or an aircraft. A river meanders broadly along the valley floor, a silver ribbon on the vast expanse. If this valley is the result of prolonged erosion, that process must have required millions of years.

Though the river in this valley occupies only a very small fraction of the width of the valley floor, it may contain a large volume of water, received from a multitude of tributaries. The valley floor is covered to depths of tens, perhaps hundreds, of feet with rock material gravitated from the valley walls and brought down by running water from the highlands. The channel is entirely in these sediments; the bedrock is far below. The channel bottom is smooth and well graded. Gradient is maintained over long distances. Although the current may be swift, the entire energy of the stream is expended in getting through sediment and transporting it. One can believe that long ago this river eroded bedrock, but those days are past. The overall scene, with its broad, quiet vistas and signs of diminishing river energy, is suggestive of old age.

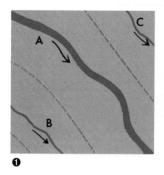 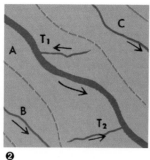 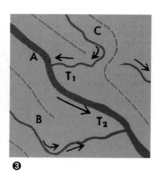

❶   ❷   ❸

Stages in Development of a Drainage System: (1) Streams **A**, **B**, and **C** are separate; (2) tributaries **T**₁ and **T**₂ develop from stream **A** in the largest drainage area; (3) by headward erosion **T**₁ captures stream **C** and **T**₂ captures stream **B**. Thus smaller streams are diverted into larger ones. Dashed lines show how divides shift.

In the broad and deep mass of sediments this river is forced into adjustments. It spreads and runs shallow rather than deep. Its energy is spent moving sediments near the surface rather than at depth, where they are more compacted, more resistant. As if to outmaneuver the resistance of so much sediment, the stream meanders over the valley floor and in some areas may resort to braiding – that is, dividing into inter-

weaving separate streams. Here and there in the channels there may be elongated islands of sediment known as channel bars. The overall scene suggests the dominance of stream deposition over stream downcutting.

Viewing the broad valley floor, one may hardly notice the valley walls. They are far away from the river, and their height is but a small fraction of the valley's width. Incidentally, though it is sometimes said that as a valley ages its walls grow less steep, this usually is not true. Because of conditions that control retreat of hillslopes, the walls of a wide valley can be just as steep as those of a narrow valley.

Drainage Systems

One of the most significant facts recognized by early geologists is that water on land everywhere, in moving to lower and lower levels, is finding its way toward the ocean. Routes followed by streams are parts of systems, in which individual streams are separated from one another by elevated areas called divides. On relatively new lands, and on old lands recently reshaped by events such as volcanic activity, uplift, or glaciation, drainage systems are likely to be makeshift, circuitous, relatively inefficient. On land eroded for hundreds of thousands or millions of years, without substantial interruptions, remarkably efficient systems are in being. On these "old lands," incidentally, streams generally meet at the same level – evidence that streamwork, with grading of channels, is advanced. On lands with weak rocks, valleys are cut and drainage systems develop faster than on lands with strong rocks, if other conditions remain constant.

A Mature Drainage System: West of Tucson, Arizona, mountains exhibit a well-organized system of valleys for transport of water to nearby lowlands. The system's texture is relatively fine, indicating that valleys are in relatively weak rocks. These are fault-block mountains – a stairlike series of tilted fault blocks.

Each stream on a landscape claims the drainage from a certain area, known as its drainage basin. As the stream descends to lower elevations and grows by the contributions of more and more tributaries, the area from which it receives drainage – that is, its drainage basin – is enlarged. Large rivers are supplied by basins covering huge areas. In the United States east of the Appalachians, most drainage is to the Atlantic Ocean, in the Midwest most is to the Gulf of Mexico, and from the mid-Rockies west it is to the Pacific Ocean.

A Drainage System Evolves

As with valleys individually, drainage patterns and associated landforms reflect sets of environmental conditions. Although these can and often do change radically, altering streamwork, it can be useful to observe how a drainage system could evolve if controlling conditions remained constant or nearly so.

Imagine a crustal block of relatively strong rock, a few hundred miles square, which has recently risen a few thousand feet. The block's top surface has little relief, and on it many small streams, mostly temporary, zigzag down gentle slopes. Relatively few of the streams are large, and these have few well-developed tributaries. Very wide divides separate major valleys, and, because efficient drainage has not yet developed, upland areas are patched with lakes and swamps. Valleys as they lengthen upward intersect, and at each intersection, in the process called stream piracy or stream capture, the valley with the steeper gradient takes the water. Thus with time there develops a more and more efficient drainage system, with streams flowing more and more directly toward the block's edges.

As development proceeds, divides become narrower, and the once-level top surface of the land block becomes a terrain of slopes. Relief (differences in elevation between ridge tops and valley bottoms) is at maximum; gravity movements are at a high rate. Although the land is being lowered by erosion and the pace of change is slowing, streams still have substantial gradient. They move through straighter, smoother channels, transporting sediment loads more efficiently. Tributaries are numerous and well developed; the

Remnants of a Divide: These monoliths in Cathedral Valley, Capitol Reef National Park, Utah, are remains of a divide almost entirely eliminated by erosion. They survive perhaps because the divide was wider than others in the vicinity or perhaps because it was capped by rock more resistant than surrounding rock. The same can be said of many buttes and mesas on the Colorado Plateau.

Meeting at the Same Level: On a long-eroded terrain with relatively efficient drainage, streams generally join at about the same level, without a waterfall, because the stream with the higher gradient cuts down faster than the other stream until they meet. Here in North Dakota is a concordant (same-level) junction. The smaller stream appears "underfit" – too small to have made such a large valley. This valley may have been made oversize by torrents from melting Pleistocene glacier ice.

Dendritic Drainage: Egypt's Western Desert, seen from an airliner, shows drainage channels cut during an earlier, humid epoch, beginning probably before the Pleistocene. Streamflow was right to left.

rock has ceased except under extraordinary conditions. Sediment is deep on the land. Creep has become the major agency of geomorphic change. Streams flow but barely. The land has been reduced almost to a plain at or near sea level.

Designs on the Land: Drainage Patterns

When valleys develop on newly made terrain, such as a coastal plain (a sector of ocean bottom recently lifted above sea level) or an uplifted lowland, the pattern formed by the valleys collectively is determined at first by the given slopes, with runoff following the steeper inclinations. As streams cut deeper, they enter zones of relatively weak rock and henceforth follow these in valley cutting. Thus with time a drainage pattern corresponds increasingly to rock composition and structure, for these determine resistance. Generally, drainage systems on weak rocks have relatively close spacing (fine texture) between streams, and systems on strong rocks have wide spacing. Because they reflect underlying geologic conditions, drainage patterns can help in the interpretation of landscapes.

Drainage patterns in a region may vary from locality to locality, and may be altered by events such as volcanic eruptions, glaciation, faulting or folding, or uplift or subsidence. However, certain basic patterns do exist and can often be recognized by casual observers, especially from the air and on maps.

In dendritic patterns, which are the most common, tributaries to the mainstream branch irregularly in all directions, usually at angles of much less than 90 degrees. These patterns are familiar on terranes of horizontal sedimentary rocks, such as the Colorado Plateau and Great Plains, and on terranes of massive igneous or metamorphic rocks that are relatively homogeneous, as in the Older Appalachians.

In trellis drainage, mainstreams are nearly parallel and are intersected by tributaries at angles near 90 degrees. Mainstreams may make right-angle turns through water gaps

system as a whole is well integrated, water is moving oceanward with relative ease.

After much-prolonged erosion, gradients on the crustal block are low and the drainage system is the scene of little activity and relatively slow change. Relief is subdued. Most of the land consists of very broad, relatively smooth, nearly level valley floors. From these floors may rise scattered highland chains or perhaps just a few isolated hills - monadnocks, inselbergs, mesas, or buttes – which have survived erosion because of superior resistance or distance from main drainage lines. Streams have shrunk, are fewer, intersect less often, and meander on floodplains. Downcutting into bed-

Radial Drainage: From an airliner over Montana a dome, probably laccolithic, is seen with small valleys running down its flanks to join a larger valley ringing the base of the dome. Valleys on the flanks are called consequent because they follow the original given slopes; the valley ringing the dome is called subsequent because it follows a weak zone made available later by erosion. Another dome is partly visible at upper right.

Common Patterns of Drainage: (1) Annular, with tributaries ringlike, from the influence of domes and basins. (2) Trellis, with linear control by folds. (3) Rectangular, with control by fault or joint structures. (4) Parallel, on moderate to steep slopes; also in areas of parallel elongated landforms. (5) Deranged, with stream channels blocked and rerouted by glaciation. (6) Dendritic, with little linear control.

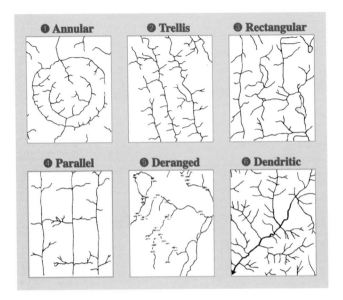

(valleys cutting crosswise through ridges). Trellis patterns are common on long, parallel rock folds, as in the Folded Appalachians, the Colorado Front Range, and the Jura Alps.

In rectangular patterns, master streams flow at right angles to one another. Side streams often are not as near to parallel as in trellis drainage and not as elongated. Valley cutting guided by rectangular joint and fault patterns is seen in many areas, among them New York state's Adirondack Mountains and Norway's fiord coast.

In annular drainage, valleys follow weak zones around a dome, such as the Black Hills Dome of South Dakota, or around a basin, such as the Nashville Basin (produced by the eroding out of the Nashville Dome). Similar drainage may develop around volcanic craters, and around sinkholes in a limestone terrane. Annular drainage often combines with a radial pattern, in which streams diverge from a high central area, as in the Black Hills. It may also combine with a centripetal pattern, in which streams converge from the sides of a basin toward a low central area, as in the Nashville Basin or in volcanic craters.

In parallel drainage, streams may follow structural features such as parallel joints and faults, as occurs also in rectangular drainage; or, streams may flow parallel to one another down a pronounced slope, as some do along the Atlantic Coastal Plain and down steep mountainsides.

A drainage pattern that appears chaotic, with little or no regularity of form, is called complex. It may occur, for example, where the terrane with its stream channels has been newly covered with lava from a neighboring volcano, and there has been insufficient time for a regular pattern to develop. So-called deranged drainage results from recent glaciation during which channels have been destroyed or modified by erosion or covered with rock debris. Such terranes have streams flowing randomly (so it seems) into and out of lakes and swamps.

Consequent Drainage on a Coastal Plain: Along the Alaskan coast near Cook Inlet, small streams follow the given inclination to the ocean. This drainage pattern is of the parallel type, and the streams are consequent.

Where Streamflow Stops

The ultimate destination for running water is the world ocean, where streamflow stops. This is known as base level, and in the broad scheme of things, over geologic time, stream activity is more or less controlled by this level. In a world without continuing tectonic and igneous activity, streams originating on land would cut networks of valleys leading to the ocean, valleys would progressively widen, the land would be lowered progressively, and in the end all

Local Base Level: In the vicinity of Estes Park, Colorado, the low, nearly flat terrain serves as local base level for streams descending from surrounding mountains. Landforms here are mainly erosional, varying according to differences in rock composition and structure. Relief in the Colorado Rockies is generally less sharp than in most of the European Alps.

Earth's lands would be at sea level and streams would not exist. But in this actual world of continuing tectonic and igneous activity, along with climatic change, streamwork is repeatedly subjected to interruption, new lands are being created, and in one way or another Earth is being renewed. Even after 4 billion years, a third of the planet's crust remains above sea level and billions of streams run down slopes. Just as it is difficult for geologists to determine with certainty what has happened here or there in the past, it is difficult to determine just what will happen in the future. We can be sure only that change will occur.

The ocean as base level (often called eustatic base level, the average level of the world ocean) obviously has little noticeable influence on local streamwork, except near coasts. Streams inland rush down slopes, pause to form ponds and lakes in depressions, and meander over plains, apparently in disregard of the ocean. Eustatic base level may be several thousand miles away, and, because of the continuing ups and downs of the crust, eustatic base level will never be reached by most streams, especially those traversing interiors of continents. During the ups and downs, eustatic sea level varies accordingly.

Inland areas where streamflow ceases or is much delayed because of the lack of gradient are known as local base levels. They can be created by any events that affect stream gradients, including prolonged erosion, sediment accumulation, tectonic or volcanic activity, and glaciation.

Interruptions Unlimited

About a century ago the brilliant pioneer geomorphologist W. M. Davis, after long observation and study, proposed that many if not most erosional landscapes pass through stages of youth, maturity, and old age, each stage being marked by distinctive landforms. Youth, he argued, is characterized by narrow highland valleys with steep gradients, swift streams, and rugged scenery. These evolve into wider valleys with gentler gradients, larger but less vigorous rivers, and more subdued topography, corresponding to maturity. Old age comes as a region is reduced to a broad lowland with minimal gradient, a few lazy streams, and perhaps scattered low hills where mountains once towered - a lowland which Davis called a "peneplain" (Latin *paene* + plain). This appealing view of landscape evolution dominated geology for a half-century or more.

Geomorphologists today generally believe that although some landscapes may evolve in that fashion, most evolve less according to advancing age than according to their tendency to reach a state of equilibrium in response to prevailing conditions. Thus a terrane of narrow, steep-walled valleys with swift streams can represent strong rock resistance rather than "youth"; a land with a fully developed, efficient drainage system may reflect a humid climate and crustal stability rather than "maturity"; and a low plain with meandering streams may owe its nature more to regional subsidence than to prolonged erosion. For most geologists today, peneplains are nonexistent or exceedingly rare.

Valley Profiles During a Stream's Life Cycle: The diagram assumes the absence of changes in regard to climate, rock type, crustal stability, and other factors affecting both the stream's downcutting and the gravity movements on valley walls. Streams are subject to changing conditions, and valley development often differs more or less from this model.

In the present view, the story of erosional landscapes – that is, landscapes in which erosion has been the main sculptor of the topography – is a story of changes in conditions that determine streamwork. These conditions are: stream discharge (volume of streamflow), which depends mainly on climate and weather patterns; elevation, affecting stream gradients; and resistance of material in and on which streams are flowing. Such conditions affect dimensions of valleys and stream channels (relatively narrow or wide, deep or shallow); amount, density, and caliber of sediments on valley floors; roughness or smoothness, straightness or crookedness, of valley and channel sides and bottoms; and kinds and amounts of vegetation, if any, growing in channels and on valley sides.

A few examples will show how change in a single environmental condition can have a far-reaching influence on stream activities and the landforms resulting.

Visualize a highland stream with a channel in relatively resistant bedrock and fully occupying the bottom of a steep-sided valley. The local climate changes from humid to more humid. Then the stream's discharge and gradient increase; sediments in the channel are transported faster and lesser amounts accumulate; downcutting accelerates; increased undercutting makes the valley sides steeper; vegetation on the valley sides becomes thicker and thus slows gravity movements; and, finally, intensified erosion tends to straighten the channel and smooth its bottom. If, on the other hand, the climate becomes dryer, stream discharge and gradient diminish; sediments accumulate more thickly on the streambed; downcutting decreases; mass wasting on valley sides slows down; further straightening and smoothing of the valley and channel sides and bottom all but cease. If the climate remains stable but the stream in downcutting comes to a stratum or zone of more resistant rock, the stream will narrow its channel, weathering and slopewash will wane on valley walls because of higher rock resistance, and the walls will steepen. If, then, the local climate cools by 5 to 10 degrees F. and the valley becomes occupied by a glacier, the entire existing pattern of conditions will be upset and the glacier will proceed with its own methods of valley erosion and landform sculpturing.

Young, mature, and old landscapes do exist. Volcanic terranes such as those of Hawaii, with rushing, zigzagging streams and sharp relief, are obviously geologically young, at most a few million years old. The more subdued relief and highly organized drainage system of the Colorado Rockies

A Young Landscape: This sawgrass prairie in Florida's Everglades is a young landscape – a marine plain which was sea bottom before Florida's uplift beginning in the Pleistocene. This land is only a few feet above sea level, except for an occasional hummock (such as the hillock in the distance).

A Rocky Mountain Landscape: Many sectors of the Colorado Rockies, as here at Independence Pass, have more or less rounded profiles, shaped by prolonged erosion of massive metamorphic rocks and granite, often on broad anticlines.

result mainly from 20 million years of erosion since the last major uplift, and thus can be called "mature." Some broad, nearly level landscapes on continental shield rocks under a relatively dry climate, as in western Australia and on the Mozambique Shield of eastern Africa, appear to have been in a nearly uninterrupted erosion cycle for several hundred million years – long enough for such landscapes to be regarded as "old" and, maybe, to be called peneplains.

Some Landscape Closeups

Considering the complexity of most erosional landscapes and the difficulty of creating categories into which landscapes can be neatly fitted, one may choose to turn away from theories and models, and instead simply glance at a few well-known lands and try to understand them in terms of specific conditions and processes prevailing there.

The Colorado Rockies primarily display – as do all mountains – forms created by prolonged erosion of an elevated land mass. These particular forms have been strongly influenced by various recognizable conditions. In some sectors, broad ridges and wide, sweeping valleys testify to streamwork on underlying fold structures, and U-shaped cross sections of major valleys record past glaciations. Narrow valleys in some sectors and the massiveness of many rock outcrops express streamwork in extensive, strong masses of granite and metamorphic rocks. Rugged summit profiles here and there tell of intense frost weathering and stream action in vertical joints and faults.

In contrast are the European Alps, which, though somewhat similar to the Rockies in rock types, uplift, glaciation, and erosional history, in many areas have a much sharper, almost chaotic relief. In most of these mountains streams have been guided by very tight folding and multitudes of

A Landscape in the French Alps: In the European Alps, the predominance of rugged profiles is due mostly to differential erosion of diversified rocks and to extreme folding and faulting. Here are the Aiguilles du Midi, a famed display of granite pinnacles shaped mainly by frost action and erosion by glaciers.

A Desert Landscape: The Wadi Rumm area of Jordan exemplifies a sand plain with not enough sand and with sparse vegetation. This terrain has been eroded down to a resistant layer, on which gradient is slight and streams almost nonexistent. The isolated hills are inselbergs. On these highlands, horizontality of the rock strata and lack of vegetation favor steep slopes.

dislocations by faulting. Much Alpine terrane is pinnacled, because of intense frost weathering in close vertical joints. The Canadian Rockies' Front Range, with predominantly sedimentary rocks, likewise has sharp relief, but this results from erosion mostly on long, regular folds, so that the highest ridges often are finlike rather than pinnacled. West of the Front Range are broad, flat summits on backs of folds.

The Basin and Range Province, in the United States, shows forms made by prolonged erosion of sunken crust and tilted mountain blocks under a dry climate. Mountain summits are often pointed, profiles ragged, slopes usually steep, testifying to pronounced variations in the nature and resistance of volcanic and other rocks, extensive faulting, and dominance of stream erosion. (Rain is infrequent, but likely to be torrential and highly erosive; weathering and gravity movements are subdued.) The modest mountains of Mt. Desert Island, Maine, are of granite, broad and rounded by virtue of their massiveness and homogeneity, and smoothed by the Pleistocene glaciers, which also oversteepened the walls of north-south valleys.

The Great Plains of the United States evidence the labors of streams carrying erosion debris eastward from the Rockies for tens of millions of years. The higher plains, in the west, are essentially merged alluvial plains; in the east are exten-

sive floodplains on mostly horizontal sedimentary strata. Here and there notable relief exists, such as the Nebraska Sand Hills, which are wind-made dunes; the North and South Dakota badlands, which show effects of erosion on sedimentary strata of varying resistance; and the Black Hills Dome, heaved up by an igneous intrusion. Notwithstanding the existence of such features, the Great Plains express most of all the ability of sheets and streams of water to spread sediments wide and far.

Volcanic landscapes exhibit their own special responses to running water. These landscapes include mountains, plateaus, plains, and seemingly every form in between. Volcanic materials – lava from flows, pyroclastics from volcanic explosions – characteristically are very diverse in composition and accumulate as diverse structures differing in resistance. Thus running water on volcanic terranes such as those

In the Basin and Range Province: In the Wild Rose Graben, in California's Panamint Valley, a part of the Mohave Desert, typical "B and R" features are noticed: bare, erosion-worn highlands, an alluvial fan (*left*), deep accumulations of rock debris on a valley floor, and – barely seen in the far distance – a "dry lake."

Following a Given Rock Structure: The cross section of a valley in the Wasatch Range north of Provo, Utah, is neatly defined by a syncline. In its downcutting the stream followed the inclinations of the limbs of the fold.

in Idaho, Hawaii, and Iceland (to name a few of hundreds) can create features unique and innumerable in variety – ridges and depressions, rock towers and waterfalls, hoodoos and honeycombed cliffs, "fins," solution furrows and whatnot.

Limestone landscapes are still another testimonial to the versatility of running water in dealing with diverse conditions. These landscapes express primarily the solubility of calcium carbonate. Sinkholes, caverns, dry valleys, haystack hills – these and other features point to solution by running water as an efficient geomorphic sculptor.

Nearly every erosional landscape, then, expresses the work of running water under a certain set of conditions. Hills and valleys, plains and seashores, deserts and glaciated lands, and all the other kinds of erosional landscapes, with their associated minor features, look the way they do largely

because of the interaction of water with rock under given – and changing – conditions. No two landscapes in the world are exactly alike. Each landscape evolves in its own way; its future cannot be exactly predicted according to any model. It is, however, ready to be understood as itself.

Some Special Valley Types

Valleys of Streams That Follow Given Surfaces or Rock Structures

A stream following the inclination of a newly made surface, such as a coastal plain, or following the inclination of a given geologic structure, such as a rock stratum, is said to be consequent, because its orientation results directly from the existence of the given slope. Examples could include streams running down any dip slope or the side of a volcanic cone. Streams flowing in the same direction as the consequent drainage but at a lower level (after the original surface has been eroded away) are known as resequent.

A stream that lengthens headward through zones of rock weakness is called subsequent, because its course has resulted from exposure of the weak zone by erosion, which eliminated original surfaces. An example is a stream that flows along the base of a ridge in fold mountains and is intersected by consequent streams descending the side of the ridge. Subsequent streams, such as some segments of the Susquehanna and Delaware rivers, follow weak zones between ridges in the Folded Appalachians.

A stream flowing contrary to the dip of rock strata is termed obsequent; for example, a stream flowing down the east side of a slope on edges of strata that dip westward. These streams, too, are common in fold mountains.

Finally there is the stream cutting into rock structures or loose material that appears not to influence its course, or whose influence is too complex to determine. Such a stream is called insequent. An example might be a meander on horizontal strata or on homogeneous igneous or metamorphic rock, or a meander in deep sediments in a swampy area.

Modes of Origin of Streams: Watercourses develop wherever there is a water supply to an inclined surface. The direction taken tends to be the direction of steepest gradient.

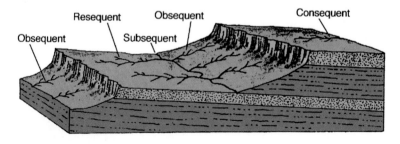

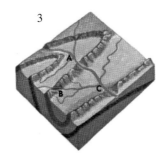
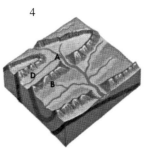

Formation of Water Gaps and a Wind Gap

(1) Streams flow on sediments that cover folds eroded during the geologic past. (2) As erosion removes sediments, folds reappear. (3) Streams cutting across resistant belts make water gaps at

A, B, C. Other streams by cutting headward along weak belts shape ridges. (4) By headward erosion a stream in a weak belt intercepts the transverse stream at **A**, leaving a wind gap at **B**.

Insequent streams are familiar on old erosional landscapes, such as the Hudson Highlands and Great Smokies, where rock structures are very complex and often difficult to trace.

Designations mentioned above are commonly reserved for streams but can be used to indicate the nature of the valleys the streams occupy.

Water Gaps and Wind Gaps

A valley that cuts across the linear structure, or "grain," of a ridge is known as a water gap. Examples are common in fold mountains such as the Folded Appalachians and Wyoming's Bighorn Mountains, in which gaps cut generally at right angles across longitudinal folds.

One kind of water gap forms as a crustal fold rises under a stream which cuts fast enough to keep pace with the uplift. This rare type of stream and its valley are called antecedent, because the stream's course was set before uplift. Examples

of streams thought to be antecedent include Washington state's Yakima River where it cuts across Umtanum Ridge, the Rio Grande where it has made scenic Santa Elena Canyon in a rising fault block in Big Bend National Park, and the Arun River where it is cutting through still-rising sectors of the Himalaya.

Much more common is the kind of gap cut by a superposed stream. This is a stream which, cutting down almost like a bandsaw into a terrane of deep sediments, reaches and slices into a weak zone – perhaps along a fault or a master joint – in an otherwise resistant buried ridge. As the region generally is lowered by erosion, the ridge, being resistant, survives as a highland with the gap cut through it. The Folded Appalachians have numerous gaps cut by superposed streams, including the Susquehanna, Delaware, and New rivers. In Wyoming similar gaps have been cut into the Big Horn Mountains by the Bighorn River and into the Owl Creek Mountains by the Wind River.

A Wind Gap: Seen westward across the Shenandoah Valley, in Virginia, the nearly level ridge of Massanutten Mountain shows the cut which is Newmarket Wind Gap. Long ago a stream which crossed the region there was captured by the Shenandoah River, in the valley, and went dry.

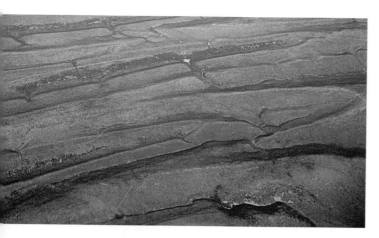

Water Gaps and Wind Gaps: A display of Appalachian folds, cut by water gaps and wind gaps, is seen from the air near Williamsport, Pennsylvania. Gaps reaching to the base of folds are water gaps; those cut only partly into folds are wind gaps. Ridges are separated by subsequent valleys. Running from lower left to right are stumps of eroded plunging anticlines.

Some geologists say convincing evidence of the existence of antecedent and superposed streams is lacking. However, the concepts of antecedence and superposition seem to be plausible ways of explaining the gaps. Here is an instance of the fact that in geology the evidence of events past is often meager or lacking, and explanations must then include some degree of supposition – and doubt.

Many streams become diverted from their valleys by capture ("piracy"), which is common in the development of drainage systems. Capture occurs when one stream by headward erosion intersects another stream at a higher elevation and diverts that stream into its own channel. Thus the upper stream's valley is left dry, with "only wind moving through it"; hence the term "wind gap." Many wind gaps are at a high elevation because the ridges themselves, being resistant, stand high above adjacent land. Like water gaps, wind gaps are common in fold mountains.

Valleys within Valleys

The energy of a stream can increase because of uplift, lowering of base level, an increase in runoff, a decrease in load, or an increase in stream volume due to stream capture. A stream thus restored to vigor is called "rejuvenated." Downcutting then accelerates. Characteristically, the stream cuts a new, narrow, steep-walled valley lengthwise along the floor of the old valley. Flanking this inner valley are terraces of bedrock at about the same level; these are remnants of the former valley floor, graded (worn nearly level) regardless of the inclination (if any) of the rock strata. Gradient increase is most often by uplift, as with the Colorado Plateau and other sedimentary plateaus during the past few million years. On the Colorado Plateau are numerous wide valleys with telltale inner valleys occupied by swift streams.

An interesting variety of these streams with restored gradient and a steep-walled valley is the incised, or entrenched, meander. The stream's recently acquired swiftness will enable it to straighten the channel, but that will take time, and, so far, little straightening has been accomplished. Among the Colorado Plateau's numerous incised meanders is the San Juan River, sculptor of the famous "goosenecks" in southern Utah. Other scenic incised meanders include the Genesee River, on the glaciated Allegheny Plateau, in New York state, and the Moselle River in the Eifel district of western Germany.

Valleys with Misfit Streams

Some valleys are occupied by streams too small or, less commonly, too large to have cut them. Such streams are called misfits. For example, if meander scars or the curves of a valley's walls indicate that the meanders were formerly wider than at present, the stream now in the valley is suspected of being underfit. Today many such streams are seen where, as in the northern coterminous United States, rivers swelled with water from melting glacier ice enlarged their valleys,

An Incised Meander: Air travelers over the Colorado Plateau see numerous narrow, meandering valleys much deepened by streams whose gradients were restored by uplift during the past few million years. In this valley, in southern Utah, an oxbow (*lower left*) has been cut off by the stream.

A Misfit Stream: Unaweep Canyon, in Uncompahgre National Forest, Colorado, was cut originally by a powerful river. Much of its water was later diverted by uplift of the Uncompahgre Plateau. Today the stream is clearly underfit. The stream is at center, a highway at left.

and then returned to near normal flows. Underfitness may result also from stream capture, drying of the climate, or any other factor that reduces streamflow or increases load.

Valleys with overfit streams – streams "too big for the valley" – are rarely seen. An overfit stream soon shapes its valley to an appropriate size, eliminating signs of overfitness.

Flooded by the Sea: Estuaries

Any drainageway through which tides of seawater reach into river valleys or coastal marshes is called an estuary. The lower ends of valleys can be converted into estuaries by a rise in sea level or sinking of the coast. Toward the end of the Pleistocene, numerous estuaries were made by the estimated 400-foot rise in sea level due to almost world-wide melting of glacier ice. Even more estuaries would have been created but for the fact that as ice on coasts melted, these terrains rose somewhat because of removal of weight.

Estuaries are common along North America's Atlantic Coast from the Gulf of St. Lawrence to South Carolina and along most coasts of Great Britain, northwestern Spain, and northern Mediterranean countries. Estuaries in which the bottoms have been eroded by glaciers to a depth below sea level are known as fiords.

Valley Cross Sections

Shapes of valley cross sections tend to be determined by rock types and structures and by dynamic relationships between stream action, weathering, and gravity movements.

It will be recalled that where a stream cuts through relatively resistant rock, such as massive granite, the valley is likely to be relatively narrow and deep, with nearly vertical walls, because weathering and gravity movements on the walls are limited and the stream is efficient in moving away the rock debris they produce. A spectacular example is Colorado's vertical-walled Black Canyon of the Gunnison River,

cut in granite, averaging 1,300 feet wide and having a maximum depth of about 3,000 feet.

Vertical walls occur also in valleys developed in narrow zones of weak rock. A master joint (that is, a long and deep joint) in bedrock is a weak zone; a zone of rock crushed by movements along a fault plane also is weak. Another example is a dike (igneous rock injected into a fracture) that is weaker than the host rock (rock into which the injection occurred) and erodes faster, leaving a trench or ravine.

An Estuary: A rise of perhaps 400 feet in sea level since the Pleistocene has drowned the Delaware Bay region between New Jersey and Delaware. Seawater entering the lower Delaware River has made it an estuary. Adjacent low areas flanking the estuary are coastal plains, barely above sea level, with typical meandering streams.

Asymmetrical Valley Walls: In New Jersey's Ramapo Mountains, a stream's downcutting is guided by an inclined joint plane. As the stream cuts deeper, it shifts diagonally downward along the plane, undercutting and steepening one valley wall (*at right*) and leaving the other wall as a gentle slope.

Deep valleys with vertical walls are seen commonly on uplands where a resistant rock stratum overlies weaker ones, as on parts of the Colorado and Appalachian plateaus. The strong stratum gives the weaker one a degree of protection and thus maintains the verticality of the valley walls.

Wall steepness will vary according to such factors as rock composition and structure, stream gradient, and climate. Where a stream is cutting down through relatively homogeneous materials and there is an approximate equilibrium between gravity movements and the rate at which the stream carries away debris shed into it, the valley's cross section tends to the V-form, ranging from narrow to broad. Wall profiles show irregularities where conditions change. In unconsolidated material, even with a high stream gradient walls are unlikely to be steeper than about 35 degrees, which is the angle of rest for many kinds of loose material.

Facing valley walls may differ in steepness for several reasons. One wall may be steeper because it has been undercut by the stream at a curve, and the facing wall will then have a gentler slope because of the lack of undercutting there and because of buildup of sediments inside the curve. Walls may differ in steepness where a stream follows a fault on one side of which the rock differs in resistance or structure from the rock on the other side. In valleys where facing walls are of the same rock, a difference in steepness may be due to a difference in exposure to ambient climatic conditions; for example, a difference in exposure to sunlight. Again, if a valley crosses inclined rock strata or an inclined fault or joint plane, the valley wall at the lower side of the incline will have the steeper slope. Asymmetrical valley walls are common in rock strata that are folded.

U-shaped valley cross sections are frequently seen in mountain regions that have been glaciated during recent geologic time. The U-shape, seen in the larger valleys, results from the severe grinding action of thick glacier ice near its base, where pressure is greatest because of the glacier's huge overlying weight.

A "Slot" Valley: The Ausable River near Keeseville, New York, follows the easy route of a major joint in sandstone. Gravity movements do not keep pace with the stream's swift downcutting. Verticality of the valley walls is due in part to this swiftness, in part to horizontality of the sandstone strata.

Potholes in Marble: These potholes are in the north fork of the Ausable River, a mountain stream, at Keene Valley, New York. Solution and abrasion, influenced by deformations in the rock, have combined to create bizarre, almost macabre forms.

Erosional Features in Valleys

Potholes and Rapids

Rock debris whirling in eddies can drill holes in the solid rock of a valley floor or wall. Called potholes, these are generally deeper than wide, and are circular or elongated. The largest are up to a few feet in diameter and may be centuries in the making. They tend to form where streamflow is turbulent because of projections from the valley floor or walls, and are made most readily in soluble rock. They are, naturally, most noticeable where bedrock is exposed. Potholes in walls above present stream level are evidence of higher water levels in the past.

Potholes generally represent an important part of erosion done by very vigorous streams. Toward the end of the Pleistocene, much pothole drilling was done by streams swollen with meltwater from waning glaciers.

The swift, turbulent, dashing flow of a stream where it suddenly increases in velocity is called a rapid or, more commonly, rapids. These can form, for example, where flow is down a stairlike incline not steep enough to make a waterfall, or where flow speeds up to get around large boulders or other obstacles, or where edges of resistant strata jut from the channel floor. Rapids may occur also where a channel has been partly blocked, as by a landslide or a lava flow. Rapids characterize streams with high gradient and uneven channel bottoms; they disappear as the stream grades its bed.

A Mix of Rapids and Waterfalls: The Great Falls of the Potomac River, in Virginia, is at the border between the Piedmont, with its ancient, resistant rocks, and the Coastal Plain, with its younger, weaker sedimentary rocks. The river descends about 40 feet in 3 miles in a series of rapids and small waterfalls.

A Niagara-type Waterfall: The stream flows over the edge of a stratum stronger than strata beneath it. Splash, aided by spring sapping, erodes edges of the weaker strata, undermining the stronger stratum and creating overhangs which periodically break off. Thus the falls gradually retreats, or "travels."

Waterfalls, Cascades, and Plunge Pools

The descent of a stream down a vertical or nearly vertical slope is a waterfall, or "fall." A series of waterfalls is called a cascade.

Some falls, such as majestic Victoria Falls in Zimbabwe, Africa, at the edge of the Rift Valley, develop where a stream falls from one crustal block to an adjacent block that has been lowered by faulting. Some falls are on edges of lava flows, as in Hawaii, Iceland, and New Zealand. More common is the falls formed where a stream on a resistant rock layer passes onto a weak layer, which erodes faster; so it is with Niagara Falls.

Among spectacular falls are those where a stream in a tributary valley reaches the edge of a major valley that has been much deepened by glacial erosion. The end of the tributary valley "hangs" above the major valley, and the water may plunge hundreds or thousands of feet to that valley's floor. Such falls are common in mountains sculptured by glaciers, the best-known examples in the United States being Bridalveil Falls and others in Yosemite Valley.

Several minor but interesting features may be associated with waterfalls. One is a plunge pool at the foot of the falls, in a basin created by impacts of the falling water and sediments. Another common feature is the cave, or alcove, behind the fall, eroded out by splash.

With time, as undercutting and other erosional action destroy the cliff beneath a fall, the fall retreats in the upstream direction and becomes a rapid; and this, too, barring interruption, is eventually eliminated by the stream's grading activity. Niagara Falls, now retreating toward Lake Erie, will – if nature has its way – become a rapid when the river has cut through the resistant layer on which it now flows and starts cutting into the softer layer beneath that. Even now, undercutting and sapping are causing parts of the resistant layer to break off and drop. Fortunately for our enjoyment, major waterfalls even if left to nature may have lifetimes of tens of thousands of years.

Lakes, Ponds, and Wetlands

Basins whose bottoms are at a level below the water table usually contain lakes, ponds, or wetlands. Depressions with bottoms that are above the water table but are impervious also may contain water, at least temporarily. Most lakes,

A Waterfall Over a Resistant Stratum: In the Appalachian Plateau's edge in northern West Virginia, the Blackwater River, like the Niagara River, goes over the edge of a resistant stratum. Below is a plunge pool, in a basin eroded out by falling water.

Influence of Rock Structures: Lake Placid, in New York's Adirondacks, lies mainly in "shear zones" – zones of rock broken by faulting. Long ago, streams cut valleys through these weakened rock masses. Later the valleys were enlarged by glacial erosion and dammed by glacial deposits.

ponds, and wetlands are fed and drained by streams; thus most can be regarded as wide segments of streams.

A lake can be regarded as a large pond. A wetland is an area seasonally wet and clogged with vegetation; it is where the water table barely intersects the land surface. These features grade into one another.

Basins in valleys have many possible origins. Some are areas hollowed out by erosion, as by a glacier or a waterfall in the past. The Great Lakes and New York's Finger Lakes – not to mention thousands of other lakes – are in valleys enlarged by glacial erosion and dammed by the debris from such erosion. Other basins are created when valleys are blocked by landslides, lava flows, and stream deposits. Many lakes in Iceland, Hawaii, New Zealand, and other volcanic regions have natural dams of lava.

Basins for lakes and ponds can occur also in fault and fold structures, volcanic craters, and sinks in the terrain. Basins for artificial lakes are made by damming a valley stream or excavating to a depth below the water table.

All these features are short-lived, geologically speaking, if left to nature. Sooner or later overflow or seepage cuts an outlet through the basin rim, and as the outlet is cut deeper and deeper, the basin can hold less and less water, and finally empties. Also, basins tend to fill with sediments from runoff, wind, and gravity movements, and with vegetation and animal remains. Lakes and ponds evolve toward becoming wetlands, and wetlands to dry ground. Preservation of lakes and reservoirs may require periodic dredging and damming.

Undercut and Slipoff Slopes

When a stream rounds a curve, it tends to undercut the streambank on the outside. Undercutting can occur in bedrock as well as unconsolidated material. It tends to produce a steep bank with overhangs, which fall into the channel. In time, the undercut slope retreats in the downstream direction.

Opposite a channel's undercut side there usually is a "slipoff" slope projecting into the channel. Here the stream is drawing away from the inside of the curve and eroding the

Lake Basins: Major types (not including those made by meteorite and asteroid impacts) are shown here. Basins left to nature are likely to be destroyed soon (geologically speaking) by erosion, but geomorphic processes continually make new ones.

Glacial gouges and kettles

Sinkholes

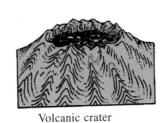
Volcanic crater

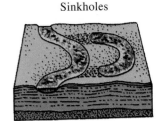
Oxbow

Graben

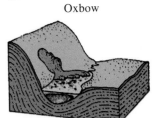
Valley blocked by landslide

Valley blocked by lava flow

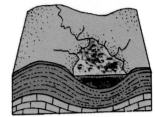
Basin in fold

bank little if at all. Instead, the stream may deposit sediments there, because stream velocity is lower at the inside of the curve than at the outside. Thus the inside is built outward, with a gentle inclination, making a "point bar."

Undercut slopes and point bars are familiar features especially on floodplains where meandering streams make their way through masses of sediments.

Rounding the Bend: Near Port Jervis, New York, the Delaware River undercuts the outside of a curve, keeping the valley wall steep. A point bar, on the inside, has been cut through.

Benches, Bedrock Terraces, and Buttresses

A stream cutting down may reach a relatively resistant rock layer which causes the channel to narrow. Thus a platform of resistant rock – a rock bench – remains, flanking the channel on one or both sides. The edge of this bench is a steep scarp, or riser, made by the stream's sidecutting. The top surface of the bench follows the inclination, if any, of the resistant stratum. Talus may cover all or part of the bench. North America's finest exhibit of benches is in the Grand Canyon of the Colorado: but benches are abundant in other steep-sided valleys also, especially on layered rocks.

As already explained, when a stream with restored gradient cuts a narrow valley within a wider one, a rock platform is left on one or both sides of the new valley. Such platforms, called bedrock terraces or strath terraces, are remnants of the old valley floor. Their surfaces are smooth even if on edges of upturned strata, because of grading by the stream; this is a characteristic that helps to distinguish these terraces from benches and alluvial terraces. If there are platforms on both sides of the new valley, they are likely to be at about the same level. Some valleys have more than one set of matching terraces, indicating that there has been more than one episode of gradient restoration.

Rock Terraces: North of Mostar, Bosnia, the Neretva River has cut deep into the mountain terrane. Renewed uplift in recent geologic time has caused an increase in the stream's gradient. Moving more swiftly, the stream has cut a narrower, deeper channel in the floor of its valley, leaving rock terraces along the valley sides.

Abode of Spirits: Pinnacles, buttes, and mesas of Monument Valley, northeast Arizona, are remnants of an ancient terrain worn down to a resistant layer. They suggest James Hutton's fine phrase: "the ruin of former worlds." Sensitive to the grandeur of nature and the mystery of creation, the Navajo have long regarded this land as sacred.

Tall, more or less rectangular, pillarlike rock masses projecting from the base of a cliff are known as buttresses. They resemble the massive structures, also called buttresses, that give lateral support to walls of Gothic churches. Rectangular or near-rectangular jointing has guided the shaping of these features, which are likely to be in strong sandstone or limestone. Examples are seen often where streams have cut deep into horizontal rock strata, as on the Colorado and Appalachian plateaus.

Mesas and Buttes

Buttes and mesas stand isolated on broad, nearly level terrains, far from valley walls, as in Monument Valley, Arizona. Buttes are generally higher than wide; mesas, wider than high. One definition says a feature is a butte if its top surface covers less than a square mile; if more, it is a mesa. Mesas and buttes may be remnants of divides mostly eroded away. They occur mostly on dry sedimentary plateaus.

Rock Towers, Pillars, and Pinnacles

Tall, slender rock masses sometimes are seen standing out from the base of a valley wall. Usually they are portions of the wall separated from the main mass by erosion in vertical joints, as has happened spectacularly in Bryce Canyon, Utah, and at Hell's Half Acre, Wyoming. Some are remnants of buttresses. Many are protected by resistant caps.

Tall, towerlike rock residuals can be of other origins too; for example, volcanic necks and sea stacks.

Natural Arches and Natural Bridges

Sapping, along with weathering and rainwash, may penetrate through a broad, high-standing, thin rock mass, creating

More Rock Terraces: As the Colorado Plateau was uplifted, streams gained gradient and began cutting new, narrow valleys within old, wider ones. Thus the Colorado River cut this inner valley in an old one near Moab, Utah. Rock terraces flanking the new valley are parts of the river's former bed.

"Sentries": Slim sandstone towers stand near a wall in Canyon de Chelly, Arizona. These were shaped mainly by running water in closely spaced vertical joints in the cliff at right.

an open arch. Blowing sand, which can be lifted as much as 5 or 6 feet above the ground, may enlarge the lower part of the opening and smooth the rock near the base. Spectacular examples occur in Arches National Park, Utah, where numerous sandstone masses are finlike because of very close vertical jointing. Close jointing resulted from stretching of the crust during the Colorado Plateau uplift.

Openings located high up in fins are appropriately known as "windows."

Natural bridges look like natural arches but are of different origin. They are remnants of stream-cut tunnels. One kind is the erosional remnant of a tunnel cut through sides of an oxbow by an incised meander. This variety is numerous on the Colorado Plateau, the most famous being Rainbow Bridge. Another type stands near the town of Natural Bridge, Virginia; this is a remnant of a tunnel cut by an underground stream in limestone, then uncovered by erosion.

Solution Furrows

Thin streams ("rills") flowing down inclined rock surfaces, especially steep ones, often make furrows ("gullies") which

Natural Arches: During the Colorado Plateau uplift, stretching made numerous vertical joints in rock strata and erosion widened them, creating rows of parallel "fins." In these, windowlike openings were made by continuing erosion. Such are the histories of Delicate Arch (*left*), in Arches National Park, and Wilson Arch, south of La Sal Junction, in Utah.

Furrows in a Cliff – I: Deep vertical furrows are seen in the side of Wuyi Mountain, southeast China. The rock is conglomerate, probably containing a natural limy cement which is vulnerable to solution and thus favors furrowing by torrential vertical streams. The furrows could be called large examples of lapiés, or karren. The term fluting may also apply.

extend straight down the surface, regardless of joints or of bedding or foliation planes. These are likely to be well developed where the rock is relatively soluble or easily decomposed. In carbonate rock the furrows are usually known by the French term lapiés (LAH-pee-ays) or the German karren (KAH-ren). They range from long, deep furrows in cliffs, separated by sharp ridges, to small, shallow grooves separated by round-backed ridges, in walls of limestone caverns and in gently sloping rock outcrops.

Visitors in Hawaii often are impressed by the deep vertical furrows, separated by sharp-backed ridges, in the basalt cliffs. Here the furrowing is favored by cliff steepness, occasional

columnar jointing, vulnerability of basalt to decomposition by moisture, and frequent torrential rains.

Shallow, closely spaced, parallel, vertical furrows on steeply inclined rock surfaces are often called flutes. They resemble fluting in columns of Greek temples of the Doric style.

Some furrows follow joints or penetrate rock along bedding or foliation planes. Furrows that extend straight down cliffs or gently sloping surfaces are likely to result at least partly from stream action; those that zigzag are likely to result from weathering and gravity movements. Solution furrows that follow joints in limestone are called grikes (a Scottish term), and the ridges that separate them are known as clints. These may be noticed on many limestone terranes. Especially fine displays are seen in Yorkshire, England, and on The Burren, in County Clare, Eire.

Niches and Caves

A stream flowing around a curve often shapes a relatively shallow recess in the undercut slope. This recess, usually called a niche, can be enlarged by rockfalls due to processes

Furrows in a Cliff – II: Deep vertical furrows are common in Hawaii's basalt lava cliffs. These result mainly from the torrential nature of temporary streams descending cliffs during Hawaii's heavy rains, and from the fact that basalt is oxidized relatively fast by moisture and thus is easily eroded.

cliffs of the Columbia Plateau are usually deeper and more common than the stream-made niches.

"Caves" differ from niches and alcoves in that caves are, strictly speaking, deep cavities with a narrow entrance and a well-defined roof. Caves form in either loose material or bedrock. Some potholes in valley walls are deep enough to qualify as caves. Caves in valley walls may be entrances or exits of tunnels made by underground streams. Some geologists use the term "cave" only for limestone caverns.

Meanders, Meander Belts, and Meander Scars

The meandering, or back-and-forth swinging of a river, in either hilly or flat country, has long been a subject of interest to both geologists and ordinary observers. Meander patterns are graceful and eye-catching; and just how meandering occurs continues to be a challenging question.

Meandering in highlands is usually explained by the rock structures the stream is following, or perhaps by other factors, such as landslides or lava flows, that have shunted the stream one way or another. Meanders on floodplains are another matter. Geologists for a century and more have been pondering just why and how these meanders develop, and a substantial literature on the subject has resulted.

A river on a floodplain – a broad, deep, nearly level mass of sediments – commonly follows a looping or sinuous course, with little or no downcutting but continued sidecutting. This meandering appears, at least superficially, to be the strategy adopted by a stream when overloaded with sediments – a phenomenon something like the tendency of a canoe paddle to move from side to side when pulled through the resisting water. However, experiments show that any stream which starts moving straight through sediments such as sand or clay on a gentle gradient will soon start weaving; laminar flow (flow within which flow lines are parallel) cannot be maintained. The reason appears to be that in any

such as sapping by groundwater, exfoliation of the rock, or frost prying. Niches made by swinging streams can be very broad. Shallow ones somewhat resemble the arched niches in cliffs left by rockfalls. Niches in finlike rock masses may be the first stage in the formation of a natural arch.

Stream-made niches are sometimes called alcoves, but this term is often used specifically for a recess developed where a spring exits from a cliff. In the West, alcoves such as those in the sandstone cliffs of Mesa Verde, Colorado, and in basalt

Where a Stream Meandered: On the Great Plains east of Bismarck, North Dakota, meander scars, oxbows, and cutoffs delineate the changing course of a stream during the past.

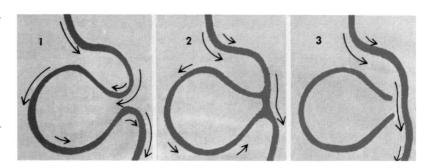

stream there is always some perturbation, or irregularity, and in any mass of sediments there are always some variations in composition and texture, and therefore in resistance. Where gradient is very steep, as on cliff faces, laminar flow can occur because of the high velocity of flow; thus parallel or nearly parallel furrows and lapiés are cut.

A meander's wavelength (the distance from loop to loop) increases with an increase in stream discharge, and may be affected also by the gradient and by the type and caliber of the sediments. The channel tends to be narrow and deep if the banks are on unusually cohesive material and if they are lined with vegetation, which also tends to limit erosion.

Meander loops travel, somewhat like loops of a rope being shaken. The bank on the outside of the curve is gradually undercut and retreats in the downstream direction. Because of friction between streamflow and the channel bank on the outside of the curve, a helical current, or swirl, develops there and carries debris from the undercutting to the other side of the channel, and deposits it on the inside of the next curve, creating the projecting mass of sediments called a point bar. As this scenario of erosion and deposition proceeds, loops migrate in the downstream direction.

The path, or zone, within which migration occurs is called the meander belt, which on long-established floodplains is commonly about a third as wide as the floodplain. Along the meander belt are curved depressions, or "meander scars," which are remnants of loops that have been cut off.

Oxbows, Cutoffs, and Oxbow Lakes

A meander in the form of a nearly closed loop is an oxbow. It results from more rapid downstream migration of the upstream side of a meander than of the downstream side.

At flood time, the stream may have the energy to break through the divide between two loops, making a channel called a cutoff. This becomes graded to the rest of the channel. When discharge moderates, the cutoff may become sealed with sediments – "clay plugs" – and in that case the stream may resume its former course. Cutoffs occur not only on floodplains but in incised meanders, where, by penetrating divides, they make the tunnels from which natural bridges evolve.

Standing water in an oxbow that was isolated by a cutoff, then sealed with alluvium (stream-deposited sediments), is called an oxbow lake. Its surface is the water table.

Formation of an Oxbow Lake: Arrows show the direction of streamflow: the longer the arrow, the faster the flow. (1) An almost complete meander circle develops. (2) High water floods across the neck of the loop, making a cutoff. (3) Deposition of sediments at low water seals the loop ends, creating the lake.

A Small Channel Bar: Near Tyndrum, Scotland, a stream rounding a curve loses velocity and portions of its bedload stall, accumulating to form a channel bar, typically with large-caliber sediments. Also here are an undercut slope, point bar, slumped channel bank, and broad alluvial terraces on glacial drift.

Deposits by Streams

When running water carrying sediments slows and loses transporting power, sediments are deposited in any of three kinds of locations: on the streambed; as channel bars or braids, at the sides of the channel; as natural levees and floodplains; and, at the stream's terminus, as alluvial fans and cones and as deltas. Sediments deposited by running water are called alluvium.

Channel Bars and Braids

Sediment deposited in a channel where a stream loses velocity may form a channel bar, elongated in the direction

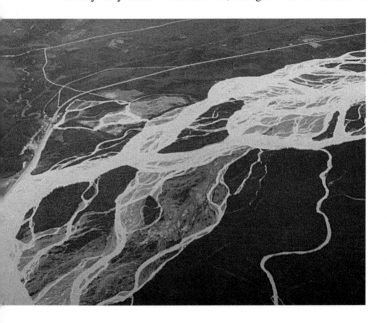

of streamflow. Often the sediment is of relatively large caliber, more difficult than other sediments for the stream to transport. Some bars may be above water when the stream is low. Bars may change in size and in form rapidly, according to changes in conditions such as volume of flow, volume of bedload, and caliber of the load.

As already indicated, where channel banks are of sand, gravel, and perhaps larger-caliber rock debris – all of which are relatively non-cohesive – the stream may resort to braiding; that is, dividing into many interweaving channels. Channels in the network are wider and shallower than those in meanders, but the gradient is higher, and some braided streams are actually downcutting. Spectacular examples of braiding are numerous in alpine valleys where glacial rock debris (including much large-caliber material) is thick, and on desert terranes, where streams have limited capacity to carry away debris from weathering and gravity movements. On some terranes, transitional processes between meandering and braiding exist.

Alluvial Cones and Fans

Sediments deposited by temporary streams where steep, narrow valleys reach a lowland often take the form of steep-sided cones. These may resemble talus cones but consist mostly of small-caliber material, which becomes finer toward the cone's edges because the velocity of the water decreases in that direction. Some alluvial cones include scatterings of talus – that is, rock fragments gravitated down from the valley walls.

Alluvial cones grade into alluvial fans, which have sides that incline more gently and which may cover a much wider expanse. Distributary channels run down from the fan's apex and, in swinging, build up the fan in the typical triangular form. Because fan material is usually more or less porous, streams usually sink into it before reaching the fan's edges; but occasionally water reappears at the base of a fan as a spring. Fans often include layers of material from mudflows and debris flows.

Dealing with Glacial Deposits: East of Fairbanks, Alaska, the Tanana River meanders and braids its way along the floor of a vast valley with deep deposits of glacial rock waste.

Braiding by a Desert Stream: Near Santa Fe, New Mexico, Tesuque Creek braids its sediments. Even in April, when the stream is replenished with water from snow melt in the mountains, the stream cannot keep much sediment moving.

An Alluvial Fan: Commonly associated with deserts, alluvial fans occur elsewhere also. This small one, in a sand quarry in New Jersey, is fairly typical, showing washes diverging from the gully made by temporary streams.

An Alluvial Cone: In a roadcut, a small intermittent stream emerging from an open fracture in limestone has deposited its sediment load to form an alluvial cone. The sediment is finer than is usual in a talus cone. However, a few small talus blocks have fallen onto the cone and slid down to its base.

are called alluvial terraces. Usually flanking the channel on both sides, they are above the stream's present level. They are remnants of floodplains formed when the stream was higher. They may now be covered with talus or vegetation.

Alluvial terraces can result from any conditions that cause the stream to narrow and deepen its channel. One such condition is uplift; another can be a change in climate. Alluvial terraces are familiar in valleys that were flooded with glacial meltwater toward the end of the Pleistocene but are now occupied by a smaller stream. Examples are common on valley floors throughout areas recently glaciated.

Cones and fans are developed most commonly on desert terranes, because the rate of sediment accumulation there is high relative to the rate at which streams carry sediments away, and also because of the contributions made by mudflows and debris flows. Cones and fans exist also on humid terranes, but are limited in size, numbers, and noticeability because of vegetation cover.

Floodplains

Floodplains are broad, nearly level surfaces on sediments, ranging from fine to coarse, that have been deposited by streams flooding over their channel banks. Floodplains are common; even in relatively narrow valleys small floodplains may here and there flank the channel. Floodplains that are miles wide, such as those of the lower Mississippi, characterize valleys in which major rivers, over millions of years, have deposited sediments received from tributaries draining hundreds of thousands of square miles.

Floodplains are numerous on many kinds of terrain but are best observed from highland summits or aircraft. Only broad views can do justice to major floodplain features such as

Alluvial Terraces

Nearly horizontal platforms of alluvium (stream-deposited sediments) extending from the base of a valley wall to a stream's channel are remnants of former floodplains. They

Development of a Floodplain: (1) The stream begins meandering and a point bar starts building up. (2) The meander loop migrates downstream, trimming spurs and widening point bars. (3) A floodplain has developed, with spurs (meander scrolls) trimmed to the width of the meander belt.

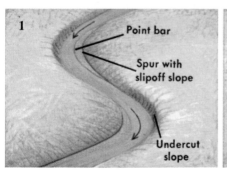
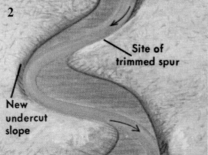
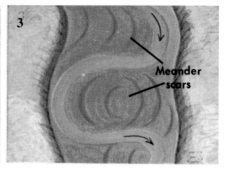

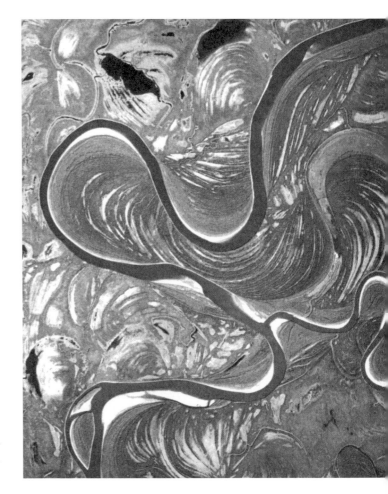

A Floodplain Exhibit: Broad valleys in Alaska, their bottoms deeply covered with rock debris from glaciation, present fine floodplain panoramas. This NASA-U.S. Department of Agriculture photograph shows meanders and meander scars, point bars (white), oxbows, and oxbow lakes.

meanders, meander belts, cutoffs, oxbow lakes, and natural levees. In the United States, aircraft afford especially fine views of floodplains along the Mississippi River from southern Illinois south, in the Midwest generally from the Canadian border south, on coastal plains, and, in Alaska, on terrains along the Yukon.

Natural Levees

A ridge of alluvium deposited along the bank of a meander by floodwaters as they suddenly spread and lose velocity is called a natural levee. Higher than the floodplain, it may rise many feet above it; it can be a mile or more wide; it is steep on the channel side, gently sloping on the other side.

Natural levees (and often artificial ones) may allow build-up of sediments in the channel, so that the river becomes more and more elevated. If the river at flood time breaks through the levee (making a new passage, called a crevasse), extensive and disastrous flooding of adjacent land may occur. Building and maintaining levees have long been a major – and often vain – measure for flood control along great rivers such as the lower Mississippi, the Nile, and the Yangtze.

An Artificial Levee: North of Tokyo, on Honshu, Japan, a levee has been built for protection of farms, residential areas, and industrial sites against periodic flooding by the river in the background of this photograph.

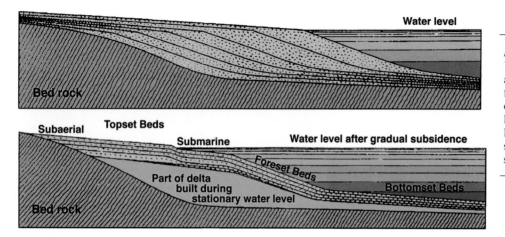

Deltas Built into Quiet Water: Delta A is built out over a stable rock platform. Deposits form a relatively simple series of layers from the shallows, at left, to deeper waters, at right. Delta B is built out over a subsiding platform; therefore its structure is notably different.

Deltas

Alluvial materials deposited by a river at the site where it enters a slow stream, a lake, or the ocean make a sedimentary landform called a delta. It can form only if sediments are ample and the slope outward from the shore is gentle, and if tides, waves, or currents are not strong enough to carry the sediments away as fast as they are deposited. Much delta-building was done during the Pleistocene by streams swollen by melting of glaciers or by heavy rains during this humid epoch.

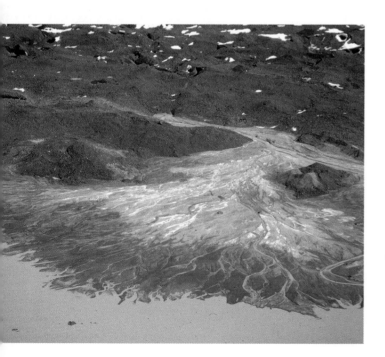

Most deltas are partly above water. The nearly level area that is above water is called a delta plain. Here the main-stream, from the mainland, divides into small streams (distributaries), which carry sediments toward the delta's edges. With time, if ambient conditions change little, sediments moved onto the delta are deposited farther and farther out; thus a well-developed delta has an advancing front, which consists of foreset beds of relatively coarse material, dipping down from the shore. Finer sediments, carried out further, form bottomset beds. A thin, nearly horizontal layer of sediments overlying foreset beds at an angle, and representing an extension of the floodplain, is known as a topset bed.

A delta plain often is patched with bodies of water, known as delta lakes, in areas where channels have been dammed by natural levees, or in depressions made by subsidence of underlying sediments.

Delta shapes vary widely because of a complex of factors. These include size of the river's drainage area, gradient, and amount and character of sediments deposited; also, shape of the coastline, nature of underlying rock structures, steepness of the slope of deposition, interactions of fresh water and seawater in terms of density and chemical composition, local climate (which influences the amount and types of vegetation present), nature of marine action (tides, currents, waves) along the shore, and stability of the underlying crust.

A Delta on a Glacier: On Alaska's Malaspina Glacier, a meltwater stream winding through masses of moraine on glacier ice finds its way to a lake on the ice and there deposits its sediment load to make an arcuate delta, complete with an intricate system of distributaries. The white color of the lake is due to suspended silt.

Delta Forms: Deltas, occurring in many forms, are difficult to classify into a few recognizable categories. Commonest are the arcuate, birdfoot (digitate), and cuspate forms.

Arcuate Delta

Birdfoot Delta

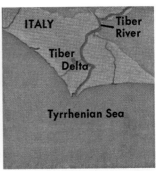

Cuspate Delta

Traditionally, deltas have been classified by shape into three broad types. Arcuate (fan, or lobate) deltas, such as those of Egypt's Nile and France's Rhône, are built of relatively coarse sediments and are subject to fairly limited wave and current action. Elongate (birdfoot) deltas, such as the Mississippi's, are characterized by fine sediments, constantly shifting distributaries, and relatively minor wave action. Cuspate deltas, such as the one built by Italy's Tiber River, involve rows of barrier beaches resulting from strong attack by waves and currents. Also recognized now is the Ganges-type delta, characterized by long, often parallel distributaries bordered by dense tropical vegetation and kept straight by strong tidal action. So-called estuarine deltas, consisting of sediments deposited by rivers as they meet seawater in estuaries, may not be deltas at all, since they differ substantially from other delta types in both development and structure.

The largest, most impressive deltas are those made by major rivers reaching the sea; they include some of the world's most extensive landforms. By recent measurements the present Mississippi Delta, for example, is about 12 miles long, 30 miles wide and some 850 feet thick. The Nile Delta has a length of 96 miles and a width of 145 miles. The dimensions of the Euphrates-Tigris Delta are 350 by 90 miles. The Huang-Ho, greatest of all, has a length of 300 miles and a width of 370. Incidentally, the Amazon has no delta; its mouth is an estuary.

An Arcuate Delta: A delta at Cook Inlet, Alaska, like others of the arcuate type, consists of deposits from narrowly spaced distributaries. Darkest areas in this photo are areas covered by vegetation; light areas, shore deposits; gray areas, ocean.

A Youthful Volcanic Landscape: Craters of the Moon National Monument covers 83 square miles of the volcanic Columbia Plateau. In the foreground here are three large spatter cones along the Great Rift; beyond are cones that produced lava and pyroclastics, the largest being Big Cinder Butte, about 800 feet high. Also in the area are pahoehoe and aa flows, pit craters, lava tunnels and caves, natural bridges, cinder crags, tree molds and casts, lava trees, volcanic blocks and bombs, and the lava plain. Volcanic activity dates back thousands of years; the most recent eruption was in 1538.

8
Features of Igneous Activity

AMONG the most interesting of landforms are volcanoes and the myriad forms associated with them. Ancient man viewed volcanic eruptions and accompanying earthquakes as acts of angry gods. It was after the Roman god of fire, Vulcan, that volcanoes were named. In the Middle Ages and even into modern times volcanoes were believed by many to be chimneys for great fires raging inside the earth – the fires of Hell, perhaps. Today, despite (or perhaps because of) the wide understanding that volcanic activity is due to a natural cause – that is, heat generated by chemical and physical changes within Earth – volcanoes continue to fascinate with their majestic explosions, spectacular and sometimes dangerous flows of lava and ash, and the multitude of strange forms taken by volcanic materials as they solidify and are reshaped by the elements.

Probably all regions of the earth have at some time known volcanic eruptions. Where these have occurred during recent geologic time – say, the past 100,000 to 500,000 years – volcanic cones, lava flows, and other volcanic forms and materials are likely to be visible at the surface and recognizable for what they are. A few examples of such terranes are Craters of the Moon, in Idaho; most sectors of Iceland, Hawaii, and New Zealand; several areas of Italy; and the vicinity of the Great Rift Valley in Africa. Where volcanic activity occurred in the more remote past, its traces have been more or less eroded away, covered by vegetation or erosion debris, or sunk beneath the sea. In northern Europe, for example, most evidences of prehistoric volcanic activity have been largely eroded away or grown over, and are hardly apparent except to informed observers.

Through Earth's history the intensity and extent of volcanic activity have fluctuated. At times the crust has been relatively quiet, with processes of weathering, erosion, and sedimentation dominating. During other intervals thousands of volcanoes and fissures have been exploding and spewing huge quantities of igneous material into the sky and onto lands and sea bottoms. India's basaltic Deccan Plateau, formed by fissure flows from the Réunion Plume over a period of several million years, around 65 million years ago, covered perhaps a million square miles to depths as great as a mile or more before erosion reduced the area to the present 200,000 square miles. The Siberian Traps, another vast lava accumulation, cover 756,000 square miles. Eruptions from the Yellowstone Plume, which built up the Columbia Plateau between 17 and 6 million years ago, covered perhaps 120,000 square miles with volcanic materials to depths as great as 3,000 feet.

Monument to a Volcanic Past: About two centuries ago, residents of Le Puy, in the Haute-Loire, France, were unbelieving yet frightened when a pioneer geologist announced that their homeland had once blazed with volcanoes. Volcanic it had been, but in prehistoric times. Eroded lava flows patch the land and volcanic necks survive as rock towers. Perched on a volcanic neck in Le Puy is St. Michel d'Aiguille Chapel.

Once a Volcanic Landscape: Highlands of the famed Lake Country of northern England are made of volcanic materials dating from the Paleozoic. During hundreds of millions of years the deep accumulations of volcanics have been altered, deformed, and eroded into their present forms.

Some scientists believe such intense, extensive volcanic activity has in some cases been triggered by collisions of Earth with asteroids and comets. The Siberian Traps and the Deccan Plateau were built up toward the end of the Permian and Cretaceous periods, respectively, suggesting that worldwide extinctions of flora and fauna that occurred at those times may have been due to billions of tons of dust, carbon dioxide, sulfur, and other harmful materials blown into the atmosphere by volcanoes. Some studies indicate that major volcanic activity may cause, not result from, rifting between crustal plates. Collisions between Earth and asteroids or comets also may result in either volcanic activity or rifting, or both.

During the past 100 million years, hundreds of thousands of volcanoes have developed and become extinct on the continents alone. During the past 2 million years eruptions of the explosive type have increased. Volcanoes active now number about 800, of which about 775 have developed in ocean basins or along fringes of continents, and only about 25 in continental interiors. Apart from volcanoes, there are areas such as Iceland where large volumes of lava emerge from the crust along fissures, spreading broadly but not forming volcanic cones. Again, in many sectors of the earth there are unnumbered bodies of igneous rock that formed within the crust and, in the future, may be uncovered by erosion and take their place as recognized landforms.

Realms of Volcanism

Volcanic eruptions occur in several kinds of crustal environments. Some magma emerges along lengthy fissures that open in the ocean bottoms or in the continents as plates move apart, or, locally, as the crust becomes fractured by tectonic forces. Eruptions occur along the midocean ridges and the

One of a Chain: St. Augustine Volcano is one of at least 80 major volcanoes in Alaska's 1,600-mile-long Aleutian Arc. Like other stratovolcanoes, this one consists of alternating layers of lava and pyroclastics, and its slopes are subject to deep gullying and gravity movements, such as the mudflow in the foreground here.

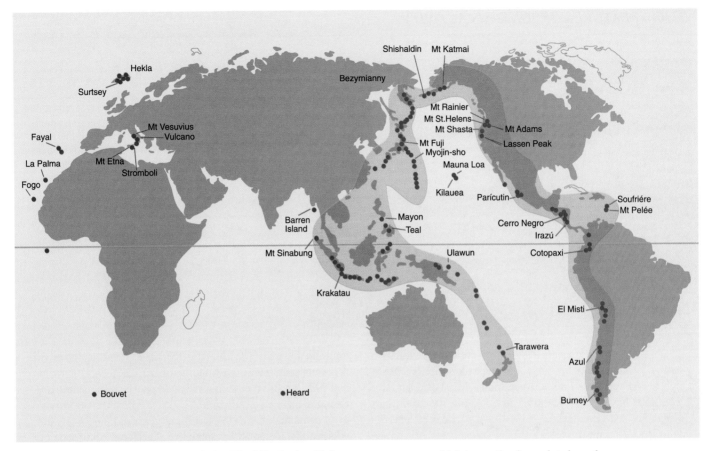

Distribution of the World's Active Volcanoes: On a small map not all volcanoes can be shown, but the overall pattern of their occurrence, which is mostly along plate boundaries, is apparent. Those indicated are volcanoes active during recent geologic time.

lava spreads sideward, in a sort of conveyor-belt motion, becoming new ocean floor. Eruptions are common in subduction zones, where one plate plunges under another, crumpling and breaking the crust, creating fractures through which magma can reach the surface. "Plumes" of very hot material rising from deep in the mantle penetrate continental crust here and there, surging to the surface through fractures. Volcanic activity can take place in polar regions as well as warmer climes; even now volcanoes are performing under the Antarctic ice sheet.

During recent geologic time (and probably earlier), most volcanic activity has occurred approximately along crustal-plate boundaries. That is true today, as witness the so-called Ring of Fire around the Pacific Ocean basin – a chain of volcanoes, active now or during recent centuries, extending south from Alaska down the western side of the Pacific through the islands of Indonesia to Antarctica, then up through western South America and Central America to western North America, thus completing the Ring of Fire.

Centered in the ring are the Hawaiian Islands, created during the past few million years by magma rising through fractures in the Pacific Plate. These islands present a geologic problem. They are oldest in the northwest, youngest in the southeast, suggesting either that the plume which is the source of the magma has remained stationary while the plate has been moving northwestward over it, or, just possibly, that currents in the mantle have broken up the plume, so that parts of it have been producing volcanoes in the different locations. In the southeast, on the Big Island (Hawaii), volcanic activity continues, from a magma chamber thought to be only about 5 miles beneath the ocean bottom.

Southeast of Hawaii are the Galápagos Islands, about 550 miles west of the South American mainland. These also have been built up from the ocean bottom during the past few million years by volcanic activity caused by a hot spot or a chain of such spots. Volcanic activity here continues.

The Atlantic region today has no ring of fire comparable to the Pacific ring, but it does have recently active volcanoes.

Remnants of a Giant Cone: Dominating about 3,000 square miles of volcanic scenery on the Colorado Plateau around Flagstaff, Arizona, is San Francisco Mountain. Its highest point, Humphreys Peak, attains 12,633 feet. Fully developed volcanoes and cinder cones were active in this area for millions of years, into the past few centuries.

Iceland, on the Midocean Ridge, was created by volcanic activity and continues to be the scene of eruptions from cones and fissures. In the West Indies Mt. Pelée and La Soufrière have been recently active. Along the Midocean Ridge are scattered islands – the Canaries, Azores, and St. Paul Rock – with active or dormant volcanoes.

In southern Europe, Italy has Mts. Vesuvius, Etna, and Stromboli, all intermittently active. From the Greek islands in the Mediterranean through the Middle East to Indonesia a scattering of volcanoes were active in recent centuries; so were cones neighboring the Great African Rift.

In prehistoric times the patterns were different. In the Oligocene and Miocene, for example, volcanic cones were erupting in areas flanking the Rockies, including parts of Montana and Idaho, Nevada, southern California, Arizona and New Mexico, and southwest Texas. In these states dry climates have favored preservation of volcanic landforms, even some that are 10 to 20 million years old, and many are still readily identifiable, thanks to scantiness of vegetation and lack of development.

In the Jurassic and Triassic periods, long chains of volcanoes were active in eastern North America from Nova Scotia to Virginia, and in western Europe from the British Isles to southern France. Cones are mostly gone, but some cone remnants and lava flows became buried under erosion debris and are now being uncovered by erosion.

A Volcanic Pile: A highland which is the eroded remnant of a mass of volcanic materials is sometimes termed a volcanic pile. An example is Ben More, rising 3,843 feet in the Grampian Mountains of Scotland.

Igneous Activity: A Closer Look

As we know from earlier pages, igneous activity is the movement of molten materials within and on the crust. Eruptions of such materials at the surface through cones or fissures are volcanic activity; movements of molten matter within the crust are plutonic activity.

The mantle consists mainly of solid or nearly solid materials of basic composition, portions of which rise in convection currents, melt or soften as pressure decreases, cool as they approach the crust, turn down, become reheated, and rise again, in repeating cycles. Hot concentrations of this material, bulbous in form, originating deep in the mantle, are called "plumes." Plume material rises against the crust and penetrates it by way of fractures or weak, spongelike zones, pushing crustal material aside and consuming some of it. Thus "hot spots" are created in the crust.

Some magma rising in the crust cools enough to solidify (freeze) there, becoming intrusive features such as batholiths, laccoliths, sills, and dikes. Other bodies, retaining enough heat to stay fluid, continue toward the ground surface, expanding as they rise and often mixing with broken crustal rock and groundwater. Finally this magma erupts into the upper world as foaming fountains or flows, called lava, or as blasts of shattered lava and rock fragments, known as pyroclastics. Lava and pyroclastics make up volcanic cones, lava flows, and beds of loose material, with countless variations of form. Associated with eruptions of lava and pyroclastics are the activities of geysers, hot springs, fumaroles, and other minor events, which too produce a variety of landforms.

"Fire" from Below: Volcanic Eruptions

Magma that reaches the surface has risen from a zone of melting within the crust or upper mantle. If it contains much crustal material it is likely to be more or less acidic; magma that is mostly of mantle material is basic. Some vents from which magma erupts are the long, deep fractures called fissures; others are intersections of fractures that act as pipes for rising magma. The top of a pipe is usually in the bottom of a crater in a cone formed of erupted materials.

As magma moves toward the surface, pressure on it diminishes, it cools and expands, and it starts to become the mineral matter that makes up lava or pyroclastics. Nearing the surface, it may encounter groundwater and convert this into steam. Finally it reaches the vent as a foaming, more or less viscous, white-hot fluid – a sort of devil's soda pop.

The nature of a volcanic eruption depends largely on whether the magma is basic, acidic, or intermediate in composition, and on whether the eruption is from a fissure or a pipe. Oceanic cones (those rising from the ocean bottom), as in

Some Forms Produced by Igneous Activity: Portions of the magma rising from the mantle or generated in the crust cool and solidify in or on the crust in varied forms. Portions penetrate crustal fractures, in some places pushing up crust and in others emerging at the surface as lava flows or lava shattered in explosive eruptions. Both magma solidified in the crust and magma erupted onto the surface become various kinds of igneous rock, and masses of these rocks become varied assortments of landforms.

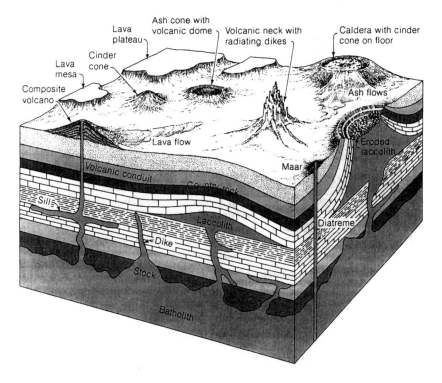

A Fissure Eruption: During the activity of Hawaii's Kilauea Volcano in 1971, this eruption occurred in the southwest rift zone. Typical were the lava fountains, forming a curtain of fire, and the spreading lava flows. Kilauea erupts every few years.

Fountaining Lava: In the 1959-1960 eruptions of Kilauea Volcano, lava fountains along the Curtain of Fire reached up to 1,900 feet. The blobs of cooling lava then fell as pyroclastics.

Hawaii, Iceland, and New Zealand, produce mostly basic lavas, from both fissures and pipes. Continental volcanoes (those rising from large land masses), such as the Cascade volcanoes of the Pacific Northwest, the Andes, and Italy's Vesuvius and Etna, produce mainly acidic and intermediate lavas, principally from pipes.

Where magma is basic and the vent is a fissure, the emerging magma becomes a relatively fluid lava with a temperature of 2000 to 2200 degrees F. and a melting point as low as 1400 degrees F. A voluminous flow can spread over tens to hundreds of square miles, filling valleys and creating a vast lava plain. Successive flows from many fissures may accumulate to form a lava plateau thousands of square miles in extent and hundreds or thousands of feet thick.

Hawaiian eruptions typify those of oceanic volcanoes. Eruptions often start with the opening of a long fissure, sometimes in the side of a cone. From this, red- or white-hot lava squirts in a line of fountains called the Curtain of Fire. As the eruption progresses, fountains can rise hundreds of feet, showering the area with blobs of lava that cool to become pumice, cinders, and other pyroclastics. After a few days, usually, the eruption settles down to quiet flows.

Some Hawaiian eruptions are from a pipe exit in the bottom of a crater in a cone. These eruptions are livelier than those of the fissure type because magma in a pipe is under greater pressure. Magma emerges as spewing gas and lava fountains, which fill the crater with a boiling lake. Repeated minor explosions blow clots of lava and cone fragments into the air, and these, cooling and solidifying in flight, fall back into the crater or onto the sides of the cone. A "lava lake" formed in the cone may flow over the crater's edge and move freely downslope, at a velocity of 30 miles per hour or so, and may travel tens of miles before becoming too cool and viscous to flow further. Some Hawaiian flows reach the ocean as brilliant "lavafalls."

Acidic lava, produced mainly by continental volcanoes and cooling to form pumice or rhyolite, is erupted from pipes in cones and occasionally from breaks in the side of a cone. Relatively viscous and stiff at its extrusion temperature of some 2200 degrees F., it cannot flow far before solidifying. Being very viscous, it tends to clog the pipe, causing pressure buildups that end in explosions, often huge. Explosions produce large quantities of pyroclastics. Coarse, relatively heavy materials – boulder-size rocks and large cindery masses – fall near the vent; finer materials, including ash and dust, fall farther away. In some eruptions lasting a few days, cubic miles of material are blown out, blanketing hundreds of square miles, and clouds of the finest particles may rise into

The World's Greatest: In Iceland, in 1783, occurred probably the greatest of all fissure flows witnessed by modern humans. Pouring out from the 15-mile-long Laki fissure, the Skaftareldahraunit, as it is called, covered some 220 square miles with pillowy basalt lava. Today the rock of this lava plain is coated with green lichen.

the stratosphere and be carried by winds around the world. Colorful sunsets worldwide are a familiar consequence, and even changes in weather can result.

Among recent major eruptions of volcanoes that produce acidic lavas have been outbursts from Mt. Pinatubo, on Luzon, in the Philippines. On June 15, 1991, after 600 years of dormancy, the mountain blew out more than 7 billion cubic yards of material. Within 2 weeks, ash and gases blown up into the stratosphere had drifted around the world, and during the following year this cloud blocked enough sunlight (3.8 per cent, according to satellite readings) to reduce average Earth temperatures by 0.5 degrees C. Eruptions continued intermittently, and the usually heavy fall rains touched off lahars. By April 1997 an estimated 9 billion cubic yards of pyroclastics had swept down the mountain and spread over the region, burying villages and farms to depths as great as 90 feet, and creating a new landscape. About 1,000 people had died directly or indirectly from Pinatubo's activity. An estimated 25 billion cubic yards of pyroclastics – makings for future lahars – remained on the mountain.

In recorded history the greatest of all explosive eruptions was that of Tambora Volcano, on Sambawa Island, Indonesia, on April 10–11, 1815. An estimated 18 cubic miles of the mountain top was blown off, and between 24 and 36 cubic miles of pyroclastics was produced. More than 88,000 people died in showers of hot pyroclastics, in tsunamis produced by earthquakes during the eruption, and, later, from illness and starvation due to noxious gases and destruction of farms. Ash rising and spreading through the atmosphere blocked so much sunlight that temperatures fell over much of the globe, and 1815 became known as "the year without a summer."

Internal pressure, due to blocking of the pipe or vent by a plug of lava, may burst the side of a cone and allow a red-hot cloud (nuée ardente) of shattered magma particles to rush forth. Heavier than air, the cloud flows downslope like an avalanche, more or less in contact with the ground, at a ve-

A Historic Mediterranean Feature: Mt. Vesuvius, rising about 4,200 feet above the Bay of Naples, Italy, originated during the Pleistocene, then went dormant until its catastrophic outburst in A.D. 79. This is the mountain as it was in 1955, before more recent eruptions. The peak at left is Monte Somma, part of the large crater that half-circles Vesuvius on the north side.

locity as great as 100 miles per hour, and may travel tens of miles, incinerating everything in its path. Such was the story in 1902 when the side of Mt. Pelée, on Martinique Island in the West Indies, blew out. An incandescent cloud swept down over the city of St. Pierre, killing all but two of its 36,000 people. (The two were prisoners in an underground cell.) A nuée ardente may result not only from a break in the cone but also as a column of gas, ash, and dust blown high above the vent loses momentum and, being heavier than air, collapses and falls, then rushes down the sides of the cone.

Numerous volcanoes erupt lavas of intermediate types, which form rocks such as andesite and dacite. Eruptions may be of basic magma at some times, acidic magma at others. Over long periods, a volcano may become either more basic or more acidic in output, depending on varying conditions in mantle and crust. A reservoir of basic magma within acidic crustal rock, such as granite or quartz sandstone, may in time absorb more and more of this rock, and lava produced in eruptions may therefore become increasingly acidic.

A volcano may become dormant for periods ranging from a few years to hundreds of thousands of years, and then suddenly begin a new series of eruptions. Mt. Vesuvius had been dormant and unrecognized as a volcano (except by a few scholars, among them the Greek geographer Strabo) for thousands of years before the catastrophic eruption of A.D. 79, which buried Pompeii under showers of pumice and flooded Herculaneum with lava. Mt. St. Helens, in Washington state, had been dormant since 1857 before it resumed activity in May 1980. Explosions blew off the top 1,300 feet of the cone and sent about a cubic mile of pulverized rock and ash into the atmosphere. The blast, with resulting rockslides and debris flows, devastated an area 8 miles long and 15 miles wide on the mountain's north side. Cities 80 miles away were blanketed with ash. The largest ash cloud reached 63,000 feet, and dust drifted around the world.

Eventually, crustal movements or solidifying magma close the fractures through which a volcano is fed. Then the volcano becomes extinct, and the cone is gradually erased by erosion or covered by erosion debris.

Volcanic Cones

A volcanic cone, or "volcano," is built up by successive eruptions of materials from a pipe. A crater is produced around the pipe by explosions, collapse of cone materials around the vent, and sinking of magma within the pipe as eruptions cease. Cone forms are determined largely by the nature of the magma.

A Volcano Awakened: After more than a century of quiet, in March 1980 an earthquake announced that Mt. St. Helens was stirring. More earthquakes followed, then a series of minor eruptions, and finally, on May 18, came the devastating blast. This photo, taken 2 years later, shows the great crater formed by the eruptions. In its center is a lava dome.

A Scoria Cone: Capulin Mountain, one of numerous cones in the Raton Basin, northeastern New Mexico, has been active during the past few thousand years. It is mostly of scoria and cinders. These readily absorb water; hence the lack of gullies on the cone's sides. The diagonal line is an auto road.

Cinder cones, usually not more than a few hundred feet high, consist mostly of layers of loose basaltic or andesitic cindery material erupted in minor explosions and falling near the vent. The sides may be as steep as 40 degrees. When young, a cinder cone is very porous and absorbs water easily; thus it tends to remain ungullied by runoff. Older cones, weathering to clay, become less porous and more subject to gullying. Cinder cones may develop as a chain along a fissure and can grow rapidly. The cinder cone of Paricutín, in Michoacan, Mexico, grew to about 450 feet during its first week, in 1943; it reached 1,000 feet in its first year and some 1,200 feet in 7 years.

The term cinder cone may apply to any cone consisting entirely or almost entirely of pyroclastics. Most such cones actually are of cindery material, "cinders" being by definition 0.5 to 5.0 centimeters (0.20 to 2 inches) in diameter. Some cones, such as New Mexico's Capulin Mountain, near Raton, are of scoria, which consists of heavier, larger, clinkery fragments. Still another type is the tuff cone, built of volcanic ash or dust welded solid by heat; examples include Koko Head, on Oahu Island, Hawaii, and numerous cones in the Galápagos Islands.

During eruptions, cones are subject to severe stresses and strains. Fractures in a cone may become conduits which lead magma to vents in the sides. Minor eruptions from these vents build up so-called parasitic cones on the flanks of a main cone. In time, the "parasite" may grow to immense size, rivaling that of its host. Mt. Etna, in Sicily, has more than 200 parasitic cones.

Basaltic lava may be erupted from minor vents mostly as fountains of gas, with clots of lava that cool and solidify as they fly through the air. This material, called spatter, cools as a clinkery rock which may accumulate to form sizable cones astride the vents. Spatter eruptions often occur along a fissure that has been invaded by substantial amounts of water, which on contact with magma changes to steam, creating pressure that helps to cause further eruptions.

Long-continued basaltic flows from a pipe, spreading widely, build what is called a shield volcano, shield cone, or lava dome. This has the shape of a shield laid on the ground, convex side up. Its great area compared to its height is due to the fluidity of basaltic lava and the relative sparseness of associated pyroclastics. The cone's sides generally increase in steepness from a few degrees near the crater, where the lava is hot and relatively fluid, to 10 or 12 degrees at the base,

Dissection of a Volcano: At Buccaneer Bay, Santiago Island, in the Galápagos, waves and currents undercut cones, causing rock and debris falls which expose layered inner structures. The cone at right is mostly of hardened tuff capped by a basaltic lava flow; the cone at left is of tuff layers and lava.

Butte, Idaho, on the volcanic Columbia Plateau, and Prospect Peak, in Lassen Volcano National Park. Earth's largest shield is Hawaii's Mauna Loa, which rises 13,677 feet above sea level and more than 28,000 feet from the ocean floor, and has a diameter of some 60 miles. Its crater is 33 miles long, 1.5 miles wide, and 600 feet deep, and its volume, 10,000 cubic miles. The largest known shield volcano in the solar system is Olympus Mons, on Mars, towering 15 miles above its surroundings and covering more than 100,000 square miles.

Most volcanoes, especially on continents, produce both lava flows and pyroclastics. During thousands or millions of years, alternating layers of lava and pyroclastics accumulate to form a large cone called a stratovolcano, or composite volcano. Pyroclastics falling near its crater may maintain a slope of 30 degrees or more, but this diminishes to 5 degrees or less toward the cone's base. Fissures radiating from the main pipe may allow lava to escape through the cone's sides. Stormwater may mix with loose material on the slopes, destabilizing it and causing rock avalanches, debris flows, or lahars. Solidified lava layers in the cone are low in permeability and allow little percolation of water; thus they become gullied by runoff.

Stratovolcanoes are the commonest kind of volcanic cones.

where the lava, having flowed far and lost much heat, reaches its solidification temperature and stops flowing. An increase in production of pyroclastics by the volcano may cause the buildup of cinder cones on the dome, and minor elongated domes may be built up along the sides of the shield by flows from fissures.

The shield form is characteristic of many oceanic volcanoes, including those of New Zealand and Samoa as well as Hawaii, Iceland, and the Galápagos. Also of shield form are a limited number of cones on the continents, such as Ferry

Mauna Loa Panorama: Several calderas, all at least 1 kilometer in diameter, appear in this north-northeastward view over the summit of Mauna Loa. A light snow covers the scene. In the background is Mokuaweoweo, 3 miles long, 1.5 miles wide, and as much as 600 feet deep. Black areas in the foreground are lava flows. Mauna Kea is in the distance, covered partly by clouds.

Among the Most Distinguished: Mt. Kilimanjaro, on the Kenya-Tanzania border, is Africa's highest peak, rising to 19,340 feet. This beautiful, famous stratovolcano, capped by snow although near the equator, has several craters, built up by intermittent volcanic activity starting around 2 million years ago. Eruptions have occurred during recent centuries.

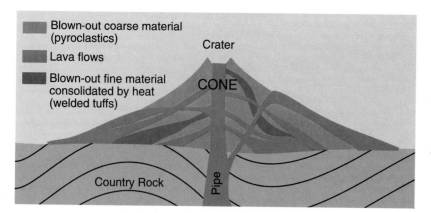

Blown-out coarse material (pyroclastics)

Lava flows

Blown-out fine material consolidated by heat (welded tuffs)

Crater

CONE

Country Rock

Pipe

Structure of a Typical Stratovolcano: The commonest form taken by large volcanic cones is the stratovolcano, built up of alternating layers of lava and pyroclastics. Its slopes often are not as steep as those of cinder cones but are generally much steeper than the slopes of shield volcanoes.

Some well-known examples active today or in the recent past are Mts. St. Helens (9,677 feet before it lost 1,300 feet in the 1980 eruptions), Rainier (14,410), Lassen (10,457), and others of the Cascade Range; also, Katmai (6,715), in Alaska; Fujiyama (12,389), in Japan; Pinatubo (5,770), in the Philippines; and Cotopaxi (19,347), in Ecuador – all in the Pacific Ring of Fire. Vesuvius (over 4,000), Stromboli (3,038) and Etna (10,700), in Italy, are prominent among the sprinkling of Mediterranean stratovolcanoes along the Europe-Africa plate boundary. Tanzania's Mt. Kilimanjaro (19,340), another stratovolcano, is the largest of many volcanoes neighboring the African Rift Valley. Mt. Hekla, in Iceland, with both shield and stratovolcano characteristics, rises only 3,280 feet above the lava plateau on which it is built up, but it is nearly 17 miles long and 1 to 3 miles wide – a respectable mountain ridge.

Principal Craters, Calderas, and Pit Craters

All volcanoes have a principal central crater – a steep-sided basin – in the top of the cone. In the bottom of the crater is the vent. The crater is formed by explosions or by accumulation of lava and pyroclastics to form a rim around the top of the cone. The crater's bottom is hardly larger than the vent. At times of inactivity, rock material slides down the crater walls into the vent and clogs it; but during active periods the vent is kept open by explosions.

With time, a crater may become enlarged by explosions and, more commonly, by collapse of its walls because of withdrawal of supporting magma from below. The enlarged crater, which can be 10 miles or more wide and more than 1,000 feet deep, is called a caldera (Portuguese *caldeira*, ket-

The "Perfect" Cone: The majestic size and form – not to mention eruptions – of Fujiyama, or Mt. Fuji, on Honshu, Japan, established it as a sacred mountain early in Japanese history. This stratovolcano is still the objective of pilgrimages, an inspiration for poets and artists, and a prime attraction for tourists. Now dormant, it last erupted in 1707.

On the Andes Volcanic Chain: Massive Cotocachi, a stratovolcano north of Quito, Ecuador, has been long dormant, erosion has furrowed its slopes, and man has developed farms and villages around its base. Many other Andean volcanoes, such as Cotopaxi, south of Quito, have been active in recent time, occasionally causing much property damage and loss of life.

tle). Its floor is much wider than the vent. Renewed activity after a dormant phase may build up new cones, sometimes nested, on the caldera's floor. Pressure beneath the floor may raise domes, called tumuli.

In the coterminous United States the best-known caldera is the one occupied by Crater Lake, 5.4 miles in diameter and nearly 2,000 feet deep, in Oregon. This is the crater of the Pleistocene volcano Mt. Mazama, in the Cascade Range. Originally about 12,000 feet high, Mazama exploded some 7,000 years ago, blowing off much of its top and raining some 18 cubic miles of pyroclastics on thousands of square miles of the Northwest. Rising above the present lake is Wizard Island, a typical cinder cone, built up by minor activity after the major outburst. Less well known than the Mazama caldera, but larger, is the 14-miles-wide Valles Caldera, 1.1 million years old, in the volcanic Jemez Mountains of New Mexico. Tanzania's Ngorogoro Crater, one of the world's most famous animal habitats, 2 to 3 million years old, is about 12 miles in diameter, with a floor area of some 190 square miles. The caldera of Alaska's Mt. Katmai,

Inside a Volcanic Cone: Irazu Volcano, in Costa Rica, rises to 11,200 feet and is the only peak in the Americas from which the Atlantic and Pacific oceans are both visible. This stratovolcano has been active in recent decades. From an edge of the principal crater one sees at right the crater lake and, on the crater walls, scars left by rockslides and mudflows. The blocky, rugged topography results from collapse of parts of the crater wall due to removal of supporting rock and lava during eruptions.

A Pit Crater Holding a Lava Lake: The Pauahi Pit Crater, on a flank of Kilauea Volcano, is partly filled by a solidified lava flow. "Frozen lava" lakes like this are common in pit craters. After the most recent eruption this "lake" cooled to ordinary temperatures in about 25 years.

A Boiling Lake: A single explosive volcanic eruption may create a crater without a cone. Around the crater, known as a maar, shattered rock forms a crater ring. Usually the crater contains a lake. This maar formed in 1886 during the eruption of New Zealand's Tarawera Volcano. Rock beneath the lake is close to magma and is therefore very hot, causing the lake to steam.

A Lake in a Caldera: Crater Lake (which might better be called "Caldera Lake"), in Oregon, occupies the caldera of Mt. Mazama, a Pleistocene volcano which blew its top about 7,000 years ago. Wizard Island, a cinder cone, now extinct, rises above the water. As with many other calderas, the bottom and sides are sealed by hardened lava and pyroclastics; thus the crater can hold a lake.

A Lava Cave: At the edge of a basalt lava flow north of Pocatello, Idaho, is this opening in the side of a lava tunnel. As the flow cooled, the lava became more viscous, and increasing pressure caused it to break through the tunnel's side. The lava is of the rugged aa type.

known as the Valley of 10,000 Smokes (from the aftermath of the great 1912 eruption), is about 2 miles wide and 3 miles long. In the La Garita Mountains in southwestern Colorado is a caldera recently reported to be 27.8 million years old, 50 miles long and 20 miles wide, produced by one of the greatest of all known volcanic eruptions. Even larger, perhaps, is 2-million-years-old Yellowstone Caldera, about 50 miles in diameter, over a hot spot and plume thought to reach down to Earth's liquid outer core. Indeed, calderas are common, having been identified even where volcanoes have been long extinct, as in eastern North America, central and southern Europe, and central Asia.

An incidental feature on or near cones, especially shields, is the depression called a pit crater, volcanic sink, or lava pit. Such features, a few feet to hundreds of feet wide, result from collapse of the roof of a depleted magma chamber or the roof of a lava tunnel. Pit craters may form in chains along zones of roof collapse. On occasion they may fill with fresh lava from a magma chamber or from overland flow. Halemaumau Crater, the "fire pit" in the caldera of Kilauea Volcano, on Hawaii Island, was formed by roof collapse over a magma chamber.

Sculptures in Basaltic Materials

While still very hot, basaltic lava flows freely, and the rock formed from it has a smooth surface, which when fresh glistens with a thin coat of volcanic glass. This rock, known by the Hawaiian name *pahoehoe* (pah-HOI-hoi), may be shaped into coils, ropes, filaments, or billows (waves), resulting from movement of lava beneath the cooling and stiffening surface. The surface is sometimes scalloped or textured like elephant skin or shark skin.

As it cools, a basaltic flow stiffens and flows less freely. Its surface breaks up, becoming blocky or clinkery and jagged. Such a surface bears the Hawaiian name aa (AH-ah) – a name appreciated when one tries walking across it. In the

Southwest, expanses of aa are *malpais* (Spanish, "bad country").

Both pahoehoe and aa may occur on the same flow. Also, a flow may carry along cinder crags, which are heaps or towers of hard lava broken off the cone. Here and there, higher portions of older land not covered by the flow rise above the flow like islands; in Hawaii these are "kipukas." Basaltic lava emerging from a vent underwater may assume close-fitting pillow forms, each a few inches to a few feet in diameter. Rounding is due to relatively rapid and uniform cooling

In a Lava Tunnel: Among other spectacular features at Lava Caves State Park, Oregon, is this lava tunnel, or tube, which is more than a mile long and in some places is 50 feet high and nearly as wide. Here, as the outside of a lava flow cooled and formed a crust, lava continued to flow inside, then subsided, leaving the tunnel.

of small lava masses as they emerge into water. Similar forms may occur along edges of flows on land.

Among the more spectacular forms of basalt are columns. These parallel pillars, often nearly vertical, take shape as a mass of lava, cooling slowly and uniformly, becomes divided by tension fractures due to contraction. Cross sections of the columns are polygons with three to eight sides, usually six (then they are hexagons). The rock being homogeneous, jointing tends to be symmetrical, forming sides of equilateral triangles, which afford the closest fit. Honeybees follow a

A Pressure Ridge: As crust on a pahoehoe flow is subjected to flow pressure, it may bulge up to form a ridge with a convex top, as here on Fernandina Island, in the Galápagos. Stretching usually causes the ridge to crack along the crest.

similar pattern in making a honeycomb, and mudcracks also tend to be polygonal, though less precisely so.

Basalt columns, occasionally seen in cliffs and roadcuts, may be bent one way or another as a result of movement in the lava during cooling or because of creep. Sometimes they are surmounted by an entablature: an irregularly jointed mass, often vesicular (pocked with holes left by escaping gases). Forming near the surface of a flow, an entablature is subjected to effects of flow movements, escaping gases, and non-uniform cooling.

As the surface of a flow cools and stiffens to form a crust, lava beneath it may flow for some distance, then break through the crust and solidify outside. The passage left vacant is called a lava tube or lava tunnel, and the exit is a lava cave. The Thurston lava tube on Kilauea Volcano, Hawaii, is a prime example of tubes, some of which are as big as subway tunnels.

Hanging from tube and cave ceilings there may be lava "stalactites," which are conical or cylindrical masses formed by lava drip as it cools and solidifies. Lava "stalagmites," which are conical or stumplike masses built up from the floor by drip, also occur. Lava stalactites and stalagmites may lengthen until they join, forming columns. Similar forms in limestone caverns are produced by a very different process: precipitation of minerals from groundwater.

In some flows, internal pressure forces lava out through a break in the crust to form a bulbous or elongated mass called a squeeze-up, or "toothpaste lava." A lava extrusion from a small tube is a "toe." An elongated hump made by fluid lava pushing against resisting stiff lava is a pressure ridge, which is fractured along the top.

Lava flows sometimes flood forests, burning lower portions of trees. Clusters of dead trees may stand like ghostly apparitions long after. If a tree burns through at the base, the ashes may be removed later by weathering and erosion, which leave a hole in the lava flow. This cavity, called a tree mold, does not show a bark imprint (the bark was completely consumed), but may show traces of knots. If a flow sub-

A Squeeze-up: Pressure in a pahoehoe flow may cause the crust to bulge up and break, allowing lava to squeeze out like toothpaste. This squeeze-up is on Santiago Island, in the Galápagos. In the distance are several kipukas.

A Lava "Toe": In this pahoehoe flow on Santiago Island, in the Galápagos, a small lava tube while still flexible was contorted by flow pressure into a form suggesting a French horn. The small opening at the end of the coil (*upper right*) was an exit for gas or a dribble of lava. The scene is about 4 feet wide.

New Pahoehoe: A new, very black pahoehoe flow from Kilauea, recently cooled, still has its glistening coat of volcanic glass. Swirls were made as fluid lava flowed up against cooling, more viscous lava that was slowing down. The glassy covering of new pahoehoe is very fragile and tinkles when walked on. Older pahoehoe is grayer and duller though still smooth.

Waves and Bubbles: On a very fluid flow on a Galápagos island, small waves are made by forward-acting pressure within the flow. The disturbance at the surface allows gas to rise from below in bubbles, which soon burst, leaving holes. The area shown is about 2 feet wide.

Pillow Basalt: About 190 million years ago, near Paterson, New Jersey, basaltic lava emerged from a vent into a lake and cooled to form the pillows shown here. Just above these are bottom portions of a row of basalt columns, formed probably above the water level, where cooling was rapid but not so rapid as it was below the water level.

Stacked Flows: Several distinct lava flows rest upon sandstone in this valley wall north of Sedona, Arizona. Lava at the edges of the flows formed columns as it cooled and contracted.

Pahoehoe Coils: Lava in a pahoehoe flow pushing lava ahead of it may cause "ropes" or "coils" to form. These occur along with pillows at a roadside near Grants, New Mexico.

Pahoehoe and Aa: In this flow at Craters of the Moon, smooth surfaces are on pahoehoe, rough surfaces on aa. Aa tends to develop as lava in a flow cools and stiffens, and then, pushed by lava behind it, breaks up into irregular fragments. "Towers" in this scene are cinder crags – material broken off cones and rafted on flows.

Dome and Spine: During February 1983 a protrusion of lava from a dome in the crater of Mt. St. Helens formed a spine about 200 feet high. This fragile mass lasted only 2 weeks before (as usual with spines) it collapsed into a heap of rubble.

sides after only partly burning a tree, the tree's lower portion may remain standing above the flow, encased in a shell of hardened lava – a "lava tree."

Long-continued eruptions may create a lava plain. This mostly level terrain is made as wide-spreading lava flows and wind-blown pyroclastics fill valleys and other depressions. Rising from the plain there may be scattered cinder and spatter cones, pressure ridges, dike ridges, and flow features such as lava trees and cinder crags. The plains can cover hundreds or thousands of square miles. Relatively large ones in the United States include those in Idaho at Craters of the Moon, in Texas in the Big Bend area, in Arizona near San Francisco Mountain, and in New Mexico in the Raton Basin. All these are youthful, little-dissected plains; older plains may show sharper relief.

Sculptures in Acidic Materials

Forms taken by materials erupted by "acidic" volcanoes differ in many respects from forms taken by materials from "basaltic" volcanoes.

As a stratovolcano ceases activity, lava hardens in the pipe, forming a plug. If activity resumes, the plug is pushed out by rising magma. A long plug may rise impressively – and threateningly – out of the crater. Such a tower, called a spine, can rise hundreds of feet, and its weight may finally crack the cone, triggering a nuée ardente ("fiery cloud") eruption.

It was the weight of such a spine that cracked the cone of Mt. Pelée in 1903, causing the catastrophic nuée ardente. Being structurally weak, with cavities left by escaping gases during the last eruption, a spine is vulnerable to weathering and erosion, and soon crumbles.

Viscous lava being pushed out of a stratovolcano's pipe may form a dome. This is a rounded mass covering or partly covering the vent, which may be in the main crater or in a crater in the cone's side. Like a spine, a dome blocking a vent can cause a violent eruption, as happened with Mt. St. Helens in 1980 and Pinatubo Volcano in 1991.

Acidic lava takes forms ranging from streaming clouds to viscous, thick flows. Some flows consist of the very gassy, frothy lava which cools to form pumice, a highly vesicular,

"On Guard": Near fountains of gases and lava on Kilauea, a lava tree stands like a guard. The wood burned away, leaving its coating of hardened lava standing upright. Being very fragile, such features are likely to crumble before long.

Scoria and Pumice: Scoria *(left)*, formed from basaltic or andesite lava flows or pyroclastics, is reddish to black, and vesicular, with relatively large holes left by escaping gases. Pumice *(right)*, formed from highly gassy flows or showers of pyroclastics, is rhyolitic and has smaller, more numerous cavities. Less dense than scoria, it is much lighter – often light enough to float on water. Cavities in pumice may be tunnel-like and elongated by stretching.

glassy rock of rhyolitic composition. Pumice is often light enough to float on water. Its cavities, left by escaping gases, are frequently tubelike from stretching; they are smaller than cavities in scoria. Pumice can also be blown out from the crater, then falls as pyroclastics over a wide area. It was showers of hot pumice that buried Pompeii.

Another rock variety occurring in acidic flows is obsidian, also called volcanic glass, formed from lava containing little gas. It is usually black, but sometimes red, green, or brown, and its fracture surfaces are conchoidal (curved). It is amorphous, having cooled too rapidly for crystallization to occur. Some acidic flows, such as the one that made Obsidian Cliff, in Yellowstone, are entirely of obsidian.

Pyroclastics: Blown Out of Cones

A blob or clot of lava blown from a vent may cool enough to solidify during flight and fall to the ground as a volcanic "bomb." Rounded or spindle-shaped, the bomb may be vesicular, even hollow, according to the gas content. Its surface often becomes cracked from rapid cooling and contraction while the interior is still hot. Bombs range from a few inches long or wide to specimens weighing tons.

Fragments of rock broken loose from the wall of the volcanic pipe or cone and blown out during eruptions are called volcanic blocks. Some are of hardened lava; others are made of the rock through which the pipe passes, which could be rock of any type – sandstone, limestone, whatever. Blocks tend to be angular but can have just about any shape, and, like bombs, may weigh tons.

Scoria (Latin, pl. *scoriae*) is a vesicular, cindery rock heavier than pumice, partly glassy and partly crystalline. It may occur not only as pyroclastics but as fragments from a crust on an andesitic or basaltic lava flow. As in pumice, vesicles are left by escape of gases and may be drawn out by stretching. More or less weathered scoria often blankets the ground

A Lava Dome: Early in the Pleistocene an immense Cascade Range volcano in northern California, now known as Mt. Tehama, blew up. Later, through a vent in the side of the remaining stump, there welled up a large mass of dacite, which hardened to make a dome over the vent. The dome is Lassen Peak, in Lassen Volcano National Park.

Two Kinds of Volcanic Bomb: The spindle bomb *(above)*, a few inches wide, formed from a pyroclast soft enough to be drawn out by friction with the air during flight. The elephant-skin bomb *(right)* preserved its rounded form during flight, but rapid surface cooling contracted the surface and fractured it into polygonal patterns. Both bombs are from Craters of the Moon.

near craters recently active. Weathering may redden scoria by oxidizing its iron content. Some scoria fragments develop a whitish crust of calcium carbonate.

Vesicular lava fragments like those of scoria but only 4 to 32mm in diameter are classified as cinders. Fragments of similar size, but not vesicular, are called lapilli (from Latin, "little stones"). In the Southwest, grape-sized nodules of obsidian are termed "Apache tears."

Some Hawaiian beaches are noted for their fine black sands. Much of this material was produced as lava flowing into the sea caused explosions of steam. These shattered the lava into tiny droplets, which cooled quickly to form glass as they flew through the air and fell into the water. Waves and currents, transporting this material along shore, have concentrated it in coves to make beaches.

Threads of volcanic glass produced by drawing out of very fluid basaltic lava during flight, or by the bursting of bubbles on a lava lake or stream, are known as Pele's hair. Pele is the goddess of Hawaiian volcanoes; offerings are still left for her at crater edges.

Angular, glassy pyroclastic particles 2.5 to 4.0 mm in diameter are classified as volcanic ash. Smaller particles, less than 2.5 mm wide, are volcanic dust.

Compacted ash and dust, sometimes with a little clay mixed in, form volcanic tuff. This may be welded by volcanic heat to make a strong rock. It may also become

A Large Spindle Bomb: The streamlined and mostly smooth surface of this heavy specimen, nearly 3 feet long, helps to distinguish it from volcanic blocks, likely to be more angular. At left are smaller bombs. All are much weathered. They lie near Haleakala Crater, on Maui, Hawaii.

Obsidian: Favored for arrowheads, knives, and decorations, obsidian is a glass, fracturing with curved surfaces. It is light in weight, varies in color, and corresponds chemically to rhyolite.

Riot of Color: Volcanic ash and dust can accumulate in beds scores of feet thick, sometimes alternating with flow lavas. Here, in Capitol Reef National Park, streams slicing through the beds expose cross sections of the material as spectacular bands of color ranging from red and green to blue, yellow, and gray.

Earth Pillars of Pyroclastics: Beds of pyroclastics often contain large blocks or bombs which protect soft underlying material from erosion; thus earth pillars may form. These specimens are near Unity Lake, Oregon, on the Columbia Plateau.

Sculptures in Volcanic Ash: Thick beds of volcanic ash in Big Bend National Park, Texas, have been shaped into bizarre forms by vigorous though infrequent rainwater streams. Volcanic bombs and remnants of old basalt lava flows, oxidized by prolonged weathering, accumulate in depressions, forming brown carpets.

A Scattering of Pyroclastics: A side of Big Crater, at Craters of the Moon, is covered with rock fragments blown out during eruptions. Angular ones are likely to be mostly blocks; rounded ones, mostly bombs. Chunks of scoria also are present.

cemented by mineral-bearing groundwater percolating through it.

Volcanic blocks cemented together by cooling lava become volcanic breccia. The mass somewhat resembles a wall of mortared stones. Tuff containing many volcanic blocks of various shapes and sizes is known as tuff breccia.

Beds of pyroclastics often contain a mix of materials ranging from bits of dust to huge blocks and bombs. This poorly consolidated mix is subject to relatively rapid erosion. As this lowers the land, pillars may form where large rocks are protecting soft material under them. Some of these features, generally called "earth pillars," are so bizarre as to deserve an alternative name: "hoodoos." In origin these differ from demoiselles and mushroom rocks.

Geothermal Regimes

In some areas volcanic activity occurs without eruptions of lava from cones or fissures. This minor activity, due to the existence of hot rock and magma underground but relatively near the surface, involves the features known as hot springs, geysers, and fumaroles, each of which produces certain landforms. Areas of such activity are called geothermal (from Greek *gaia,* earth, and *therme,* heat). In the coterminous United States, the most prominent geothermal area is in the Yellowstone Caldera. Notable geothermal areas occur also in Iceland and New Zealand. With increasing world-wide demands for energy, such areas are of much commercial interest, which at times conflicts with efforts by conservationists to preserve geothermal features in their natural state.

Hot Springs and Mudpots

In geothermal areas, meteoric water (water from precipitation) percolates from the surface down through systems of fractures to depths of hundreds or thousands of feet and there

A Geothermal Area: This land of steam and sulfurous vapors, near dormant Lassen Peak, is Bumpass Hell, named for a cowboy named Bumpass. After breaking through the crust and scalding his legs, he was asked where he had been. "In hell!" he said.

A Roadside Hot Spring: Yellowstone has thousands of hot springs. This is one of the modest ones. White material surrounding it is siliceous sinter, a silica-related chemical deposited by hot springs; it derives from rhyolitic rock underground.

A Yellowstone Mudpot: Mudpots are relatively small, shallow, usually circular depressions containing boiling mud. They are a form of hot spring, but with relatively little water. Pots with muds of varying colors are known as paint pots.

becomes heated by contact with hot rock or magma. Water then rises convectionally, mixes with descending water, and may finally emerge from a vent at the surface as a hot spring, boiling or simply warm. Pressure on the water having been relieved, the dissolved volcanic gases and vapors form bubbles and escape.

Hot springs usually contain much silica, which is deposited as siliceous sinter (geyserite) to form rims, platforms, and terraces around pools, or domes or towers astride vents. Deposits may be colorful with sulfur and compounds of iron or other materials. Over long periods, hot-spring areas can become covered with sinter, which kills most vegetation and creates a fascinating though bleak scene.

If subsurface rock is limestone, as in the northwest sector of Yellowstone, a hot spring will deposit travertine, which is a dense, finely crystalline form of calcium carbonate. Large, colorful travertine terraces are seen in Yellowstone at Mammoth Hot Springs; there the favorite is Minerva Terrace, tinted brown, red, orange, and green by algae that live in the warm pools. Nearby is Liberty Cap, a 37-foot tower of travertine astride a vent.

Small hot springs that contain mainly a boiling mud are called mudpots, or paint pots if they are colorful. The mud churns and bubbles as gases and vapors escape. Water entering the basin is not sufficient to fill it and spread material beyond it; but highly active mudpots can pile up material into a cone, called a mud volcano.

Hot springs are relatively rare, existing only where underground rock structures allow circulation of groundwater

Travertine Terraces in Turkey: One of the world's finest exhibits of travertine terraces is seen at Pamukkale, in Anatolia, Turkey. Here, groundwater that has dissolved calcium carbonate beneath the surface emerges from a slope, and the mineral is deposited to make pure-white terraces.

Made by a Hot Spring: Yellowstone's Liberty Cap is of travertine deposited by hot water emerging from a natural pipe. Flow kept this open as deposits grew higher, forming a tower. Eventually flow ceased but the tower remained. For scale, note the human figure *(bottom, left)*.

against hot rock and provide conduits to the surface. Thousands of springs occur in Yellowstone, where volcanoes became extinct during the Pleistocene. Not so well known but impressive are the hot springs and colorful terraces at Thermopolis, Wyoming. Other notable hot-spring areas include Iceland and New Zealand.

Geysers: Intermittent Hot Fountains

Geysers are hot springs from which a fountain of water and steam is intermittently ejected, reaching heights up to 175 feet. Such behavior has prompted many explanations, of which the following seems most plausible:

Water from rains and melting percolates through interconnecting fractures to depths of several thousand feet, where it accumulates in cavities and is heated by nearby magma. Hy-

drostatic pressure (the weight of overlying water) and constrictions in the "plumbing system" temporarily prevent the water from changing to steam even though it is superheated – that is, above the boiling point. Finally, the water gets so hot that steam bubbles do start forming. They push up against water higher in the system, causing some of it to rise and flow out at the surface. This movement relieves pressure on the superheated water below, allowing it to flash into steam. The steam thrusts water above it violently upward, and it spouts forth at the surface as a geyser. When the water has been mostly ejected, pressure is reduced, the eruption ceases, and the system starts filling for its next act.

The term geyser comes from the Icelandic *geysir*, "gush," which is the name of Iceland's most famous feature of the type. Probably the world's best-known geyser, however, is Yellowstone's Old Faithful, which has been entertaining

Old Faithful: This famed Yellowstone geyser is not faithful enough to set a watch by; but it shows some regularity. The average interval between eruptions has increased from about 62 to 79 minutes during the past century. The longer the play of water in one eruption, the longer the interval that follows.

OLD FAITHFUL

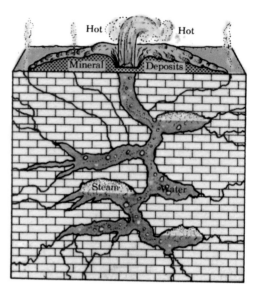

A Landform Made by a Geyser: Yellowstone's Grotto Geyser erupts gently, depositing minerals near the vent. These take shapes according to such influences as vigor of eruptions, wind directions, and erosion by rainwash.

A Geyser's "Plumbing": Geyser action results from heating of water by hot rock in passages underground, sudden conversion of water to steam, and ejection of water at the surface by steam pressure. Geyser performances are intermittent.

tourists since 1870. At the turn of the century Old Faithful erupted almost hourly, in 1950 its interval averaged 62 minutes, in 1970, 66 minutes, in 1997 it was 79 minutes (minimum 45 minutes, maximum 105 minutes). Its decline is likely due to the Yellowstone region's numerous earthquakes and to assorted materials tossed into the vent by thoughtless visitors.

Until recently, the world had 10 areas where geysers were active; now, because of well-pumping for geothermal energy, few areas have active geysers. Iceland's geysers have been much reduced; all in New Zealand are quiet. In the United States, geysers in northern California and Nevada have ceased. Yellowstone National Park still has about 250 active geysers. Russia's Kamchatka Island also has many. Geysers that survive today are threatened by commercial development, earthquakes, and tourists who toss trash into geyser vents to see what will happen.

From Escaping Gases: Fumaroles and Solfataras

Fumaroles are small vents that emit only volcanic gases – mostly steam – and vapors. Vents may be scattered, or in a

Fillings of Fumarole Tubes: Gases and vapors rising through fumarole tubes deposit minerals on the sides of the tubes. The minerals are chiefly sulphates of various metals. After surrounding soft material has been eroded away, the mineral deposits, being harder, may remain as towers. These are The Pinnacles, in Crater Lake National Park, Oregon.

A "Splash" on Granite: What looks like a splash of paint on this granite outcrop in Wyoming's Bighorn Mountains is a pattern of quartz intrusions into a system of fractures. The quartz is whitened by gases and moisture it contained when it cooled.

chain along a fissure. The steam is mostly from meteoric water; some may be juvenile (that is, from magma). Vapors consist usually of sulfur along with chlorides of iron, aluminum, potassium, or other materials, diffused and suspended in gases; these materials are deposited on the surface as vapors emerge and cool. Steam from fumaroles condenses to become rivulets and sheets of hot water, which spread mineral materials around the vents. Vents that produce sulfur mainly are called solfataras.

Minerals, such as sulfur and sulfates of various metals, that are deposited in a fumarole pipe may accumulate to form a hard core. After activity has ceased and surrounding soft materials have been eroded sway, the core may remain as a pillar. The Pinnacles, near Crater Lake, Oregon, were formed in this fashion.

Fumaroles are common in areas of geologically recent igneous activity. The United States has many in the volcanic Cascade Mountains and in Yellowstone. Mt. Lassen has fumaroles; so does The Geysers, east of Cloverdale, California. Fumaroles developing after the 1912 eruption of Mt. Katmai, Alaska, gave the Valley of Ten Thousand Smokes its name. Among active fumarole areas in Hawaii is Sulfur Banks, near Kilauea Volcano. About 200 fumaroles have been counted on the smoking slopes of Mt. Etna, in Sicily. Other notable fumarole areas are in Central America, South America along the Andes, Iceland and New Zealand.

Created Underground: Plutonic Landforms

Igneous rock masses that form within the crust are known as plutons. Their emplacement can occur before, during, or after volcanic activity, and also where there is no volcanic activity at all. Cooling and solidification of the magma may take only minutes for fine, hairlike intrusions, and hundreds of thousands of years for plutons with dimensions measured in miles. Wherever volcanic features exist, there are plutons underground. The crust contains vastly more plutons than the surface might suggest.

Plutons take forms according to factors such as the magma's composition and volume, its nearness to the surface, the rate of cooling, and the composition and structures of the country rock. Magmas range from basic to acidic, yielding

A Cluster of Laccoliths Exposed by Erosion: The La Sal Mountains, in southern Utah, have been carved from a group of laccoliths intruded into sedimentary strata in Miocene time. Stumps of strata are still visible in some locations. The nearby Abajo and Henry mountains, also in Utah, are of similar origin.

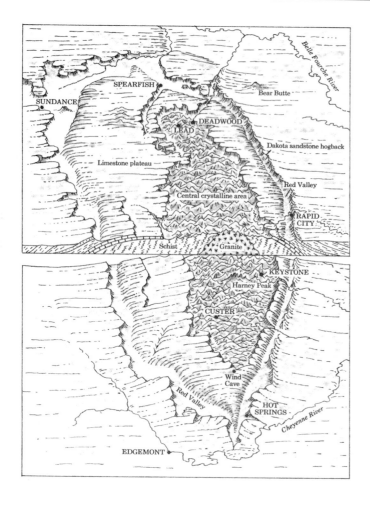

Anatomy of a Batholith: Long ago, the batholith from which the Black Hills of South Dakota have been sculptured rose in the crust like a great bubble, heaving up sedimentary rocks that covered it. Erosion has stripped off the sedimentary strata, except for stumps around the batholithic mass, and this mass itself has been dissected to make a terrain of ridges and valleys.

the familiar rocks diabase, diorite, and granite, among others. The country rock (rock into which magma is injected) may be of almost any kind, and it may alter the magma somewhat by being absorbed into it. Country-rock characteristics that influence pluton forms include stratification (if any), strong-weak patterns, attitudes of strata relative to one another, cavities in the rock, and joint patterns. Accordingly, magma cools in such forms as veins, pillars, walls, plates, tables, and bulbs.

As magma forces its way through crustal fractures underground, it usually produces no noticeable landforms unless the intrusion is near enough to the surface to make it bulge up. When, in time, a pluton is uncovered by erosion, it becomes a landform. If the uncovered mass is more resistant than surrounding rock, it will erode more slowly and sooner or later become a positive relief feature, such as a ridge or a dome; if less resistant, a negative feature, such as a ravine or a basin. Most granite plutons are more resistant than the country rock (especially if the country rock is sedimentary) and therefore become positive features after exposure. Plutons of basic rock usually are chemically less resistant than granites and metamorphic rocks, and in those settings usually erode to form negative features. Basic plutons are, however, stronger than most sedimentary rocks, notably shale

Carved from a Batholithic Dome: New York's Adirondacks have been shaped mostly from a dome of igneous rocks, much of it anorthosite, and remnants of Precambrian metamorphic rocks that once covered them. The view is from Mt. Marcy to Gothics Peak *(near background, center)* and Giant Mountain *(far background, center)*. Anorthosite is very resistant, and wide spacing between joints restricts weathering and favors rounded, smooth profiles. Some valley sides were oversteepened and many smoothed by Pleistocene glaciers.

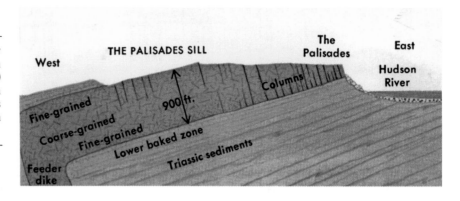

The Hudson Palisades in Cross Section: Magma rising from depth forced its way up through a fracture *(left)* and formed a 900-foot-thick sill between tilted sandstone strata. Erosion has exposed portions of the sill, the eastern edge of which now fronts the river.

and limestone; thus erosion tends to leave these plutons standing higher.

A pluton when exposed at the surface not only becomes a landform but influences adjacent topography. For example, where erosion is lowering a terrane of horizontal shale strata and has uncovered a relatively resistant layer of diabase emplaced between the strata, the diabase will somewhat protect the stratum beneath it; thus a highland may form, thanks to the protective cap. Eventually erosion may remove this cap and expose the underlying strata, which will then start eroding down to surrounding levels. Many landforms owe their existence and form to processes like this, in which the original "cause" of the feature disappears, leaving the observer to wonder how the feature was produced in the beginning.

Some plutons may be easily recognizable for what they are. Thus on a terrain where volcanic activity ceased perhaps 5 or 10 million years ago and cones have been erased by erosion, there may be a scattering of jagged rock "stumps" up to a few hundred feet in diameter: these are likely to be volcanic plugs or necks. (Although parts of volcanoes, these were formed below the surface and therefore qualify as plutons.) On old volcanic terranes sharp-backed ridges of hard igneous rock rising above beds of pyroclastics are likely to be dikes, and a hill in the form of an elongated dome can be suspected of being a laccolith.

Many plutons, especially big ones miles wide, can easily be confused with other large rock masses, perhaps of metamorphic or even sedimentary rock, shaped by erosion into pluton-like forms. Large plutons may require identification by determining the rock type, investigating the body's relationship to surrounding rock masses, and testing with seismological equipment. In pluton country, as elsewhere, an observer is well advised to carry a guide to the regional or local geology.

Pluton Closeups

The largest plutons are batholiths (Greek *bathos*, depth; *lithos*, stone). These are mainly of granitic rock, with exposed portions 40 square miles or more in area. Batholiths are so deep that they have no detectable bottoms. With time, a granitic batholith rises in the crust because it is less dense than the basic rock of the lower crust, which surrounds it and buoys it up.

A batholith does not conform to the stratification, if any, of surrounding country rock. Often it is more resistant and thus its exposed portion is a highland, in some cases with stumps of arched and eroded country rock strata projecting up around its edges.

Well-known batholiths in North America include the Coast Range Batholith, in British Columbia, nearly 1,000 miles long and 100 miles wide; the Idaho Batholith, 10,000 square miles; the Black Hills Batholith, 125 by 60 miles; and the Sierra Nevada Batholith, 400 miles long and 40 to 80 miles wide.

Somewhat resembling batholiths is the stock, but by definition this is less than 40 miles in area. Some stocks are parts of batholiths. Often of basic rock, such as diorite, the rock may on occasion approach granitic composition. Like batholiths, stocks often are more resistant than the country rock, and their exposed upper portions may stand as highlands. Among stocks in the coterminous United States are those in Utah's La Sal Mountains and in the Crazy Mountains of western Montana. Cabezon Peak, in New Mexico, is a stock with interesting columnar forms. In the East, Mt. Ascutney, New Hampshire, is reportedly a stock.

A stock-like mass sunk in the center, like a saucer, is called a lopolith (Greek *lopos*, shell). The La Plata Mountains, in Colorado, are said to include examples.

A slab or plate of igneous rock formed by an intrusion between layers of country rock is a sill. Very common in

A Sill to Help the Romans: This segment of England's Great Whin Sill, near Housestead, Northumberland, carries on its back the famed Roman Wall, built by the Roman emperor Hadrian during the second century for defense of conquered territory against the Scots. The enclosure at lower left in this view was one of the guard stations manned by Roman soldiers.

On the Hudson Palisades: The Hudson River flows along the base of the Palisades, the edge of a diabase sill between sandstone strata, emplaced during the Jurassic or early Cretaceous. This view is southeast from the Palisades across the river. Polygonal patterns on the diabase surfaces are cross sections of columns, mostly 3-, 4-, and 5-sided.

volcanic areas, a sill when exposed in cliffs and roadcuts suggests a slice of ham in a sandwich. Sills can be horizontal or at an angle, according to the attitude of the strata they have intruded. Often a sill is more resistant than country rock and so, when exposed, it can protect underlying horizontal strata from erosion; thus a mesa can form, as have many on the Colorado Plateau. (Protective caps are not always sills; they can be of any relatively resistant rock.) A tilted sill when exposed by erosion can form a hogback ridge or part of one.

The north-south part of New York's Hudson River Palisades is the eastern edge of a diabase sill more than 900 feet thick, emplaced between late Jurassic or early Triassic sandstone strata. Emplacement occurred during igneous activity that accompanied separation of the North American and African plates around 190 million years ago. In Glacier National Park, Montana, a 200-foot-thick sill is exposed in the face

The Mt. Sopris Laccolith: Among various scattered intrusive bodies south of Glenwood Springs, Colorado, along the western fringe of the Rockies, is Mt. Sopris. Like many other laccoliths it is domal, but when seen from the side it looks like a long ridge. In the foreground is the Colorado River with one of its minor floodplains.

The Packsaddle Mountain Laccolith: Named for its humps, this intrusive body near the highway south of Alpine, Texas, still shows on its flanks portions of limestone strata it intruded and heaved up millions of years ago.

of The Garden Wall, in the Many Glacier area; it resulted from igneous activity accompanying the rise of the Northern Rockies. Limestone strata next to the sill were turned to marble by heat and pressure.

Prominent in northern England is the Great Whin Sill, an intermittent ridge of diabase that runs scores of miles southwest from the Farne Islands to the Pennine Hills and thence southeast to Middleton-in-Teesdale. Along the northern part of the sill the Romans built Hadrian's Wall for protection against invaders from Scotland. (The Scots came anyway.)

Like sills, laccoliths (Greek *laccos*, cistern) are intrusions between strata of country rock, but they are lens-shaped, with a convex top and a flat bottom. A laccolith can be up to 5 miles in diameter and from a few feet to several hundred feet thick. It is fed by relatively acidic, viscous magma from a feeder dike probably below its thickest point. Magma intruding between strata near the surface causes them to arch up, forming a dome. Erosion of the dome exposes the upper part of the laccolith and leaves a ring of truncated country-rock strata around its edges.

Numerous laccoliths flank the Rocky Mountains, examples being Crow and Elkhorn peaks, South Dakota; Mt. Sopris, Colorado; the Henry Mountains, Utah; Little Sundance Mountain, Wyoming; the Highwood Mountains, Montana; Navajo Mountain, Utah; and Packsadddle Mountain,

southwest Texas. In the world of experts, an argument continues as to whether Devils Tower, that wondrous vertically grooved, or columned, feature in Wyoming, is a laccolith or a volcanic neck; it has some characteristics of both. An Indian legend says the grooves are claw marks of a giant bear that tried to reach a maiden who took refuge on top.

An intrusive sheet or wall-like mass that breaks through country-rock layers, usually at a high angle, is a dike. A similar mass injected through a joint in unlayered rock also is a dike. Dikes of igneous rock can be granitic or basic. Often they extend out from a larger pluton such as a stock, or from a volcanic pipe as "dike swarms," as in the Spanish Peaks area of southern Colorado. Dikes can be up to 6 miles

The La Plata Mountains – a Plutonic Mix: In the Rockies west of Durango, Colorado, are the La Platas, where intrusions into colorful shales and sandstones created sills, laccoliths, dikes, and stocks. These bodies now uncovered by erosion are in a zone that may have been the source of magma for volcanic activity that created portions of the San Juan Mountains. This view is from about 20 miles to the west.

A Trench Cut in a Dike (left): Near Upper Jay, in New York's Adirondacks, the Ausable River has cut into a diabase dike, which is dark and less resistant than the anorthosite host rock. Diabase outcrops often have the blocky, steplike aspect seen here.

Dike and Sills (right): An outcrop of granite gneiss in New York's Hudson Highlands shows layers of host rock invaded by granitic material from below. The light-colored vertical mass is a dike; the light-colored horizontal layers are sills.

thick and 100 miles or more long, as in Africa near the Great Rift, but most are much smaller and readily identifiable when exposed.

On terrains eroded deeply enough to expose plutons, a dike often appears as a long, jagged ridge or chain of ridges rising above less resistant rock, such as shale or a soft sandstone. A dike in country rock that is more resistant, such as a diabase dike in massive granite, usually erodes out to leave a steep-walled trench. In highlands of granitic or other strong rock, such as the Sierra Nevada, most of the Rockies, Older Appalachians, and Adirondacks, numerous streams follow diabase dikes as lines of lesser resistance; but on the same terranes dikes of quartz, very resistant, stand out as ridges.

Dike ridges are numerous on the Colorado Plateau and on relatively soft sedimentary and volcanic rocks along the Pacific Coast from Washington south. They are especially

Safe for Kings: Volcanic necks were good locations for castles not simply for the view but because they were natural fortresses. On one of them stands Edinburgh Castle, in Edinburgh, Scotland, seat of Scottish kings.

Dike Ridges: Cathedral Valley, in Utah's Capitol Reef National Park, includes many basaltic dikes intruded into sandstone. Dikes form ridges as surrounding sandstone erodes away. Ridges shown extend 100 feet or more into the background.

History of a Volcanic Neck

(1) Volcano becomes extinct. Magma "freezes" hard in throat of cone.
(2) Soft material of cone erodes away, exposing part of hard core.
(3) Cone is eroded low, but core survives as a tower, flanked by dike ridges.

 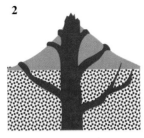 

well exposed and recognizable in arid localities such as California's Mohave Desert and the Big Bend country of Texas. In an interesting turn of nature, the sill whose edge forms the Palisades of the Hudson River curves west at its northern end, becoming a dike cutting through the sandstone strata rather than between them.

As a volcano becomes extinct, any magma remaining in its pipe solidifies, and as the cone is eroded away this mass, being more resistant than the loose cone materials, may survive on the landscape as a towerlike form rising from the ruins of the cone. Such features are known as volcanic necks. Though part of a volcano, which is an extrusive feature, a neck is generally regarded as intrusive, because it solidified underground.

Some necks erode to ragged or jagged forms, because the magma from which they formed was still gassy or was not compact when it froze in the volcano's pipe; thus when exposed it erodes unevenly. Other necks are more massive, more compact, more regular in form.

Necks are likely to be seen where erosion of a volcanic area is well advanced but has not yet destroyed the most resistant volcanic features. An impressive scattering of necks – some 200 – appear in the Hopi Buttes area of Arizona. Monument Valley has about 20, dominated by Agathla Peak, which is about 1,000 feet in diameter and rises more than 1,200 feet above its surroundings. More famous is New Mexico's 1,400-foot Ship Rock, south of the town of Shiprock, a structure of tuff breccia that stands like a ghost ship on a sea of sand, its dikes radiating out like tentacles.

In the eastern United States, hardly thought of as a volcanic area, volcanic necks of Triassic age have been identified in Connecticut, New Jersey, and Virginia. Castle Rock in Edinburgh, Scotland, which holds the royal castle, is a volcanic neck. So is the rock tower of volcanic breccia crowned by an 11th-century chapel, St. Michel d'Aiguille, at Le Puy, in the Auvergne volcanic area of France. The Kaiserstuhl (King's Chair), near Freiburg, Germany, is another neck of historical interest.

A Desert "Ship": Rising about 1,400 feet above the desert, spectacular Ship Rock, south of Farmington, New Mexico, dominates its surroundings. Best known of North American volcanic necks, it consists of a large central neck flanked by smaller ones. Ridges radiating from it are crests of dikes partly uncovered by erosion. Other, less imposing necks are in the vicinity.

A Community of Faults and Folds: A Route 14 roadcut
south of Palmdale, California, in the San Andreas fault zone,
reveals intense deformation. Most noticeable, left to right, are
several small folds, a large syncline with its left limb broken by
a vertical fault, an anticline with a fault along its axial plane,
and an open downfold with its left limb broken by a vertical
fault. The gently undulating ground surface shows little evidence of deformations beneath it.

9
Rock Structures in Landforms

A Chaos of Folds: Seen in this cliff, shaved by a valley glacier, are edges of shale strata originally deposited horizontally, later subjected to the intense folding typical of some portions of the European Alps. This scene, about 25 feet wide, is in the Combe des Fonds, near Ferret, in the Swiss Valais Alps.

FROM ancient times, man has been impressed by mountains thrusting boldly up into the sky, land masses that look broken and deformed, rock strata that have been tilted, folded, and jumbled. Primitive man appears to have viewed such features as the handiwork of giants. Aristotle and other Greek scholars, who lived on a terrane of greatly deformed rock masses (deformed thanks to the collision of northern Africa and Europe), attributed such features to great winds bursting forth from Earth's interior, volcanic eruptions, and earthquakes. In the Middle Ages and later, such explanations became linked with the majestic details of the Creation, Noah's Flood, and activities of Satan. In the fifteenth century Leonardo Da Vinci, observing fossils of marine shellfish in Italy's limestone mountains, speculated that these terranes were formerly portions of sea bottoms, but he fell short of perceiving that they had been uplifted. In later centuries pioneers in geology realized that crustal deformations, including uplifts, had occurred in the past as natural events, but, unaware of the vastness of geologic time, they attributed such events to long-ago catastrophes rather than to ordinary crustal processes. A detailed, coherent understanding of crustal movements, of structures they produce, and of their influences on landforms, came only with the twentieth century.

Structures: Original and Acquired

All rock is made of materials arranged in some pattern; that is, all rock has structure. As rock is formed, minerals that constitute it become arranged in layers, clusters, or some other pattern that becomes basic to the rock's nature; such structures are called primary. Examples include layers in sandstone and interlocking crystals in granite. After forming, most rock sooner or later becomes subject to deforming processes that produce new structures, called secondary; these are seen in jointed, faulted, folded, or warped rock

masses, such as mountains raised by folding and depressions formed by sinking of crustal blocks. Secondary structures result from continuous shifting of mass in the crust associated with plate movements, isostatic adjustments, and activity in the mantle.

Rock masses exist in relation to other rock masses, forming with them combinations which are structures with their own distinctive characteristics; for example, a highland made by thrusting of a mass of old rocks over new ones, or a hogback formed by the edge of a resistant stratum projecting above others in tilted strata. Any variety of structures can exist in Earth scenery.

Since rock structures are fundamental aspects of Earth's crust, most regions have been classified according to the

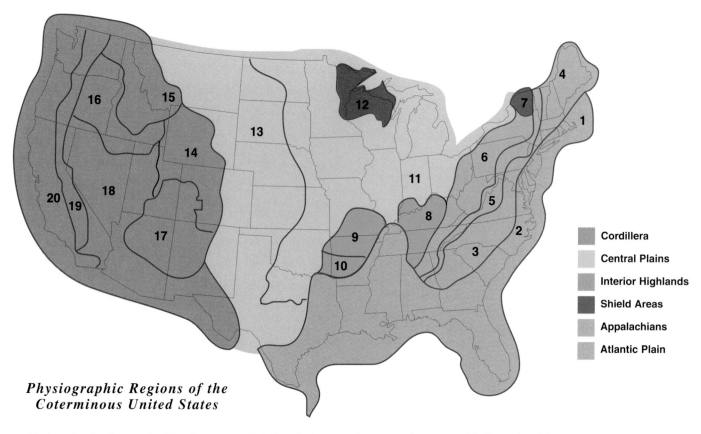

Physiographic Regions of the Coterminous United States

Each region is characterized by the prominence of certain sets of processes or conditions, e.g. igneous activity, mountain building, deposition of sediments, or erosion of certain kinds of geologic structures.

1 Atlantic Shelf: sloping submarine plain of sedimentation from shoreline to 600-ft. depth

2 Coastal Plain: low, hilly to nearly flat, belted and terraced plain on soft sediments

3 Blue Ridge-Piedmont: complex mountain structures eroded low on east, remaining higher on west (Blue Ridge Mts.); rounded summits, 3,000–6,000 ft.

4 New England: glaciated equivalent of Blue-Ridge Piedmont; Blue Ridge replaced by sharper peaks of White and Green Mts.

5 Ridge and Valley: long, parallel mountain ridges eroded from regular folds

6 Appalachian Plateau: generally steep-sided, deeply dissected; sandstones and limestones predominating; 2,000-4,000 ft. on east, lower toward west

7 Adirondacks: complex mountains on ancient crystalline rocks; summits up to about 5,000 ft.; wide areas of moderate altitudes and relief

8 Interior Low Plateaus: open, well-dissected domes and basins; low escarpments separating more level stretches

9 Ozark Plateaus: high, hilly; no stratified rocks

10 Ouachita Mts.: parallel ridges and valleys eroded from folded strata

11 Central Lowlands: low, rolling plains; mostly veneer of glacial deposits, including lake beds and lake-dotted moraines; unglaciated southwestern part showing flat sediments little deformed or dissected

12 Superior Upland: Hilly area of erosional topography on ancient igneous and metamorphic rocks

13 Great Plains: erosion in north exposing stratified and igneous rocks related to Rockies buildup

14 Central and Southern Rockies: broad anticlines eroded down to crystalline basement rock, which preserves ranges and peaks to 14,000 ft.; broad synclinal basins separating anticlines; extrusive volcanic rocks forming dissected plateaus in northwest portion

15 Northern Rockies: complex mountains with narrow intermontane basins

16 Columbia Plateau: high, rolling plateaus on extensive volcanics, deeply canyoned

17 Colorado Plateau: high plateau on sedimentary rocks, deeply canyoned

18 Basin and Range: mostly isolated fault-block ranges separated by wide desert plains; many ancient lake plains and alluvial fans

19 Cascade-Sierra: Sierras, in south, eroded from huge fault blocks in granitic rocks; Cascades, in north, being warped volcanics surmounted by high volcanic cones

20 Pacific Border: young fold-fault mountains flanked on east by extensive river plains in California section, on west by structurally active shelf area

nature of rock structures that underlie them, as well as other characteristics. Accompanying is a map of the coterminous United States showing both structural and physiographic characteristics of the various regions. Journeying from region to region, the observer familiar with these classifications is prepared to see certain forms dominating in each region and to understand, at least in part, their origins.

Perspectives on Deformation

Rock masses everywhere are subject to deformation: faulting, folding, shearing, compression, or extension, despite their hardness. Crustal blocks and slabs are raised or lowered, bent, folded, warped (twisted), or broken and displaced. Resulting rock forms range from the "micro" to the "mega," from the size of a fingernail to mountains miles high. Some landscapes show deformations more or less plainly; these terrains are likely to have high relief, with but a thin cover of sediments and vegetation, and bedrock at or near the surface. But on flat or gently rolling landscapes, effects of crustal deformation often are obscured; here the surface is on a nearly horizontal rock platform or on deposits of sediments spread by water, wind, or glaciers, as on a plateau, a floodplain, a desert, or an alpine plain. Subsurface structures may be noticeable only where exposed in cliffs or roadcuts or when encountered in drilling.

Since patterns of deformation frequently are complex, have been more or less removed by erosion, and are concealed to some extent by overlying materials, tracing out the full story of deformations that have influenced a given landscape may be difficult if not impossible even for an expert observer. However, some bits of the story usually are noticeable, and the pleasure of recognizing them and thinking about what they may mean can be well worth one's time.

On some landscapes, major topographic features and many minor ones were produced directly by crustal movements. Such features might include, for example, a basin where the crust sank, a mountain formed where a block rose up, or a ridge made by an upfolding of strata. Prominent examples in the coterminous United States include the numerous highlands and basins of the Basin and Range Province and the California Coast Ranges.

More common are landscapes that consist mainly of features produced not directly by crustal movements but rather by erosion of rock masses that were earlier deformed or displaced by such movements. As degradation of the land proceeds, patterns of deformation are uncovered. Since these are patterns also of varying resistance, the topography produced by erosion has strong rock masses generally standing higher than weak ones. (There are exceptions, as we shall see.) Such are all topographies in the United States east of the Rockies.

A Panorama of Structures: In this sector of a relief map drawn by the eminent geographer Erwin Raisz, dominant characteristics of four physiographic provinces appear. From left to right: *Appalachian Plateau* – on horizontal strata, with rolling topography, steepsided major valleys. *Ridge and Valley (Folded Appalachians)* – long, parallel, steep-sided ridges, which are erosional remnants of regular folds. *Blue Ridge-Piedmont* – rather complex mountain structures, resistant rocks intricately dissected. *Coastal Plain* – low, flat to hilly, terraced, on soft sediments, and with parallel consequent streams.

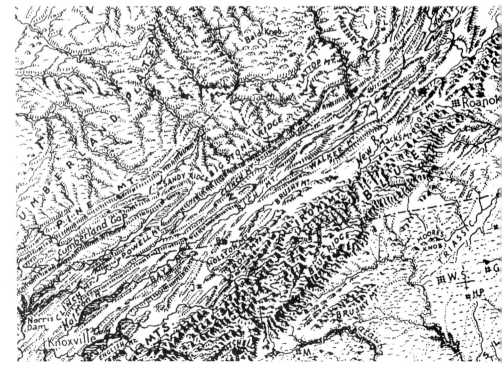

Rock Under Stress and Strain: How Deformations Occur

That rock masses can be broken and crushed when sufficient force is applied is familiar to everyone. It can be difficult, however, for persons lacking a scientific background to imagine the magnitude of forces required to heave up mountains; perhaps it is even more difficult to imagine how solid rock can be warped, folded, stretched, or dilated. In truth, such processes are quite as common in Earth's crust as breaking and crushing.

Deformation is accomplished by compressive (squeezing) stress, tensile (pulling-apart) stress, and shearing (cutting) stress. Like a cloth spread out on a table, a rock mass can be deformed in any of these respects when subjected to different stresses – that is, forces operating in different directions. Susceptibility to deformation is determined not only by the original nature of the rock, including composition, foliation, and layering, but by the extent to which it has been weathered, jointed, or faulted.

Susceptibility to deformation has limits. First is the rock mass's elastic limit: the amount of strain the rock can sustain without being permanently deformed. An example is seen in lands that were depressed by Pleistocene glacier ice and are now "rebounding." Second is the plastic limit: the extent to which particles in rock can flow (move independently of one another without becoming separated from the mass); an example is the flow that may occur in rock metamorphism. Third is the rupture limit: the force required to break the rock by compression or shearing.

Elastic, plastic, and rupture limits can vary widely even between specimens of the same rock type. However, a few generalizations can be risked. All rocks tend to be weakened by weathering and fracturing. Granite, massive basalt and andesite, and metamorphic rocks are usually strong. Rhyolite can be strong if well baked. A sandstone or conglomerate cemented with quartz will likely be stronger than one cemented with limy material. Shale, usually relatively soft, can be resistant but tends to split easily along bedding planes. Massive limestone can be quite resistant, though subject to solution; shelly limestone tends to be weak.

In rocks generally, resistance to crushing is high and resistance to tensile stress low. When strata are folded, typically there is tension in the tops of anticlines (upfolds) and compression in the bottoms of synclines (downfolds); thus in tops of anticlines fractures open up, and in bottoms of synclines crushing occurs.

Deformation of a rock mass can occur by movement along bedding or foliation planes, just as playing cards in a deck slide against one another when the deck as a whole is bent. The slipping of one stratum against another often produces a polish and shallow grooves, together called slickensides. These are smoother in the direction of movement than in the reverse direction; thus they can indicate for the geologist the direction in which a rock mass has moved.

Under some conditions, grains in rock may creep. Movement may occur along glide planes within individual mineral crystals. If solution of some material occurs in the rock mass and it precipitates out later, with recrystallization occurring, new crystals may assume patterns that change the shape of the rock mass as a whole.

Movement of particles in rock being deformed tends to be from locations of higher to lower pressure. Movement accelerates if the rock is heated, because then the space between atoms increases and atoms can move more freely. Movement may accelerate if the rock is invaded by fluids that act as a lubricant. Chemical changes due to heat, pressure, and infiltration by other materials may be involved. Recrystallization of component minerals may occur because of pressure. Finally there is the factor of time. With sufficient time, slow-moving particles of rock can migrate; with sufficient time, crustal forces can be nearly irresistible.

Horizontal Structures in Rock Masses

The simplest type of rock structure influencing topography is a horizontal or nearly horizontal layer. This may form when the rock originates, as when limestone forms on a sea bottom. It may result from deposition, folding, or faulting. Horizontality of strata is extremely common; good examples are seen in locations as varied as the Colorado and Appalachian plateaus, the Edwards Plateau in Texas, and the Salisbury Plain in southern England.

Highlands that are underlain by horizontal layers, stand more than 450 feet above surrounding terrain, have at least one side that is steep, and cover hundreds or thousands of square miles qualify as plateaus. Prolonged erosion of such highlands produces a landscape of mesas and buttes, with broad flat areas between them. A mesa (Spanish, "table") thus formed is wider than high, commonly is capped with a resistant, nearly horizontal stratum, and has steep sides. A butte (from "butt," of a tree) is similar but pillarlike – that is, higher than wide. Some plateaus, notably lava plateaus and uplifted, eroded portions of granitic mountain cores, do not consist of sedimentary rocks but may have some horizontal layering.

Remnants of highlands on sedimentary layers have flat, sometimes horizontal tops, and the sides show layer edges. Early in the degradation process the valley sides tend to be stairlike, as in the Grand Canyon of the Colorado, with benches on the edges of stronger layers, usually sandstone or limestone, and gentle slopes on the edges of weaker layers, usually shale. During degradation of these highlands, steepness of the upper portions of the valley sides tends to be maintained. The reason is that edges of weaker strata erode faster than edges of stronger ones, which thus tend to project out farther. The overhangs occasionally break off and fall, exposing underlying weaker strata to further erosion; thus the upper part of the valley side remains steep as the side retreats. Meanwhile talus builds up at the foot of the cliffs, protecting underlying rock and allowing it to form a pedestal around the highland's base.

Under a dry climate, as in the Southwest, there is little vegetation to stabilize cliffs on a mesa or a butte, and little running water to erode them. These conditions favor steepness. But under a humid climate, as on the Appalachian Plateau, in Spain's Central Basin, and on the Salisbury Plain, running water is more effective in beveling cliffs, more vegetation is

"Flattop": Typical of innumerable mesas on the Colorado Plateau is this one near Springdale, Utah. Mesas usually are capped with a resistant layer, which is visible here. Mesa caps may be of either sedimentary or igneous rock.

present to stabilize the slopes, and undercutting is less intensive; thus the highland's sides tend not to be precipitous. There are, of course, exceptions, among them the lofty sandstone tablelands (tepuis) of eastern Colombia, northern Brazil, and southern Venezuela. In the Venezuela sector is famed Angel Falls, the world's highest waterfall, with a drop of 3,212 feet.

Fold Structures
Ordinary Folds

Folds are essentially undulations, or bends, in rock. They are most noticeable in layered rocks, sedimentary or metamorphic, but may occur less noticeably in volcanic and plutonic rock masses. Folds range in size from tiny waves and swirls in a specimen of metamorphic rock to folds that form mountain ridges miles high and tens of miles long.

Folds are described and measured in terms of several basic elements. The axial plane, or axial surface, is a plane that divides the fold as nearly symmetrically as possible. The fold axis is the line formed by the intersection of the axial plane with any stratum in the fold. The limbs, or flanks, of a fold are its sides. The crest is the upper limit, or top, of this landform.

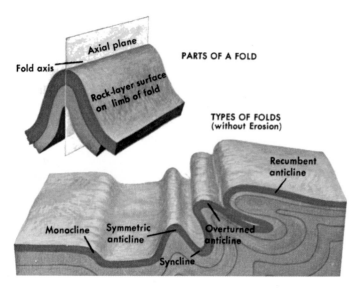

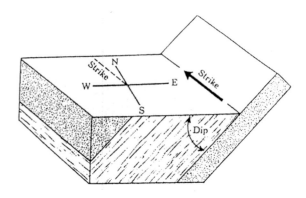

Anatomy of Folds: At left: Diagrams show simplest elements and types of folds as studied and described in structural geology. Wherever deformed bedrock layers are seen in cross section, some variation of one or more of these fold forms may be observed. At right: How dip and strike are determined.

A fold is described also according to the strike and dip, or attitude, of its strata. The strike is the compass direction of a line formed by the intersection of the stratum with a horizontal plane. Dip is the angle between the horizontal and a line drawn on the stratum perpendicular to the strike.

Folds that are perfectly symmetrical are rare if they exist at all. From point to point any given fold is likely to depart more or less from the "regular" in terms of strike or dip, degree of openness or tightness, or some other characteristic. Many folds are very intricate and, in addition, have been complicated by faulting. Rock exposures in roadcuts, cliffs, and other such situations usually show only portions of folds – either edges or part of a limb; most of the structure is hidden. Given these common conditions, the tracing of folds in three dimensions across country is ordinarily a task for geologists with knowhow and special equipment. However, folds as the nonprofessional notices them are always interesting, not only for their forms – waves, chevrons, loops, zigzags, S's and other letters, warps, domes, cones, and just about every other conceivable geometric shape – but for the clues they give to processes and conditions that have shaped surrounding topography.

Warps, Cuestas, Domes, and Basins

Regional uplift of horizontal strata is never quite uniform throughout a region; it involves local deformations of strata into warps, domes, and basins, all involving more or less folding. These features tend to develop mostly in association with uplifts of nearby mountains.

Profile of an Upwarp: Upwarps are not likely to be noticed as such by ordinary observers; they are scarcely noticeable even on very broad plains. Low on the horizon here, about 60 miles to the north, is New Mexico's Zuni Upwarp, with a relief of 500 feet.

A warp is a slight flexure of strata upward or downward, or into a slightly twisted form. Often it suggests what happens to an unseasoned knotty plank left in the sun. Some degree of warping is often associated with other kinds of folding. Warps may be responsible for certain topographic forms etched out by erosion, such as long, curving ridges, shallow depressions, or cuestas – those hills or ridges that have a long, gentle slope on one side and a short, steep slope on the other. Warps covering hundreds or thousands of square miles are hardly noticeable except to geologists.

Prominent upwarps in the Southwest include the Monument Upwarp, about 2,500 square miles in extent, rising northwest of Monument Valley, and the Zuni Upwarp, involving some 22,500 square miles, in west-central New Mexico. Among downwarps in the Southwest are the San Juan Basin, covering nearly 10,600 square miles, straddling northern New Mexico and southern Colorado, and the somewhat smaller Uinta Basin, in northern Utah. The Mississippi Valley is thought to have originated as a downwarp perhaps 25 to 20 million years ago.

One of the simpler fold types is the dome. This is essentially an upfold that is circular or elliptical. Some domes are created by upward pressure from igneous bodies rising in the crust. Domes made by rising batholiths can be very high and wide, such as South Dakota's Black Hills Dome, extending about 65 miles east to west and 125 miles north to south.

Navajo Mountain, in southeastern Utah, and Bear Butte, in South Dakota, are among numerous laccolithic domes flanking the Rockies.

Domes made by regional uplift of horizontal sedimentary strata are called structural, or tectonic. Middle Dome, near Harlowton, Montana, is an example. The Appalachian Plateau has the Nashville Dome and Cincinnati Dome, each in limestone, about 90 miles in diameter. The Wealden Dome, in southeast England, is similar. Covering hundreds of square miles, such domes can be traced on a topographic map.

As domes on stratified rock rise, the increase of elevation accelerates stream action. By the time the dome has risen to substantial height, the original top has been eroded off and the interior has become more or less exposed. By now, erosion has likely reached relatively weak layers within the dome, and rapid erosion of these layers has created a central basin. Minor streams follow consequent valleys down the sides of the basin toward the center, joining mainstreams which flow out of the basin through valleys cut through its sides. Other consequent streams run down the outer slopes of the dome, often joining subsequent valleys in edges of weak strata ringing the dome. So it is with the Nashville, Cincinnati, and many other domes.

Where a dome has been forced up by an intrusion, often the intrusive rock is stronger than the covering host rock. When erosion reaches the intrusive mass, this erodes less

---

A Pair of Cuestas Seen from the Air: Somewhere over the Colorado Plateau an airliner passes over cuestas which rise above surrounding terrain like ocean waves about to break.

Cuestiform Topography – a Ground View: Edges of upper ends of cuestas are seen in this unidentified view, probably in Capitol Reef National Park, on the Colorado Plateau.

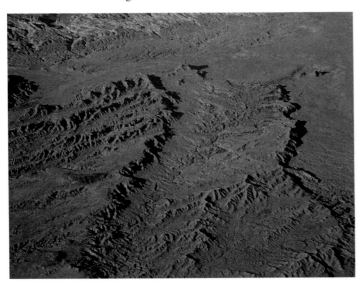

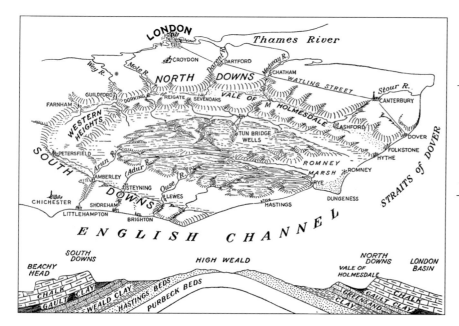

The Wealden Dome: A drawing from an old textbook shows how erosion of the dome left a nearly complete circle of escarpments around a central basin. Many domes, such as the Nashville in Tennessee and the Cincinnati in Ohio, have been eroded out similarly.

A Salt Dome *(right):* In this simplified drawing a dome in a humid region has broken through sedimentary rocks and reached the ground surface, there to be eroded rapidly and form a basin. On dry lands, where solution is subdued, salt domes may stand as highlands, as does Jebel Usdum, in Israel.

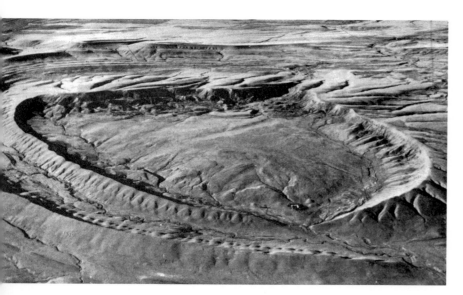

A Deeply Eroded Tectonic Dome on the Great Plains: Southeast of Harlowton, Montana, is Middle Dome, dissected to form a ridge circling a basin. Near the bottom in this view is the erosional remnant of an earlier circling ridge. The triangular features – "flatirons" – on the outsides of the ridges are stumps of strata that formerly arched over the dome.

rapidly than the covering rock, and thus it becomes a highland flanked by stumps of weaker layered rocks, which form hogbacks around its base. These stumps of sedimentary rocks are seen around the periphery of the Black Hills and the laccolithic domes mentioned above.

Another member of the fold family is the salt dome. This is produced as a mass of salt, heated and compressed 3 to 6 miles deep in the crust, flows upward, pushing overlying strata up. The salt commonly forms a plug a mile or so in diameter. In humid regions, solution processes acting on the plug may create a broad depression in the top. In desert regions the plug may rise far enough to form a mountain – for example, Jebel Usdum (Sodom), near the south end of the Dead Sea. (On Jebel Usdum is the "pillar of salt" into which Lot's wife, of Biblical fame, was changed when she looked back longingly toward Sodom, the sinful city.) Salt domes are numerous along the Gulf Coast of the United States and the North German Plain in Europe.

Members of the fold family include also basins separated by uplifts or arches. Such basins may be approximately circular or elongated. Having few if any outlets for running water, they tend to accumulate sediments. These structural basins are very numerous in terranes on sedimentary rocks. Examples include the Denver Basin, in Colorado; the Pow-

der River basin, between the Black Hills and the Bighorn Mountains; and the Raton Basin, in northeastern New Mexico. Often similar in appearance, but different in origin, are basins created by faulting, most notably the Great Basin in the Southwest.

Half-folds: Monoclines

An extensive stratum that is mostly horizontal may include local areas where a single flexure interrupts the dip, forming a kind of step. Such flexures are monoclines, common on terranes of predominantly horizontal strata, on flanks of folds, and on fault scarps (sides of crustal blocks uplifted by faulting). The dip slope often is interrupted by secondary (subsidiary) flexures of many types and by fault scarps, and thus may have an irregular profile when viewed along the

"Appalachian Folding": Relatively symmetrical and rather open alternating anticlines and synclines characterize much of the Folded Appalachians. This example appears in an east-west roadcut through Kanouse Mountain, northern New Jersey.

Roadside Folds: A roadcut near Monterey, Tennessee, on the Appalachian Plateau, reveals an open syncline and an anticline of chevron form. The bedrock is limestone; the black seam is the edge of a coal layer. At right is a fault.

fold axis. A monocline may extend long distances across country in broad, sweeping curves.

On the Colorado Plateau, monoclines are the most characteristic structures, ranging from 10 to 300 miles long, with dips from 10 to 85 per cent and elevations in thousands of feet. Among them is the East Kaibab Monocline, 222 to 30 miles wide, with 2,000 to 5,000 feet of relief, extending 150 miles from the San Francisco volcanic field, in Arizona, to Bryce Canyon, Utah. Another is the Echo Cliffs Monocline, spanning 1,300 square miles and rising up to 500 feet above its surroundings.

Up- and Down-folds: Anticlines and Synclines

The commonest fold types are anticlines, which are tent- or arch-shaped upfolds, and synclines, which are downfolds, rather like inverted anticlines. These folds generally alternate in series, with axes more or less parallel. Some folds lean one way or another; often the axis is inclined. Folds may be relatively tight, with limbs close and nearly parallel, or relatively wide, with limbs at a very wide angle to one another. Often the fold is not symmetrical; the limbs may be at different angles to the horizontal and may be interrupted by secondary folds or by faults.

Cross sections of folds often appear in cliffs and roadcuts. Folds that intersect the ground surface are likely to have been truncated by erosion. Usually the rock is sedimentary or metamorphic, but folds can occur in volcanic rocks too.

Nearly symmetrical alternating anticlines and synclines characterize the Folded Appalachians. Cross sections are seen in numerous roadcuts by travelers going east or west. In most of the European Alps, deformation is extreme, and many rock masses – especially layered ones – have been so disrupted by faulting as to resemble marble cake.

Overturned Folds and Isoclines

In some folds one limb has been rotated more than 90 degrees from the horizontal and both limbs are inclined in the same direction. These folds are called overturned. If the fold has been rotated so far that its limbs are nearly horizontal, it is called recumbent. Where the limbs have been so compressed as to be nearly parallel, they are an isocline.

Folds That Plunge

In many folds the axis inclines from the horizontal, so that the fold "dies out" in the direction of inclination. These folds

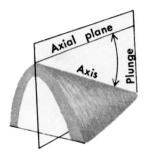

A Plunging Anticline

are called plunging. If both ends of the fold plunge, the fold is said to be doubly plunging; then the feature is essentially a dome. Spectacular specimens of plunging folds in the Appalachians are seen from the air over east-central Pennsylvania.

Homoclines

Any rock unit in which the strata all have the same dip is called a homocline. Often it is a limb of a fold that has been eroded away or is not visible as a whole. The Dakota Sandstone hogbacks in the Front Range of the Rockies between Colorado Springs and Boulder, Colorado, are homoclines; so are the lofty hogbacks of the Canadian Rockies in Jasper and Banff National Parks.

Landscapes on Folds

When a landscape on folds is geologically youthful and little eroded, ridges are likely to be on anticlines and tops of homoclines, and valleys usually are in synclines and at the foot of homoclines. However, as erosion cuts more and more rock off the folds, the topography will increasingly reflect variations in resistance. Higher relief will be more and more on stronger rock, lower relief on weaker rock. Meanwhile, consequent streams running down sides of folds will cut valleys across the folds, and subsequent streams will cut valleys along the base of the folds. Eventually the original topography is erased and a new topography replaces it. This new relief may be somewhat like the original, with anticlinal ridges and synclinal valleys; quite as often it is "inverted," with synclinal ridges and anticlinal valleys or, as usually happens, it may have a mix. In the Folded Appalachians, where time has been ample for inverted topography to develop,

Appalachian Ridges and Valleys: In north-central Pennsylvania, as seen from an airliner, erosion of two broad plunging anticlines has produced a pattern of sharp-backed ridges, with slopes characteristically steeper on the insides of the breached folds. Most valleys cutting across the ridges are water gaps, but at least one (*low right center*) appears to be a wind gap. In the upper left corner here is the town of Lock Haven, with the west branch of the Susquehannna River to its right.

Large Homoclines: Precipitous dip slopes on sawtooth ridges of the Queen Elizabeth Range, in the Front Ranges of the Canadian Rockies, are homoclines.

An Anticlinal Mountain: North of Lake Minnewanka, in the Front Ranges of the Canadian Rockies, the cross section of a ridge exposes a large, open anticline. During erosion a resistant layer has tended to preserve the anticlinal form.

A Synclinal Mountain: In the De Smet Range, in the Front Ranges of the Canadian Rockies, the cross section of a ridge exposes a large, open syncline. At top are minor, irregular folds nested in the large one. A normal fault appears at right.

homoclinal ridges and ridges on parts of eroded synclines are common. Many splendid folds appear clearly in the Canadian Rockies, as in Mt. Kerkeslin, a synclinal mountain, and the Mistaya Valley, cut in an anticline.

During development, a landscape on folds may be strongly influenced by factors other than variations in rock resistance. Renewed tectonic activity may raise or lower the land. Volcanic eruptions may cover it with lava or ash. Glaciers may grind the terrain, shaving ridges, gouging out valleys, deranging streams. Low areas will become covered by erosion debris. Also, at any given time during development a ridge may be on weak rock because the last remnants of a

strong cap rock were eroded off just recently, and a valley may be in strong rock because removal of overlying weak rock has been only recent. Thus, as time goes on, a landscape created by tectonic events increasingly shows the influences of other geologic processes.

Made by Fracturing: Joints

All rock masses of substantial size are jointed – that is, divided by relatively smooth fractures or by planes along which fracturing is likely to occur. Joints may be formed as molten rock, such as a basalt flow, cools and solidifies, con-

A Large Anticlinal Valley: Sides of the Mistaya Valley, in the Canadian Rockies, are scarp slopes, indicating that the valley has been cut into an anticline. The valley was deepened and widened by valley glaciers during the Pleistocene. Peyto Lake, in an ice-gouged depression, has the turquoise hue typical of lakes fed by streams from glaciers. The hue is due to rock flour carried into the lake by the streams.

Flatirons in Arizona: Erosion of folds in colorful rock strata produces triangular forms along the Comb Ridge Monocline, at the eastern edge of the Monument Upwarp between Mexican Hat and Bluff, Utah.

tracting as it does so. They can be produced as the crust is stretched ("tension fractures") or compressed. Most joints are planar; some are curved. Joints may be inches or thousands of feet long, and may define blocks, sheets, and other forms with either straight or curved surfaces. They may be at any angle to the horizontal, and together they form sets, or systems, often of great variety and intricacy. Joints may be invaded by fluids, such as quartz solutions, which deposit minerals within them.

Joints may be tight or open. As with strata, they are described in terms of strike and dip. In jointing there is no displacement parallel to the fracture; thus joints differ from faults, in which displacement does occur.

Joints give access to air, water, tree roots, and other agents that tend to decompose or disintegrate rock. Streams find joints and follow them as weak zones in which to cut valleys. A high-standing rock mass with predominantly vertical joints tends to erode to form rows of forms such as buttes, pillars, or needles, and a similar mass with predominantly horizontal joints will probably erode to make broad, flat-topped features such as mesas. Where joints are parallel to a slope, this is likely to be relatively unstable, prone to landslides; but where joints dip in the opposite direction, the slope may be relatively free of slides. On terrains that have strong relief, broad expanses of bedrock parallel to the slopes are in many instances joint planes; in others, fault planes. Generally, erosional landscapes show the influence of jointing quite as much as the influences of faulting and rock composition.

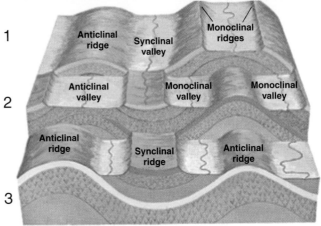

Erosion of Folds: (**1**), Original folds partly eroded; (**2**), original folds eroded away except for the syncline (topography now inverted); (**3**), original topography completely eroded away.

Made by Fracturing: Faults

A fault is a fracture along which one rock mass has moved relative to another. Crustal movements that produce such fractures are occurring continually, in most parts of the world, to the accompaniment of earthquakes. The great majority of earthquakes are too slight to be noticed without the aid of seismographs; most movements amount to only a fraction of an inch at a time. The 40-foot rise of a crustal block in Alaska during the 1964 earthquake was extraordi-

A Joint System: A granite mass on a highland of Mt. Desert Island, Maine, is divided by relatively close joints into blocky, mostly rectangular patterns. Weathering is widening the joints and blunting sharp rock edges.

Influence of Vertical Joints: Here, in New Jersey's Ramapos, joints divide granite gneiss into rectangular blocks. Blocks loosened by weathering fall from this small cliff, leaving smooth vertical surfaces which are joint planes.

A Vertical Fault: In Arizona's Apache Mountains, west of Carizzo, a roadcut exposes a vertical fault in sandstone. Rock strata at left were displaced upward relative to the strata at right. Note the effect on the strata of compression (*at left*) and of tension (*at right*).

A Reverse Fault: Here the hanging wall has been raised relative to the footwall; thus the fault is of the reverse type. This scene is on the east side of the Sandia Mountains, below Sandia Crest, New Mexico.

nary. But over geologic time, tiny increments of displacement can amount to tens of miles.

Fault movements that make mountains are as common as folding events, and faults are as influential on topography as folds are. Faults can be inches or hundreds of miles long, and may reach all the way down to the mantle. Fault movements have created many a mountain range, plateau, and crustal depression. They occur also beneath the ocean bottoms, sometimes generating huge waves (tsunamis, sometimes called – improperly – tidal waves) which travel hundreds or thousands of miles and work havoc on coastal areas.

The crust in or near a fault may be quiet for years or thousands of years, then suddenly become active again. One of the prime tasks in geology is to determine exactly the conditions that produce fault movements and, by detection and measurement of these conditions, to predict earthquakes and thus save lives and property.

Faults are most noticeable in cliffs and quarry walls as a break in the continuity of a rock stratum or other unit; the segments of the unit flanking the fault don't line up. It is the edge of the fracture that we see. Faults occur also, of course, in unstratified rocks of all kinds. Often the fault is not clear-

ly defined, but is rather a zone inches to miles wide in which the rock shows numerous small fractures or is crushed, forming what is called fault gouge (pulverized rock) or fault breccia (shattered rock). A rock surface on a fault plane may show slickensides, a kind of grooves and polish produced by abrasion during movement.

All faults have certain geometric elements. They have strike and dip, as folds do. The fault plane is the surface along which movement occurs. The rock mass that lies over the fault is called the hanging wall, and the mass lower than the fault is the footwall. The vertical displacement in a fault is the throw. The intersection of the fault with Earth's surface is the fault line, or fault trace.

Faults are of great variety, including many combinations, but those likely to be perceived and traced by a nonprofessional observer are limited to a few basic types:

Vertical faults are those in which movement is approximately straight up or down, perpendicular to the horizon. Such faults, attributed to the direct action of gravity, may be called gravity faults.

Normal faults are gravity faults in which the hanging wall has moved steeply downward relative to the footwall. The

Ramapo Fault, which extends from New England through New York and New Jersey to Pennsylvania, is a normal fault; so is the fault or faults bordering Utah's Wasatch Mountains on their western side.

In a reverse fault the footwall drops steeply relative to the hanging wall. A thrust fault is a reverse fault in which the fault plane makes an angle of 45 degrees or less with the dis-placed bodies. An overthrust fault is a low-angle thrust fault in which displacement may extend over many miles at a low angle to the horizontal.

In a strike-slip fault, such as California's notorious San Andreas feature, the movement is mostly horizontal along the trend of the fault. The San Andreas is traced at the sur-face for over 500 miles, from north of San Francisco to the Gulf of California. Total horizontal displacement since the Jurassic Period has been as much as 350 miles; present rates of about 2 inches per year have been measured. Uplift in the Palmdale sector, in southern California, is about 0.5 to 0.7 inch annually. The San Andreas Fault is not a single fracture but a fault zone; it is a chain of near-parallel, connecting fractures marked by crushed rocks, offset streams, ridges built up by pressure, and flanking trenches.

Strike-slip faults often are called transverse or, more expressively, wrench or tear faults. In faults of the dip-slip type, movement is mainly down the dip – that is, vertically down the inclined surface.

Topography on Faults
Fault Scarps

The common topographic feature produced by fault move-ments is the fault scarp: a cliff which is the exposed side of

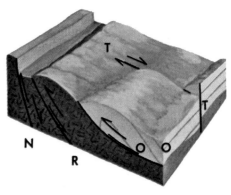

Fault Types Classified as to Relative Motion: **T**, tear or wrench (strike-slip); **N**, nor-mal (gravity); **R**, reverse (thrust); **O**, overthrust.

a block raised relative to an adjacent block. It can be inches to miles high, at any angle with the horizontal. A single raised block can involve several kinds of faults and several fault scarps. Scarps can pass into folds, and folds into scarps.

Ordinarily, fault scarps are created in small increments by earth movements over long periods. The age of a scarp, which is the length of time the scarp has been subject to ero-sion, normally increases toward the top. This, on a high scarp, may be thousands or millions of years older than the bottom. Thus a valley cut into a scarp becomes older and usually wider in the upward direction, often assuming the profile of a wineglass, with a wide top and a narrow bottom. If uplift has occurred relatively recently, the valley may not reach to the bottom of the scarp.

As a fault scarp ages, its valleys widen. Spaces between them assume triangular forms, called facets, and the scarp

Fault Movements and How They Affect the Land Surface

Fault plane
Upthrown block
Downthrown block
Landsliding

Normal fault

Reverse fault

Up
Down

Rift zone

Strike-slip fault

Thrust plane

Overthrust fault

Evidence of Movement Along a Fault: Granite gneiss exposed in a roadcut in New York's Hudson Highlands shows slickensides – grooves and polish due to abrasion by one rock mass moving against another. Grooves appeared when covering rock was blasted off during roadbuilding.

face may become a row of alternating upright and inverted triangles. These become smaller as erosion proceeds. After uplift ceases, the original face of the scarp including the facets will be gradually eroded away and a new face will be formed, this one expressing variations in rock resistance. The new face is unlikely to bear much resemblance to the original. Thus, in localities where faulting ceased very long ago, the original fault scarps are gone, although eroded stumps of the original highlands may survive. On the other hand, where uplifts have continued, an original fault scarp may have ascending rows of fault scarps behind it, like tiers of seats in an amphitheater, each scarp representing an episode of uplift.

In the West, examples of fault scarps of various ages are legion. The west-facing Wasatch Range fault scarp between Salt Lake City and Logan, Utah, is relatively young, showing clear-cut facets and wineglass valleys. East-facing scarps of the southern Sierra Nevada show clear-cut facets in some localities, all-but-obliterated facets in others. The east-facing scarp of the Teton Range, in Wyoming, is well worn, but some facets are still recognizable. In the Basin and Range Province in Arizona, some original fault scarps have been entirely eroded away, as in the Santa Catalina Mountains, near Tucson. In California north of Los Angeles the northern slopes of the relatively young, still-rising San Gabriel Mountains show tier upon tier of fault scarps.

Tilted Blocks

The commonest kind of highland produced by faulting is the tilt block. The side view (along the fault plane) often suggests the form of a sinking ship, bow up and stern under water. Typically the foreslope, which is a fault scarp, is steep and the backslope is long and gentle. Tilt blocks are rock masses that partly sank between adjacent blocks which were being moved in opposite directions by tensional forces in the crust.

Tilt blocks are notably common in the Basin and Range Province, where faulting occurred during the rise of the Rockies. Utah's Wasatch Range is a huge tilt block with the fault scarp facing west; the Sierra Nevada of California and Nevada, 400 miles long with summits 10,000 feet above lands to the east, has east-facing scarps. In central England,

On a Famous Strike-slip Fault: In this view of a segment of the San Andreas Fault, right-angle bends in streams show that the land mass at lower right has moved diagonally downward toward the left. Note also that ridges on opposite sides of the fault do not match. The horizontal displacement is the "offset."

A Pair of Very Young Fault Scarps: In Death Valley, California, near the edge of the great alluvial fan at the bottom of Copper Canyon, these two low but clearly defined scarps are evidence of relatively recent movements along a fault.

A Young Fault Scarp: Lime Ridge, south of Salina, Utah, displays wineglass valleys. These grow younger downward, becoming little gullies toward the scarp base. Strike-slip movements have displaced lower ends of the valleys toward the right.

Landforms Produced Directly by Fault Movements: These are the four basic forms. Often faults occur in complex combinations. During fault movements erosion works to modify the forms, and may eventually alter them radically.

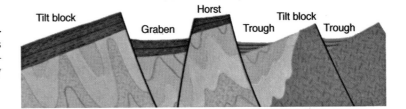

Step-faulting: Faults often occur in a series, with displacements forming "stairs." So it is in the Whetstone Mountains, southeast of Tucson, Arizona. On these peaks as shown here scarp slopes are at left, dip slopes at right, with fault troughs separating them. The backs of the tilt blocks are cuestas.

A Tilt Block on a Volcanic Terrane: Big Bend country in southwest Texas was the scene of intense volcanic activity during and after the Oligocene. Faulting meanwhile and later tilted numerous blocks. Here is Bee Mountain, with fine columnar jointing and thick talus against its sides.

a tilt block extends north to south from the Cheviot Hills to Manchester, with a prominent scarp forming the west side of the Pennine Hills, most scenically at Crossfell Edge.

Fault Troughs

Elongated depressions produced directly by faulting are called fault troughs. A common type follows a fault between two tilt blocks, one side of the trough being the fault scarp of one block and the other side the backslope of another block. These troughs are familiar in the Basin and Range Province and other localities of block faulting. A chain of troughs may appear as steps on a mountainside.

Another variety of fault trough is the graben (German, "ditch"), in which the depression is bounded by two approx- imately parallel normal faults, with facing fault scarps. Germany's Rhine Graben is 180 miles long and 20 to 25 miles wide, lying between two tilt blocks which confine the Rhine River: the Vosges block, with its fault scarp facing east, and the Black Forest block, with a fault scarp facing west. In Africa, the Great Graben east of Lake Victoria, in the African Rift Zone, is 300 miles long and up to 35 miles wide; also in the Rift Zone is the Nyasa Graben, 360 miles long and about 25 miles wide, with a maximum depth of 2,226 feet and a total vertical displacement estimated at 8,000 feet. Farther north is the segment of the Rift Zone known as the Jordan Valley, holding the Sea of Galilee. In Siberia is the crescent-shaped graben that holds the world's deepest lake: Baikal, the "sacred sea," 395 miles long, formerly about 40 miles wide and a mile or more deep, but now shrinking.

North America's Most Scenic Graben: This view in Death Valley is northward from the mouth of Desolation Canyon. At left are the Panamint Mountains, with snow-covered Telescope Peak, and at right is the Amargosa Range. Barely visible in the far background are shallow salt ponds, relics of Pleistocene Lake Manly.

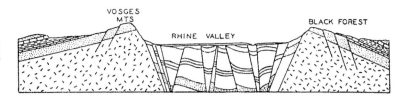
Death Valley (not really a valley – it was not cut by a stream) is in many ways representative of large grabens. It is North America's lowest terrain, reaching 282 feet below sea level. Bounded by normal faults, with eastward-dipping strata of the Panamint Range on the west and the Amargosa Range on the east, the graben is 5 to 20 miles wide and some 220 miles long. Facets still show on scarps that form the valley sides, and continued faulting has produced new fault scarps along edges of the valley floor. This floor, on long-accumulated sediments, at lowest is 11,331 feet below Telescope Peak, the highest summit of the Panamints. The bedrock floor of the graben is believed to be about 8,000 feet beneath the sediments; the overall displacement in the graben is probably about 19,000 feet. During the humid period toward the end of the Pleistocene, Death Valley held a large body of water, Lake Manly, some 90 miles long, 6 to 10 miles wide, and 400 to 500 feet deep – a veritable inland sea, where only a few shallow, scattered salt ponds remain.

Rifts

Rifts are fissures along which fault movements have been horizontal and, in some instances, vertical also. Along a rift there is likely to be much breccia or "fault gouge" (rock debris produced by fault movements) along with minor depressions, sometimes occupied by bodies of water called sag ponds. In the United States the best-known rift is the chain of fissures which are the surface expression of California's San Andreas Fault.

The Great African Rift Valley is not a valley but rather a chain of fault-produced depressions of various kinds, *en echelon*, extending 1,800 miles from the Jordan Valley southward to Mozambique. The term rift valley is better used for a rift that has been occupied and enlarged by a stream to make a true valley.

Looking Along a "Rift Valley": The chain of fault troughs known as the African Rift Valley has majestic scenery. This view is southward along the east-facing scarp of a major branch of the rift valley in Tanzania. The scarp rises several thousand feet above the rift's floor. At left is Lake Manyara.

Overthrusts

Some faulting occurs at a low angle, with one of the opposite rock masses moving over the other. Such faults are called overthrusts. Occurring in zones of intense compression, they usually form highlands and are most frequently seen in or near fold mountains. The overthrusting brings relatively young rock masses over older ones, reversing the usual order in the geologic column. In large overthrusts the vertical distance between the top of the overthrust block and the top of the block beneath can be miles.

A Basin and Range Profile: This province is a vast depression, covering 189,000 square miles, bounded by high mountain ranges on the west and east. Tilt blocks rise from the depression's floor to form "basin ranges."

Sierra Nevada   Basin Ranges of Fault Block Type   Wasatch Range

A Sag Pond: Sunken areas along rifts often are below the water table and thus hold lakes or ponds. This one is in the San Andreas fault zone; the view is northwest toward San Francisco. Hummocky topography has been produced by fault movements.

A Klippe in Switzerland: Some of the world's most famous mountains are klippen. Such is the Matterhorn, an eroded remnant of a rock mass overthrust in the collision between the European and African plates. During the Pleistocene the mountain's upper part was shaped into a "horn" by adjacent glaciers. (Courtesy Corbis Images, 1999)

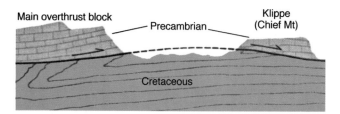

The Lewis Overthrust: Precambrian rocks moved eastward (*to the right*) over the thrust surface, on much younger Cretaceous strata. Erosion cut through the overthrust slice, leaving the eastern portion, known as Chief Mountain, isolated from the main mass. Thus klippen are created.

The Lewis Overthrust in Montana is a 200-mile-wide mass which moved 15 miles eastward. The Cumberland Mountain Overthrust, in the Folded Appalachians in Virginia, Kentucky, and Tennessee, is 25 miles wide and 125 miles long; the thrust was westward. More spectacular overthrusts are noted in the European Alps, a zone of extraordinarily intense folding and fracturing. Here are many nappes (French *nappe*, "sheet"): bodies of rock moved over adjacent bodies by either thrust faulting or recumbent folding. The Pre-Alps, between Lake Geneva and Lake Thun, are a pile of outliers of nappes that were thrust tens of miles northward from their place of origin onto the Swiss Plain.

During overthrusting, erosion proceeds and the overthrust block is gradually reduced in area and thickness. Erosion may hollow out part of this block, revealing the underlying block; the opening is called a window. An example is the Cades Cove area in the Great Smoky Mountains in North Carolina. Large windows have been identified in the Engadine and High Tauern sectors of the eastern European Alps.

Klippen and Horsts

Erosion may completely cut through an overthrust block or a nappe. The cut-off portion may then become a type of outlier called a klippe (German, "cliff"; plural, klippen). In time, erosion may slice up the block to make several klippen. Chief Mountain, in Montana, is a klippe from the Lewis Overthrust. In southeastern New York state, north of Monroe, are several klippen of Precambrian granite gneiss; they are remnants of a sheet that glided from the Hudson Highlands westward over Paleozoic rocks.

A highland crustal block with fault scarps on all sides is called a horst. Less common than tilt blocks, horsts tend to form where the crust is stretched, so that blocks can rise or sink isostatically; thus it is in the Basin and Range Province,

Fault Scarps vs. Fault-line Scarps: Fault scarps are produced directly by fault movements. Fault-line scarps are produced by erosion of rock masses displaced by faulting.

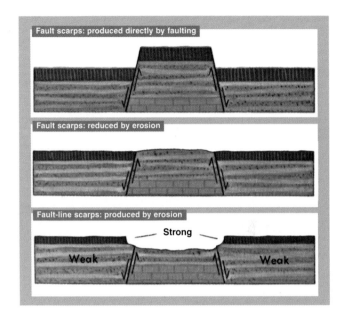

where the Ruby-Humboldt Range, in Nevada, is a major example. Horsts of similar origin occur in the rift zone in East Africa. Some horsts may form also where compression causes much breakage and uplift, as in Europe's Alps.

Topography on Faults: An Overview

An ideal landscape with strong relief produced directly by block faulting during recent geologic time would have an angular, blocky appearance, with relatively steep, often smooth hillslopes or mountainsides. These would be fault scarps, with wineglass valleys, slickensided rock surfaces, and triangular facets in the making. Ridges here and there would be offset; that is, a segment or segments would be shifted horizontally to one side or the other. Fault troughs between ridges would probably have staired or stepped bottoms and would likely zigzag, often beginning or ending at a blank wall. Where streams had occupied these depressions waterfalls, rapids, and ponds would exist, indicating that there had been insufficient time for streams to cut and shape true valleys. Some streams also would be offset. Weathering and gravity movements would not have been prolonged enough to round off edges of fault blocks and build substantial talus slopes. Along the fault line there might be springs, also ponds lacking surface outlets. Over all, the land would look rather chaotic, "unfinished," little modified by erosion.

Relief on fault blocks, like relief on folds, is produced very gradually and is subject to accelerating erosion as blocks rise. The original relief is altered slowly, becoming replaced by relief shaped mainly according to variations in rock resistance. Rock slopes become smoother or rougher depending on the attitudes and varying resistance of exposed strata. Streams tend to straighten their courses, rapids and waterfalls slowly vanish, true valleys develop, divides narrow and shrink, overall contours become gentler. Again, as with land-

Long-term Erosion of a Tilt Block: Stages in erosion are shown left to right. The diagram assumes no substantial interruption of the erosion process by major factors such as climate change, uplift or subsidence, or volcanic activity.

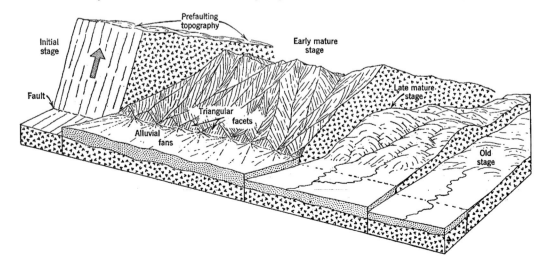

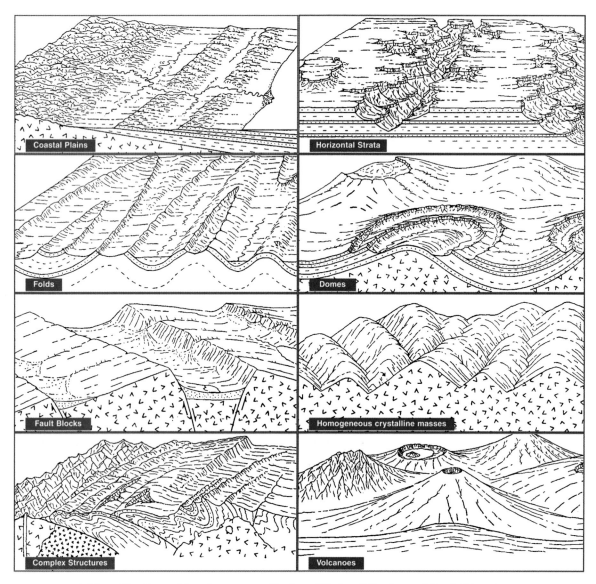

A Classification of Landscapes on Various Kinds of Rocks and Rock Structures

scapes on folds, prolonged erosion guided by lines of weakness may produce inverted topography, with highlands standing where lowlands used to be, and vice versa. Along fault planes, original scarps erode away and new scarps (fault-line scarps) replace them on either side of the fault plane. Gradually an erosional landscape takes form.

Topographic details due completely to faulting are uncommon. Usually some folding has occurred, complicating the task of tracing deformations. Further, many sorts of interruptions of "normal" development can take place, such as renewed faulting, volcanic activity, climate change, or glaciation. However, some landscapes correspond to models

rather well. In the United States, prime areas of topography produced by relatively recent faulting, during the past 20 million years or so, are all in the West, ranging from parts of the Basin and Range Province westward through the Sierra Nevada to the Pacific Coast and north to California's Klamath Mountains. In Basin and Range sectors, land lowering and block faulting appear to have resulted from stretching of the crust, with collapse of parts of it, due to upthrusting of the mountains on the east and west. The Sierra Nevada are carved from a tilt block, or blocks, with the fault scarp facing east. Parts of the Klamath Mountains are faulted and uplifted lava flows.

A Compound Fault Scarp: West-facing slopes of Utah's Wasatch Range are in different stages of erosion. Near Ogden, original facets are still well defined; valleys barely reach the base of the front scarp. Beyond, higher, are ranks of other, older scarps, each rank representing a period of uplift. Nearly horizontal lines on the front scarp show levels of Lake Bonneville, ancestor of Great Salt Lake.

East of the Mississippi all topography is erosional, including even the highest portions of the Great Smokies and New England Upland. Although erosional, these landscapes show the influence of rock arrangements produced by ancient faulting. So it is where the Ramapo Fault in New Jersey and southeastern New York separates high-standing Precambrian rocks in the west from lower Mesozoic rocks in the east. In the Adirondacks, particularly the east portion, evidence of ancient faults is plentiful, including fault-line scarps, zones of crushing, outliers of Paleozoic rocks in Precambrian zones, and resequent grabens – that is, depressions eroded out along old fault lines – such as those holding Saranac Lake and Lake George.

Observing Topography on Folds and Faults

Knowing in advance whether a landscape is tectonic or erosional is basic to understanding it. The map of physiographic provinces of the coterminous United States helps. Textbooks, regional guidebooks, and road guides generally provide this information, as they do also for glaciated, karst, and other landscape types.

If the terrane has been indicated as tectonic, each ridge, valley, basin, and cliff can be suspected of being the result of crustal movements – subject, of course, to inspection for specific signs of such movements. Clues may include abrupt changes of inclination, steep slopes, surfaces little weathered and eroded, and streams that follow zigzag, rough-bottomed depressions, some of which hold ponds. Bedrock surfaces may exhibit slickensides and smoothing due to abrasive contacts with other rock masses. In cliffs and roadcuts where cross sections of folds or faults are exposed, it may be apparent that these deformations have influenced the topography – ridges, ravines, flat areas – above them.

If, on the contrary, the terrane is primarily erosional, the influences of crustal deformation will be less evident. The landforms will likely show evidence of advanced weathering and erosion, probably with considerable chemical alteration and rounding of rock surfaces. In cliffs and roadcuts one may see cross sections of upfolds or downfolds beneath ridges, upfolds or downfolds beneath depressions, raised or lowered blocks beneath either high or low surfaces. Valleys will have well-developed, graded stream channels, and valley systems will be well organized, valley leading to valley, and drainage consistently downward with little if any ponding.

The identification of a landscape as either tectonic or erosional is complicated by the fact that all tectonic landscapes are at least partly erosional, because erosion was happening even as the tectonic features were being produced.

A Panorama of Fault Scarps: Northeast of Los Angeles rise the San Gabriel Mountains, consisting of relatively young fault blocks. Still rising, they are bounded by faults (including the San Andreas), with much deformation, displacements, and crushing. This view is south from Acton, California.

In the Italian Dolomites: At Pordoi Pass, as in some other parts of Italy's Alps of dolomite, predominantly vertical jointing and relatively swift solution have com- bined to produce a tower topography. At lower elevations, where gradients are gentler and the joints are more wide- ly spaced, the relief is smoother and gentler.

Among the Mountains of Kerry: The Gap of Dunloe, 1,500 feet deep, was cut by a tongue of a Pleistocene glacier through a water gap between Purple Mountain (*left*) and Macgillicuddy's Reeks, in the Killarney area of County Kerry. The view is southward from Dunloe Castle. The Mountains of Kerry, Eire's highest, are geologically related to the Folded Appalachians.

10   Mountains: Landforms Supreme

OUR understanding of mountains began with the perceptions of our early forebears. The massiveness of mountains, their height above surrounding land, the often-rugged topography, storms sweeping the summits, difficulties in climbing or living on mountains, their immovability and seeming permanence – such aspects of mountains understandably prompted ancient man and later mythmakers to regard these features as works of all-powerful gods, spirits, or giants, or as embodiments of such beings.

The widely held ancient belief that mountains are a kind of sacred wilderness is seen in the story of Moses and his ascent of Mt. Sinai to receive the word of his people's deity. Among the prescientific but inquiring minds of ancient Greece (a region much disrupted by the collision between Europe and Africa), mountains were regarded as having been created violently – by great winds from Earth's interior, volcanic eruptions, earthquakes – and among ordinary people mountains were thought to be, as was Mt. Olympus, the abode of

gods. In the Middle Ages (as for many people even today), mountains were regarded as made by divinely ordered earthquakes and floods, they had a religious meaning, and, incidentally, they were places of peril – hardly proper for visits by mere mortals.

In the fifteenth century Leonardo Da Vinci, observing fossils of marine shellfish in Italy's limestone mountains, speculated that these terranes were former sea bottoms, and they were, but he fell short of perceiving that they had been uplifted by natural Earth forces. In later centuries pioneers in geology came to understand that crustal deformations, including uplifts, had occurred in the past as natural events, but, unaware of the vastness of geologic time, they attributed such events to long-ago catastrophes rather than to ordinary crustal processes. A detailed and coherent understanding of crustal movements, of structures they produce, and of mountains came only with the twentieth century.

Mountains have been powerful environmental influences

upon humankind and upon our fortunes. Symbols of strength and durability, they dominate their surroundings. Mountains are barriers to travel, tending to isolate social groups and to retard intercultural exchange, but also they have defined paths for communication and have served as bulwarks for defense. Volcanic mountains can be monsters of terror and destruction; yet slopes on extinct or dormant volcanoes have long been favored for farming. Steep-sided, landslide-prone mountains discourage most kinds of agriculture, forcing farmers to resort to terracing and contour plowing. Mountain chains condense moisture from winds, creat-

Mountains and Man: Until a century or two ago, mountains were considered hostile to man and were visited by few. Today they are frequented and cherished by countless climbers and skiers, and by many who just like to sit and look. This is Champex, Switzerland, a resort on a cirque lake in the Valais Alps.

ing "rain shadows" – that is, zones of aridity – to leeward. Rain on mountains gathers into streams that become water supply – and, sometimes, floods. Minerals produced by metamorphism, igneous intrusions, and other natural processes in and on the crust during mountain-building are basic materials for civilization's multitudinous needs. Mountains are rich in attractions for people seeking recreation, scientific information, and communion with nature.

Mountains — An Overview

In popular understanding, a mountain is any land mass bigger than a "hill." What is called a mountain in some localities would be called just a hill in others. According to a standard geological definition, a mountain is any highland that rises abruptly and impressively 1,000 feet or more above its surroundings, has a relatively limited summit area (this distinguishes it from a plateau or a large mesa), and consists essentially of bedrock, which is exposed or only thinly mantled with soil. For present purposes, this definition will do.

Comparatively speaking, mountains are but slight wrinkles on Earth's surface. The elevation above sea level of the world's highest mountain, Mt. Everest, at 29,028 feet, is only about one sixth of the crust's average thickness and one seven hundredth of Earth's radius. Only 8.4 per cent of land rises as much as 1 kilometer (3,168 feet) above sea level and 1.4 per cent 3 kilometers (9,821 feet) above it.

Mountains are places of extremes not only in height but in other respects. Generally, the air gets colder as we move higher and higher in the atmosphere; and sun rays penetrate clearer, thinner air more easily. With low temperatures, frost action in mountain rock often is intense, and chemical weathering may be at a high rate because the rock is exposed. With much fragmentation of rock by frost action, with steep slopes, and with little vegetation cover, gravity movements – notably rockfalls, landslides, and avalanches – are relatively

Mountain Stumps in the Desert: In deserts are seen the bare bones of geology. Egypt's Western Desert between Safaga and Abu Sinbel shows old mountain structures dissected by erosion during a humid past, later polished by blowing sand. Patterns low in the picture are ridge-and-valley forms, with a plunging anticline pointing diagonally down to the right.

Alpine Panorama: Flights over southern Europe on a clear day often provide superb mountain views. Looking north, the highest summits here (*left to right*) are Mt. Blanc, the Matterhorn, and Monte Rosa.

common and massive. Winds are strong. Precipitation tends to be plentiful because air rising against the mountainside becomes chilled and its moisture condenses on the mountain (thus, incidentally, leaving the lowland to leeward dry). Low mountain temperatures may allow the formation of glaciers, with their very high potential for erosion. Streams may be small and most of them may be only temporary, but because slopes are steep, erosional action is vigorous. The higher a mountain reaches above its surroundings, the steeper its slopes are likely to be and the more rapid the rates of erosion and denudation. However, at or near the summits of very high mountains, such as Everest, and mountains in polar regions, moisture stays frozen; therefore weathering and stream erosion are subdued.

Terrains of mountain height are moderately rare; most people must travel a long distance to see a sizable mountain. But mountains have been plentiful during Earth's history. Continental mountain chains and systems shown by relief maps show only relatively high elevations that exist now. Continent interiors are laced and patched with modest highlands whose rocks are relics of very ancient mountains. Continent foundations consist mainly of such rocks, which since

the mountain-building events have been further deformed, metamorphosed, intruded by igneous materials, eroded low, perhaps raised somewhat once more by tectonic movements, and in many instances covered by sediments.

In general, mountains do not undergo erosion without interruptions. Not only during plate collisions but after the thrusting ceases, mountain masses are subject to the isostatic process. As erosion removes material from the mountain block, the block becomes lighter and rises higher in the crust. This cycle of degradation and renewed uplift may occur many times during the life history of the mountain mass. After millions of years, the mountain roots may have risen enough to be exposed at the surface. Despite renewed uplifts,

Old and Complex: The Great Smokies are eroded roots of mountains a billion years old. Rocks are mainly gneisses and schists, with complex structures. Original folds of sedimentary rocks eroded away long ago; root rocks have undergone repeated deformations and uplifts. The topography is erosional. North Carolina's 6,684-foot Mt. Mitchell (*not shown here*) is the highest peak in the United States east of the Rockies. This view is southeast from Newfound Gap.

Stumps of Foreland Folds: On the eastern side of Colorado's Front Range, Dakota Sandstone strata that once arched westward over granitic and metamorphic core rocks have been eroded down to stumps. At Boulder, north of Denver, these stumps point impressively skyward. To the south, in Garden of the Gods Park at Colorado Springs, similar stumps are nearly vertical, like giant fins.

however, erosion wins eventually. The greatest and most resistant of mountains are not eternal, whatever the poets say. Given geologic time, the destiny of all mountains is to be all but leveled.

Some mountains, worn low, become buried under sediments and are later exhumed by erosion. Such mountains often are cut by superposed streams. Other mountains are cut by antecedent streams – streams which existed before the uplift and, in downcutting, kept pace with uplift.

Mountains as seen today are in various stages of degradation. The oldest known mountains, those of the continental shields, with ages of 2 to 4 billion years, have been reduced to cores of granite and metamorphic rocks. Such rock masses, usually with little relief, form the cratons, in continental interiors. Mountains 1 to 2 billion years old, such as the Older Appalachians and Scotland's Grampians, are now mostly chains or clusters of root rocks with relief mostly less than 3,000 feet. These may be parts of cratons, or chains of highlands near continental borders. Mountains thrust up a few hundred million years ago, such as the Younger Appalachians and Eire's Mountains of Kerry, are familiar as deeply eroded folds of sedimentary strata with relief up to a few thousand feet. The Colorado Rockies and the Swiss Alps, with present relief dating back 50 to 60 million years, are masses which, standing high with steep slopes, have been subjected to intense erosion and in most sectors have been truncated down to their core rocks. Finally there are the "really young" mountains, some only a few million years old, along present plate borders, ranging from modest specimens such as California's San Gabriels or San Bernardinos to the towering Himalaya, all rising relatively fast and still possessing much of their original upper rock masses.

Four Types of Mountains According to Mode of Origin: (1) mountains of volcanic materials; (2) mountains from folding and uplift; (3) mountains consisting of an uplifted fault block or blocks; (4) mountains from vertical uplift. Often mountain masses are combinations of these types.

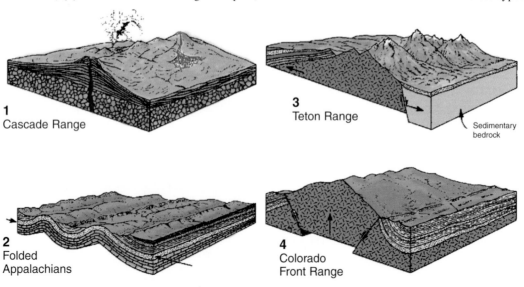

1
Cascade Range

3
Teton Range

Sedimentary bedrock

2
Folded
Appalachians

4
Colorado
Front Range

Complex Mountains in the Rockies: East of Colorado's Front Range the original sedimentary rocks have been eroded away, and mountain cores of granite and metamorphic rocks are exposed. Here, at the Dillon Reservoir, mountain forms are broad and massive. Smooth profiles, the valley with a U-shaped cross section, and the hollow (a cirque) in the mountainside (*right*) are evidences of glaciation.

A Diversified Lot

Mountains are a much diversified lot, but for simplicity here they are put into a few broad categories. The world's largest and most impressive mountain ranges originate as collisions of plates push crustal rock up into giant folds and overthrusts along the zone of collision. These so-called fold mountains grade into the complex type, which also result from plate collisions but are more deformed, with tight and complicated folding, much faulting, and multiple igneous intrusions. Fold mountains themselves may evolve into the complex forms as, through ages, they are stripped of sedimentary rocks by erosion, further deformed by earth movements, and intruded by igneous materials. In a third class, the fault-block mountains, some are groups of fault blocks rising isostatically between plates or parts of plates as these drift apart during mountain-building; others are blocks produced by faulting as mountains of any kind undergo aging.

Large, high-standing bodies of igneous rock – volcanic cones, exhumed batholiths and laccoliths, volcanic piles – also are sometimes called mountains although they differ in origin from fold, complex, and fault-block mountains. Mauna Loa and Kilimanjaro, Utah's Henry Mountains and Navajo Mountain, the Cascade Mountains in the Northwest – such scenic, massive features deserve the name of mountain. However, in this book we examine them in more detail elsewhere as features of igneous activity.

Still other kinds of highlands are often called mountains. Among these are massifs, such as the Beartooth "Mountains" of northern Wyoming and Montana, and the Massif Central of south-central France. High plateau terranes that have been cut into peaks are often regarded as mountains, examples including parts of the Colorado Plateau and, in Australia, the Great Western Plateau and the Great Dividing Range, with its Blue "Mountains." Massifs and plateaus have their own place elsewhere among these pages.

A Mountain Terrane with Little Deformation: In parts of the Montana Rockies, as here in Logan Pass, rock strata are nearly horizontal. Formed from sediments in the marine sector of the geosyncline, they were uplifted with relatively little folding and faulting.

The Birth and Nature of Mountains

The building of a mountain chain by a plate collision is known as orogeny (Greek *oro*, mountain, and *-genes*, -born). Orogeny has been occurring for 4 billion years or more and has been fundamental in the construction of continents. The entire process takes millions of years, is very complex, and is not yet fully known, but a brief summary can be risked.

A typical mountain-building event occurs as an oceanic plate, driven by seafloor spreading, collides with a continental plate. Being basic in composition, and therefore denser and heavier than the predominantly granitic continental rock, the oceanic plate begins subducting it, dragging crust downward as it goes, forming a broad depression-and-trench called a geosyncline. This may be thousands of miles long and 5 to 7 miles deep.

The seaward part of a geosyncline is called a eugeosyncline. Here marine sediments and sometimes terrestrial sediments (those from nearby land) may accumulate to depths of miles. Landward from the trench, molten material surges up through fractures to build up a chain of volcanoes, called an island arc. Farther landward, continental rock and sediments under extreme pressure are metamorphosed, folded, broken up, and invaded by igneous materials, and may be raised as so-called complex mountains. Still farther landward in the geosyncline, in a depression known as a miogeosyncline, accumulating sediments from the continental plate are compressed, heaved up in folds, and pushed forward as overthrusts, becoming what are called foreland fold mountains. These have granite and metamorphosed rocks as foundations.

Mountains can be created by collisions between masses of continental crust. In these events the opposing rock masses

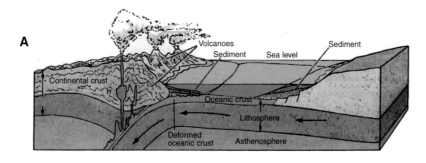

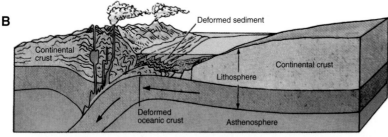

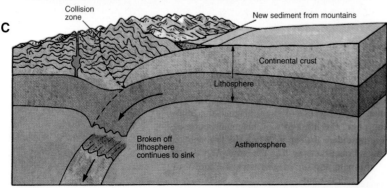

Mountains Formed by Collision of Two Masses of Continental Crust: (A) Subducting oceanic lithosphere compresses and deforms sediments at edge of continent (*left*). Sediments at edge of approaching continent (*right*) remains undeformed. (B) Collision begins. Sediments at edge of approaching continent (*right*) start being deformed and become welded onto already-deformed continental crust. (C) Collision is complete. Descending slab of lithosphere breaks off and continues to sink. The two continental masses are welded together with a mountain range of deformed sediments in the zone of collision. (From Skinner and Porter, *The Dynamic Earth*, © 1995 John Wiley & Sons, Inc., used by permission.)

are of about the same weight and resistance; therefore, neither mass subducts the other, and the energy of the collision is spent in extreme deformation and uplift.

Uplift of young foreland-fold mountains varies widely but often is about ¼ inch per year. The rate of uplift, commonly about 8 times the rate of lowering by erosion, tends to decline with age. Locally, tectonic movements with earthquakes can cause uplifts of many feet within hours. The height to which mountains can rise tends to be determined by the strength of the foundations. The higher any mountain mass rises with respect to sea level, the deeper the mountain roots penetrate into the lithosphere, even to its base.

Foreland Folds vs. Complex Mountains

Foreland-fold mountains consist of relatively regular folds, some of them plunging, with occasional thrust faults and strike-slip faults. Faults increase in number and degree of displacement toward the center of the geosyncline. As the mountains erode and the folds are truncated, ridges form on upturned edges of resistant strata and valleys form in less resistant zones. The higher the mountains, the steeper are the slopes, the more vigorous the erosion, and the sharper the relief. Erosion of relatively regular, open (wide) folds tends to produce a rolling or billowy kind of landscape; tight, overturned or recumbent folds and numerous faults tend to yield rugged relief – narrow summits, precipitous slopes. Well-known foreland-fold mountains include the Folded Appalachians, Europe's Jura Alps and Carpathians, and the Musgrave and Macdonnell Ranges in central Australia.

Formed near the center of a geosyncline, where compression is likely to be extreme, complex mountains are truly complex. Recumbent and overturned folds, isoclines, and faults of the vertical and thrust types are common. Metamorphosed rocks and granite that has intruded between or through the folds may be exposed. According to variations in rock structure and resistance, the topography may be flat-topped, hog-backed, or towerlike, with long, curving ridges, abrupt escarpments defined by thrust faults, plateau-like surfaces, and clusters of variously shaped hills. Among and along the sides of these highlands there may be intrusive features, such as laccoliths and stocks, uncovered by erosion. The general aspect is variegated and irregular. Among prime examples of complex mountains are the Middle and Southern Rockies in the United States, most of the European Alps, Russia's Ural Mountains, the Himalaya in India and Nepal, and the Tien Shan on China's western border.

With time, as foreland-fold mountains are worn down, sedimentary upper portions are stripped away, roots of intruded and metamorphic rocks become exposed, and more faulting, folding, and uplifts occur; thus these mountains become complex. The Rockies are essentially foreland folds in Cana-

Fold Mountains with "Flatirons": Erosion of rock strata parallel to the slope often produces triangular facets suggesting old-fashioned hand laundry irons resting on their heels. Flatirons are common on foreland-fold mountains, as here in the Sawbuck Range of the Canadian Rockies, north of Banff. In the foreground is the Bow River.

A Typical Appalachian Profile: Seen from east or west, most of today's Folded Appalachians are long, nearly parallel ridges *en echelon*, rising to about 2,000 feet, with nearly horizontal crest lines interrupted by occasional water and wind gaps. This view is east from Camelback Mountain, southwest of Stroudsburg, Pennsylvania. Note V-shaped wind gap in the distant ridge.

da but, as we move southward, they show more and more a complex character.

As they age, complex mountains continue to be lowered, though often with occasional uplifts of their surviving parts – that is, the masses of granite and metamorphic rocks which are their roots. Among the aged complex mountains in the United States are the so-called Older Appalachians, including the Green Mountains of Vermont and the Great Smokies of North Carolina and Tennessee. Among other major complex mountains are the Grampians, in Scotland.

Some old complex ranges are worn very low and may become covered by sediments; then the region is uplifted and erosion of the old mountain masses is renewed. Among such highlands in the United States are New York's Hudson Highlands, Vermont's Green Mountains, the Wichita Mountains in Oklahoma, and Missouri's St. Francis Mountains. In these highlands elevations have been much reduced, the pace of erosion is relatively slow, and most slopes are gentle, especially where rock resistance varies little. Where the highlands have been recently uplifted somewhat or have been glaciated (as were the Hudson Highlands), some relatively sharp relief may exist.

The Folded Appalachians

The Younger, or Folded, Appalachians, extending from Newfoundland to Alabama, are foreland folds dating from the collision of the North American and Atlantic plates around 350 million years ago, in the mid-Paleozoic. At their highest they may have reached elevations of 15,000 feet or so, but because of erosion they now do not exceed 3,500 feet of elevation or about 2,000 feet of relief (that is, height above surrounding terrain). During the past they have been eroded almost to sea level several times, then raised again. In the northern sector, including New York, New Jersey, and Pennsylvania, there was some faulting and tight folding, but folding generally was in the form of relatively open, rather regular anticlines and synclines. These trend mostly northeast-southwest, with incidental faults, which are mainly vertical. In the southern sector, from Maryland south, folding and faulting were more intense, and thrusting resulted in westward displacement of large masses such as the Cumberland thrust sheet.

Seen from the air, the Folded Appalachians are mostly long, nearly parallel ridges, like canoes *en echelon*, cut by numerous water and wind gaps. The ridges, mostly on relatively strong sandstones and conglomerates, alternate with valleys which are usually in shale and limestone. Plunging folds, as in eastern Pennsylvania, appear as V- or S-shaped topographic forms. Faults, especially in the south, are responsible for many offsets (horizontal displacements) of ridge segments. Roadcuts for east-west highways expose cross sections of anticlines and synclines, revealing that most of the ridges are on truncated limbs of synclines.

Closely related to the Folded Appalachians are the Mountains of Kerry, in southwestern Eire. Like the Folded Appalachians they date from the Paleozoic and consist of fairly regular folds displaced in some sectors by thrust and horizontal faults. The resemblance of "Kerry" rock types and structures, as well as fossils, to those of the Folded Appalachians indicates that the Kerry rock masses originated as part of the Folded Appalachians and later, in the Jurassic Period, moved eastward as part of a crustal plate while the modern Atlantic Ocean basin was opening up. Other highlands apparently likewise related to the Appalachians exist in Scandinavia, Scotland, and northwestern Africa.

The Rocky Mountains

The Rocky Mountains extend from Mexico to the Yukon Territory, in Canada. They originated as fold mountains some 130 million years ago, during the Cretaceous Period, in the Rocky Mountain geosyncline (which dates back to the Precambrian). Since the Rockies in the United States are far from known plate borders and seem not to result from a collision of plates, some geologists suppose that long ago some platelike mass beneath the lithosphere moved east from the Pacific region at a low angle and caused the uplift. Today, after several major cycles of erosion and uplift (most recently during the Pliocene and Pleistocene epochs), the Rockies have hundreds of peaks exceeding 10,000 feet, the highest in the United States being Mt. Elbert, in Colorado, at 14,433 feet, and in Canada (in British Columbia) Mt. Robson, at 12,972 feet.

The Rockies are a complex of ranges varying widely in structure, rock types, age, erosional stage, and topographic forms. All have been deeply eroded and degraded, but more or less uplifted and deformed again during late Cenozoic tectonic activity. Thus, in various sectors, present scenery is tectonic as well as erosional. During the Pleistocene even the southern Rockies had valley glaciers. All the Rockies show features of glacial erosion and deposition, such as valleys with U-shaped cross sections, truncated spurs, and glacier-produced rock debris on valley floors. Glaciers now survive in the Rockies only from Colorado northward.

The Northern Rockies, from Montana and Idaho to the Yukon Territory, are mainly of the foreland fold type, on sedimentary rock strata. In the front (eastern) ranges strata have been strongly folded and thrust-faulted and thus present rugged, spectacular scenery with enormous hogbacks, precipitous slopes, and, incidentally, multitudes of glacial cirques. Such are the mountains in the eastern portions of Banff and Jasper national parks, in Alberta. To the west, nearer the center of what was the geosyncline, strata are often nearly horizontal and the faults vertical, so that many summits are relatively broad, even flat-topped in some instances. So it is in western sectors of the Canadian Rockies and in Montana's Glacier National Park.

The Middle Rockies – those of northern Colorado, Utah, and Wyoming – are more complex. Sedimentary strata which once arched over the mountain mass have been mostly stripped off, and now survive only as "stumps" on opposite sides of the mountain mass – for example, the scenic Dakota sandstone hogbacks north and south of Denver on the eastern side of Colorado's Front Range, and the hogbacks at Glenwood Springs on the western side. Between the stumps, forming the main bodies of the Rockies, are exposed granite and metamorphic bodies which are the mountain cores and roots. In the Middle Rockies and near them are a scattering of volcanic necks and intrusive features, such as Devils Tower, Bear Butte (a laccolith), and the Black Hills batholith, all now uncovered more or less by erosion.

The Southern Rockies extend from the northern end of the Laramie Range, south of Casper, Colorado, into Mexico. They are complex mountains consisting mostly of linear ridges with cores of Precambrian igneous and metamorphic rocks. Offering a highly varied mix of rock types, structures, and scenery, they are in general the highest of the Rockies, with at least 46 peaks exceeding 14,000 feet.

In the Southern Rockies erosion is farther advanced than in the north. Mountain roots of granite and metamorphic rocks

The Canadian Rockies – Simplified: In the east, flat-lying sedimentary rocks lie under the Great Plains, layer on layer. In the Foothills, rocks are broken into steeply dipping slices. The Front Ranges are rock slices severely folded and faulted, uplifted, and eroded so that layers once deep beneath the Plains are now at the surface, and, in valleys, older rocks lie on younger ones.

The Main Ranges are of sedimentary rocks not strongly folded but lifted high. Erosion has stripped off younger rocks; flat-lying older rocks are seen high in the peaks. The Western Ranges are of faulted and folded younger rocks. The Rocky Mountain Trench, the Rockies' western boundary, is filled with sands and gravels, and occupied by major rivers such as the Columbia and the Fraser.

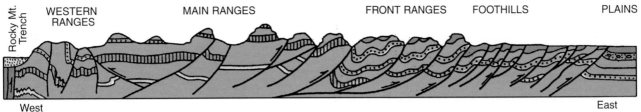

Rocky Mt. Trench — WESTERN RANGES — MAIN RANGES — FRONT RANGES — FOOTHILLS — PLAINS

West   East

"Marble Cake": A drawing by Albert Heim, an early investigator of Alpine geology, visualizes the structure of part of the Santis Mountains in northeastern Switzerland. Dotted lines indicate parts of folds probably removed by erosion, and incidentally suggest the difficulties faced by geologists as they attempt to trace the histories of landforms.

are exposed widely. From southern Colorado to Mexico, volcanoes active in recent times increase in numbers. Bordering the old mountain structures are numerous laccoliths and stocks, and also highlands consisting largely of ancient volcanics, notably the sprawling San Juans, in southern Colorado, and the rugged Chisos Mountains, in the Texas Big Bend country.

Among the most scenic of the Southern Rockies is the Sangre de Cristo Range, extending from southern Colorado into northern New Mexico. This is a mix of sedimentary and metamorphic rocks with volcanics and intrusive masses. It has been much faulted and deeply eroded but still features many high peaks, including Sierra Blanca, at 12,003 feet.

Foreland Folds and Complex Mountains in Eurasia

The European Alps, the eastern part of the Alps-Himalaya system, are being raised by the collision of the Eurasian and African plates. Most of the Alps qualify as complex mountains. They extend about 660 miles from the Mediterranean coast between France and Italy to northern Croatia. They are of Cenozoic origin, of widely varied rock types, with erosion still in a highly active stage and the relief mostly very sharp and rugged. Numerous peaks reach over 12,000 feet, the highest being Mt. Blanc, at 15,771 feet. Two centuries of geologic explorations in the Alps have revealed a dumbfounding complexity of folds, faults, and displacements, such that the familiar comparison of the rock structures to layers in a marble cake hardly begins to suggest the reality.

In the Jura and Chartreuse regions of southern France the Alps are foreland folds. Actively rising until the past few million years, they reach to about 6,000 feet, with a relief up to 4,000 feet. They still have a blanket of sedimentary rocks which show, approximately, the original tectonic forms. As in the Folded Appalachians, the folds are rather regular, though here and there displaced by thrust and horizontal faulting.

The Himalaya, the greatest of complex mountains, extend eastward 1,800 miles from the Hindi Kush Mountains of Afghanistan to northern Myanmar. They are being raised by the collision of the Eurasian and Indian plates, beginning during the Cenozoic. Much of the scenery is of sedimentary strata and includes tectonic as well as erosional forms.

Alpine Hogbacks: In Switzerland's Valais Alps, nearly vertical limbs of truncated folds tower impressively. Precipitation is frequent, and meltwater easily works down between tilted layers during daytime, then freezes at night, peeling off the layers.

Mountain "Needles": Very high mountains often have sharp, jagged profiles because of intense frost weathering, close jointing due to stresses and strains associated with mountain-building, and lack of vegetation that would soften contours. This view is from a cable car on the north side of Mt. Blanc, in the French Alps.

At 29,028 feet (a satellite measurement) the Himalayas' Mt. Everest is the world's highest peak. Himalaya peaks are so high – many are over 20,000 feet – at least partly because the opposing plates, which are both of continental rock, are similar in weight and resistance; therefore, collision with extreme crumpling and uplift occurs rather than subduction.

The Himalaya have been rising at about a half inch per year (a rate of some 30,000 feet per million years) faster than erosion, even with intense glaciation, is cutting them down. They are so high that, according to some geologists, gains in elevation must soon cease or slow down. Gains are limited not only by erosion but also, perhaps, by sagging or melting of crust beneath the mountains and by the tendency of mountain rock to spread slightly under its own enormous weight.

The Andes, another system of complex mountains, extend about 4,500 miles from Panama to Tierra del Fuego. Raised by the collision of the Nazca and South American plates, the Andes include many volcanoes as well as mountains of sedimentary and metamorphic rocks. The highest peak is the volcano Aconcagua, at 22,834 feet. Recently, fossils of whales and other marine animals that date back about 15 million years have been found in southern Chile at an elevation of 5,000 feet, indicating an average uplift rate of 4 inches per 1,000 years in this sector of the Andes.

Where the Ultimate Mountain Rises: Mt. Everest (*background*), in the Himalaya, rises above a veritable chaos of lofty peaks. The strong relief and ruggedness of this topography testify to a high rate of erosion, with intense frost weathering and alpine glaciation, but even more to the energy of the collision between two continental plates, the Eurasian and the Indian.

The Ural Mountains, at the border between Europe and Asia, extend about 1,640 miles southwestward from the Kara Sea to western Kazakh. They were thrust up by the collision of European and Asian plates 600 to 300 million years ago. As with the Himalaya, the opposing rock masses were both continental, and much of the energy of the collision was expended in deformation and uplift. Recently, to the amazement of investigators, seismic tests indicated that the Urals' roots extend down 90 to 120 miles, to the mantle. Though relatively old, the Urals still include many summits at 3,000 to 4,000 feet, and Mt. Narodnaya, the highest, reaches 6,214 feet.

Erosion of Complex Mountains: The Alaska Range, among Alaska's younger mountains, includes rocks of varied composition and structures. West of Teklanika, in Denali National Park, erosional processes work on tilted sedimentary strata. Contours are softened by vegetation-covered rock debris produced by weathering and glaciation. The stream braids its way through thick glacial outwash.

Around the world there are numerous other ranges of complex mountains – too many even to list here. However, a few major ranges can be mentioned as examples.

The Alaska Range extends about 600 miles in a semicircle from the Alaskan Peninsula to the Yukon boundary, varying from 30 to 120 miles in width. It rises to 20,300 ft. in Denali (formerly known as Mt. McKinley), North America's highest peak.

From Poland southward into eastern Romania are the historic Carpathian Mountains, partly complex, with Mt. Gerlachkovka, its highest summit, at 8,700 feet. Separating France and Spain are the Pyrenees, with Pico de Aneto at 11,168 feet. Along the western border of Sinkiang Province, in western China, are the towering Tien Shan, with Pobeda Peak rising to 24,406 feet. New Zealand's Alps include Mt. Cook, at 12,349 feet. Eastern Australia has in New South Wales its Great Dividing Range, with the Alps portion rising to 7,316 feet in Mt. Kosciusko, Australia's highest.

Fault-block Mountains

Mountains that consist of fault blocks are called fault-block mountains or, simply, block mountains. Many originate where crustal uplift and stretching, late in the orogenic process, produce gravity faults and these allow blocks to move up or down. Large blocks that stand high qualify as mountains. Formation of these features may occur near or at the edge of a geosyncline where crustal stretching is being caused by uplift and folding; thus mountains of the Basin and Range were created in association with the rise of the Rocky Mountains.

Block mountains may take form also as erosion reduces a much-faulted mountain mass and the fault blocks move isostatically along the fault planes to restore balance. That can

Rank on Rank: Some fault-block mountains show remarkable symmetry. These ridges, in Arizona northwest of Tucson, evidence a series of mostly parallel faults. Seen from the side, the ridges would appear stairlike or might suggest the teeth of a saw. In deserts, clear skies and sparsity of vegetation often allow remarkably detailed views of geologic forms.

Evolving Fault-block Mountains: The Lost River Range, northwest of Mackay, Idaho, is a horst. It has relatively young fault scarps in the foreground and older ones beyond. Valleys that do not reach to the bottom of the slopes are evidence of recent uplift.

occur in complex mountains as they age and in faulted cores of fold mountains as cores become uncovered.

Fault-block mountains usually appear as tilted blocks, less frequently as horsts and klippen. They may be either heaps or clusters of blocks or isolated features – that is, monadnocks or inselbergs.

Often the nature of block mountains is apparent in their profiles, which are usually angular. Tilt blocks generally have short, steep scarp slopes and long, gently inclined backslopes; they are commonly compared (as we recall) to sinking ships, bow up. In horsts, all sides are steep scarp slopes; they can resemble mesas but are, of course, different in origin. Klippen, being products of prolonged erosion, often have gentle slopes. In uplifted fault blocks, rock layers are usually older than those at about the same elevation in adjacent blocks, which were lifted less.

Relatively young fault scarps on block mountains usually rise abruptly from their surroundings and show tell-tale wineglass valleys, separated by broad triangular facets on which there are slickensides. Edges of rock layers exposed in the scarp face may show offsets due to faulting, and may show folds as well.

Extending out from the base of many fault scarps is a ped-

iment: a relatively smooth surface on bedrock dipping away from the scarp at a very low angle, often with a slightly concave longitudinal profile. Pediments are erosional features of arid regions mainly. Usually they are mostly covered with erosional debris, except near the scarp base.

Where erosion debris at the bottom of a fault scarp has been carried away by creep and running water as fast as it has been produced, degradation will have reached below the bottom of the original fault scarp and started forming a fault-line scarp. The upper and lower faces together form what is called a compound scarp, such as the east face of Wyoming's Teton Range.

As a scarp ages and erosion wears it back, valleys in the scarp widen and the facets gradually disappear. Retreat of

A Compound Fault Scarp: The Teton Range, stretching more than 200 miles along Wyoming's western border, reaching 13,766 feet in the Grand Teton (*here at upper right*), is a tilt block with an east-facing fault scarp rising steeply above Jackson Hole, a graben (*foreground*). The scarp has been deeply eroded, but facets and wineglass valleys are recognizable. Valley glaciation has shaped some peaks into horns. Erosion has lowered the surface of Jackson Hole below the bottom of the original scarp; thus the lower part of the present face of the mountains is a fault-line scarp. The entire mountain slope is a compound scarp.

A "Broken" Mountain? In the Pindos Mountains east of Livadia in Thessaly, Greece, the highway passes between two huge blocks which appear to be fragments of an original rock mass divided by faulting – not unusual in Greece, a very active sector of Earth's crust. The fault block known as Mt. Parnassus, famed in myth, towers in the background.

the scarp carries it back from the original fault line. Resumption of movement along the fault may thus create a new, low fault scarp in front of the old one, along the old fault line. Thus a series of movements can produce several ranks of fault scarps, the oldest in the rear, deeply eroded, and the youngest in front, with its original features nearly intact. After movements cease, erosion continues, gradually erasing facets on all scarps and steadily lowering the mountains. Eventually the original topography will be replaced by an erosional topography, perhaps inverted.

Some Fault-block Mountain Terranes

Fault-block mountain terranes are encountered on all continents. Some are located in zones where plates are colliding today, others where collisions occurred in the geologic past. Often they do not stand alone but are parts of complex mountain ranges. A few can be mentioned here.

All fault-block mountains in the coterminous United States are west of the Mississippi, except for portions of the Adirondacks. Best known are the Sierra Nevada of California, sculptured from a huge block, or series of blocks, mostly of granite, dating from the Jurassic. Uplift and deformation have resulted from the glancing collision of the Pacific Plate with the North American Plate, starting in the Jurassic some 200 million years ago. Extending north to south about 400 miles, the peaks rise as high as 10,000 feet above the Great Basin on the east and more than 14,000 feet above sea level. The fault scarp faces east; its backslope dips gradually westward. The original fault scarp is deeply eroded in most sectors but in others – west of the Owens Lake graben, for example – well-defined wineglass valleys and triangular facets survive. Back of the front scarp rise older, more eroded scarps rank on rank.

Northwest of the Sierra Nevada block are the sprawling Klamath Mountains, extending about 100 miles in Oregon. Mainly of metamorphic rocks, with some granite intrusions, the Klamath mass appears to be a part of the Sierra Nevada block broken off and moved 60 miles northwest.

The Wasatch Mountains of Utah are sculptured from another huge block. This extends 130 miles north to south and rises to about 12,000 feet, nearly 8,000 feet above Great Salt Lake. Wineglass valleys and facets on its west-facing scarp are clearly visible, especially between Salt Lake City

Long-eroded Fault-block Mountains: New Mexico's Peloncillo Mountains consist mainly of Tertiary tuffs and lava flows. In this step-faulted portion of the mountains, west of Lordsburg, tilt blocks such as this one are much worn.

and Logan. The existence of high, massive mountains on each side of the Great Basin suggests that as the mountains rose, the land between them sank to compensate.

The Great Basin, though a sort of supergraben, has scattered highlands ranging from 7,000 to 10,000 feet in elevation. These are tiltblocks, many being inselbergs while others are grouped into short north-south chains. These, as they appear on a relief map, were compared by one geomorphologist to an army of caterpillars crawling toward Mexico. The basin with its fault blocks began forming in the Oligocene Epoch in association with the late phase of Rocky Mountain orogeny. Many of the blocks have been long eroded and are half-buried in their own debris, which lies as much as 5,000 feet thick in basins where there are through-flowing streams. Beneath sediments in some sectors are ancient lava flows.

In central England a great tiltblock extends north-south from Manchester to the Cheviot Hills. Its prominent scarp forms the west side of the Pennine Hills, with dramatic scenery at Crossfell Edge. Other fault-block terranes – to mention just a few – occur in eastern Germany, eastern Africa along the Great Rift, and the Gobi and Hong Kong sectors of China.

Volcanic Mountain Ranges

In time, lava and pyroclastics produced by a chain or cluster of volcanoes and fissures may, despite erosion, accumulate to depths of hundreds or thousands of feet. Even after the cones have been obliterated by erosion, a large mass of material may survive – large enough to qualify as a mountain or a mountain range. Plentiful examples are seen in volcanic areas such as Hawaii, Iceland, and New Zealand. In some instances the volcanic mass becomes involved in geosynclinal or other processes that deform it, intrude it with plutonic rocks, mix it with sedimentary or metamorphic rocks, and elevate it. If the range has originated mainly by volcanic activity or consists mainly of volcanic rocks, it can be called a volcanic mountain range. Such are the Cascade Mountains in the Northwest and the San Juans in Colorado.

A volcanic range is likely to be irregular in relief and profiles, exhibiting a very wide variety of forms. Irregularities occur not only because the lavas and pyroclastics as originally produced are varied in form but because they vary in resistance and thus erode unevenly; the same is true of other rocks, sedimentary and metamorphic, that may become mixed in. During and after episodes of volcanic activity, faulting produces further variations in resistance and profiles. Faulting may open up new conduits for magma rising toward the surface, so that an old range may acquire a scattering of new, active cones and an assortment of intrusive bodies such as stocks and laccoliths.

As hot spots in the crust cool or move to new locations, conduits for magma close up and volcanic activity ceases; then erosion dominates the scene. Cone forms are gradually erased, and the rock masses become shaped according to their resistance to the elements. Meanwhile, as often happens, the entire mass or parts of it may be uplifted, so that erosion accelerates, producing much debris, and this accumulates on the mountain slopes and on surrounding lowlands. In time

Scenery on Uplifted Volcanics: In the Pacific Northwest, the Columbia Plateau was built up by lava flows and pyroclastic eruptions, accompanied in the western sector by the development of the Cascade Range. These mountains consist of stratovolcanoes and old metamorphic rocks capped by volcanics. This view of the Cascades is east from Vista Station, Crown Point, Oregon, over the Columbia River.

An Ancient Volcanic Domain: The rocks of New Hampshire's Presidential Range, in the White Mountains, consist of volcanics metamorphosed into schist. Volcanic activity occurred here during the mid-Paleozoic, when the Avalonian and North American plates were in collision. The cones were eroded away long ago. In this view south from Mt. Jefferson, Mt. Washington, at 6,280 feet the highest peak in New England, is in the background at left. On this alpine terrane vegetation is sparse and felsenmeer is widespread.

erosion will likely win over uplift, and the mountain range may be reduced to near base level, with a low, gentle relief that suggests little of the volcanic past.

Present-day volcanic mountain ranges are in various stages of evolution. The northern islands of Hawaii show few cones or cone remnants; here volcanic activity ceased hundreds of thousands of years ago. Cones are seen increasingly toward the south, with Hawaii (the "Big Island") the site of still-active or recently active cones such as Mauna Loa and Mauna Kea. Weathering, gravity movements, and stream erosion are relatively rapid on all the islands because of the moist climate and the weakness of mixed volcanic rocks. Thick accumulations of debris on ocean bottoms around the

islands indicate that during the past huge blocks of lava and pyroclastics broke loose from mountainsides and slid into the ocean. Today the Koolau Range, on Oahu, displays a large gap left by a slide a few thousand years ago.

The largest assemblage of volcanic mountains in the coterminous United States is the Cascade Range, extending from northeastern California across Oregon and Washington, and for a short distance into British Columbia. The Cascades include relatively old plutonic, sedimentary, and metamorphic rocks but are mostly an uplifted accumulation of more or less eroded flow lavas and pyroclastics, with a scattering of cones. The highest cone in the United States sector is Mt. Rainier, at 14,410 feet; it is heating up, and seismic tests suggest a possible eruption in coming decades. In British Columbia the highest cone is Mt. Waddington, at 13,260 feet. Other well-known cones of the range include Mts. Shasta, Lassen, and St. Helens, to name only three.

Portions of the Cascades date from Oligocene times, when fold mountains of sedimentary rocks arose here and some

A Coastal Range of Volcanics: The Koolau Range, rising steeply above the east coast of Oahu Island, Hawaii, is an eroded mass of old volcanic cones, lava flows, and pyroclastics, uplifted and subject to rapid erosion because of rock weakness and humid climate. In prehistoric time a huge portion of this side of the range broke loose and slid into the ocean, spreading debris thickly over the ocean floor.

A Volcanic Land Exhumed: In North Wales and the English Lake District, Paleozoic volcanic and plutonic rocks were lifted to form mountains, worn low, covered by sediments, lifted again, and exhumed. In Wales the highest of these volcanic relics is Mt. Snowdon, 3,560 feet. In the side of Mt. Snowdon here is a glacial cirque, with large talus cones and a lake.

volcanic activity occurred; most of the volcanics have been produced between the Miocene and Recent epochs. Uplift during that span has been perhaps 2,700 feet. To be distinguished from the Cascade Mountains is the Columbia Plateau, to the east, created during the Miocene by lava flows and ashfalls, up to a mile thick, and by uplift later.

To the south, and rivaling the Cascades in magnitude, is the San Juan Range, in southwestern Colorado. This is on a broad domal uplift covering some 6,400 square miles, with Uncompahgre Peak, at 14,309 feet, reaching highest. In some sectors the rocks are sedimentary; elsewhere they consist of colorful pyroclastics and eroded lava flows produced by volcanic activity from the Miocene to the Pleistocene. Eroded to low relief by Pliocene time, these mountains have since been uplifted 2,000 to 2,500 feet. In some sectors the slopes, on clayey pyroclastics, are very steep, and landslide hazards are severe. Glaciated during the Pleistocene, the San Juans display cirques, U-shaped valleys, and other features of glacial origin, truly spectacular when seen from the air.

In the Texas Big Bend country, the Chisos Mountains and other highlands are in part a deeply eroded, roughly profiled pile of volcanic materials, with some other rocks, dating back 10 to 20 million years.

Spectacular examples of irregular forms in old, long-eroded volcanic mountains are seen in the Basin and Range area of southern Arizona. Here most or all of the cones and much other volcanic material were eroded away long ago.

The Andes Mountains, in South America, include both massive volcanic cones and mountains built of volcanic and non-volcanic materials during the recent geologic past. Mountainous Iceland and much of New Zealand consist of volcanic rock, some in the form of recent fissure flows and cones still active or dormant. Chains of volcanoes in the western Pacific, notably in the East Indies, are intermittently active and growing in size.

In Europe, a few solitary mountains such as Vesuvius and Etna, in Italy, maintain their height and form by relatively frequent eruptions. Highlands built of volcanics dating from prehistoric times are seen in North Wales, Scotland, Northern Ireland, and the English Lake District; also, in Germany, France, the Czech Republic, and other regions of eastern Europe, and in Turkey and lands to the south in the Mideast. Africa has its chain of recently active volcanoes in the Rift Valley region.

On a Land of Lava: In parts of Iceland, mountains of volcanic rock rise abruptly from plains. The higher mountains are capped with snow and ice. Here an outlet glacier from the Vathajokull Ice Cap descends onto an outwash plain.

Plateau Topography: Sedimentary plateaus usually have nearly horizontal strata and canyon-making by swift streams, especially near plateau edges. So it is with the Colorado Plateau. This view from Deadhorse Point State Park, Utah, includes a glimpse of the Colorado River, vast reaches of horizontal rock strata, and, beyond, the La Sal Mountains, carved from laccoliths.

11 Plateaus: Raised and Built-up Tablelands

FROM the beginning, plateaus have been significant features on the stage of human history. Standing high above surroundings, they have been somewhat separate worlds, with their own climates, ecosystems, potential for defense, and scenery. Some plateaus are cold places harsh for agriculture and general living, habitable mostly for people who are individualistic, strong, accustomed to hardship. Such is the Plateau of Tibet, which averages 16,000 feet above sea level and is frigid and bleak. Many plateaus are of gentler nature, thankfully cool in summer, generally friendly toward agriculture and other human activities; such are the Appalachian Plateau, with elevations rarely over 4,000 feet, the Deccan Plateau of India, averaging less than 2,500 feet, and the East African Lake Plateaus, mostly less than 3,700 feet. The concept of plateaus is so familiar that in everyday language the word plateau signifies a high, level place or stage where one pauses before going on to greater heights.

In geology, according to a widely accepted definition, a plateau is a land area larger in extent than a mesa, rising some 1,500 feet or more above its surroundings, having more flat areas than strong relief, and having also at least one steep side. If the terrain is almost completely flat, it may be called a plain, or a plateau plain.

Plateaus may be on any kind of rock or combination of rocks. The Appalachian Plateau, east of the Mississippi, is entirely on sedimentary rocks, and the Colorado Plateau, west of the Rockies, is on sedimentary rocks except for a few patches of igneous varieties; both plateaus are in the sedimentary category. The Massif Central, in France, largely on eroded, uplifted mountain cores of plutonic and metamorphic rocks, exposes both sedimentary rocks and volcanics; such plateaus could be called compound, or complex. The Columbia Plateau, in the northwestern United States, is almost entirely on lavas and pyroclastics; thus it is in the category known as lava, basalt, and volcanic plateaus. Still another kind is the ice plateau, on polar lands such as Greenland and Antarctica.

All plateaus are elevated lands. Sedimentary plateaus are

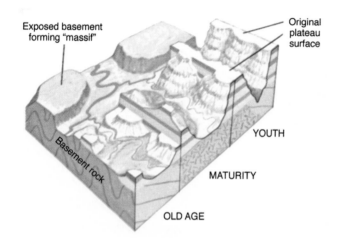

Exposed basement forming "massif"

Original plateau surface

Basement rock

YOUTH

MATURITY

OLD AGE

Dissection of a Sedimentary Plateau: The original surface of a sedimentary plateau is removed gradually by erosion. If local conditions remain about the same for a long period, the topography may evolve through stages suggesting youth, maturity, and old age. Basement rocks (at bottom in diagram) may become exposed after prolonged erosion; thus massifs are developed.

on generally horizontal layers, so that landforms of the plateau tend to have flattish, level tops. Compound plateaus, being on a variety of rocks, with tilted layers (if any), are likely to exhibit an unevenly eroded top surface, with landforms ranging from broad, flat-topped forms to needles and towers. Volcanic plateaus, made of lava and beds of pyroclastics, are likely to be layered. Many plateaus, such as the Colorado and the Appalachian, have been much dissected, and the separate sectors, with their varying characteristics, are themselves sometimes referred to as distinct plateaus.

Plateaus are common throughout the world, but are so extensive that they cannot be seen as wholes except on a map or in a satellite view. Sedimentary and volcanic plateaus tend to have topographic forms which are characteristic and may be observed with some understanding by casual observers during travels across country. Compound plateaus, with wide variations in landforms according to the nature of the local rocks and their structures, are more challenging.

Sedimentary Plateaus

Plateaus on sedimentary strata usually are made by vertical uplift of moderately deep sediments in a geosyncline. Those formed in the interior of the depression are called median plateaus; those that rise near the depression's edges, outside the foreland folds, are foreland plateaus. The plateau type of uplift, being very extensive, is called epeirogenic (Greek *epeiros*, mainland, or continent; and *genes*, kind). Vertical uplift involves a terrain of thousands of square miles; it contrasts with orogenic uplift, which is less extensive but involves strong folding that makes mountains.

Canyon-cutting in a Plateau: Streams near the edge of a plateau tend to have steep gradients, partly because of the nearness of the lowland and partly (in the case of sedimentary plateaus) because of undercutting in weak strata underlying strong ones. Here is part of the canyon of the Genesee River, cut into Appalachian Plateau sandstone in Letchworth State Park, in northwestern New York state. On this terrain the Genesee is an incised meander.

"Mountains" on a Plateau: The rugged highlands known as the Boston Mountains, in northern Arkansas and eastern Oklahoma, are actually part of the Ozark Plateaus Province. These scenic highlands are mostly of sandstones and shale.

Some sedimentary plateaus lie between mountain ranges; thus the Colorado Plateau, which is of the median type, lies between the Rockies and the Wasatch Range. Other plateaus are flanked on only one side by a mountain range; so with the Appalachian Plateau, which is of the foreland type, lying west of the Folded Appalachians. Plateaus are lifted less than adjacent mountains, and their sedimentary strata remain horizontal or, in most sectors, nearly so.

In early stages, the uplift that makes a sedimentary plateau tends to consist of gentle warping and broad folding, with formation of low domes and monoclines, but as uplift progresses there is likely to be considerable faulting, so that scarps form around the plateau edges. Crustal fractures may allow magma from the mantle to rise into the crust of the plateau and cool to make plutons, or to erupt at the surface and create volcanoes. Volcanic activity has occurred on the Colorado Plateau but not on the Appalachian Plateau.

As uplift of a plateau begins, erosion accelerates, and by the time a few hundred feet of elevation has been gained, dissection is well under way, with dendritic drainage predominating. In early stages of the erosion cycle the plateau may be virtually a high plain, with widely spaced streams cutting valleys which deepen toward the plateau edges, often forming spectacular relief features there. With time, drainage density (closeness of spacing between streams) increases and the topography becomes one of mesas and buttes separated by broad valleys. In late stages of erosion the plateau surface will consist of broad plains, usually on relatively resistant strata, with streams meandering among scattered outliers.

Among plateaus of the United States that are on, or mostly on, sedimentary rocks, the Colorado and Appalachian plateaus are the most extensive. In the so-called Interior Plateaus Province, the highlands are very modest in height and barely deserve the name "plateau"; their principal features are the deeply eroded Cincinnati and Nashville domes. The Ozark Plateaus Province, comprised of several subplateaus, is a broad uplift with elevations up to some 2,800 feet and relief as much as 1,000 feet. Projecting above level areas in the east are the St. Francois Mountains, a complex of Precambrian igneous rocks. In the south, erosion has shaped the edge of the sedimentary plateau area into relief features known as the Boston Mountains.

Compared to compound plateaus, which are likely to be irregular in surface forms, sedimentary plateaus are likely to be simple in form and readily understandable. Thus an overview of the Colorado and Appalachian plateaus can suggest characteristics of many if not most sedimentary plateaus elsewhere in the world.

The Colorado Plateau

Uplift of the Colorado Plateau began during the Cretaceous and continues today. The crust beneath the plateau is very thick, some 25 miles, suggesting that during uplift much mass may have moved into this zone from beneath the neighboring Great Basin. Covering some 310,000 square miles, the plateau is mainly on sandstones and shales that span geologic time from the Precambrian to Recent. Portions of Arizona, New Mexico, Utah, and Colorado are included. In the northeast and east the plateau is bordered by the Rockies, and on the west and south by the Basin and Range Province. Elevations above sea level average around 8,000 feet in the southern sectors and 6,000 feet in the north. The plateau lacks mountain ranges, but some parts, such as the Aquarius Plateau at 11,000 feet, and Grand Mesa and the Kaibab Plateau at about 9,000 feet, rise so high that when

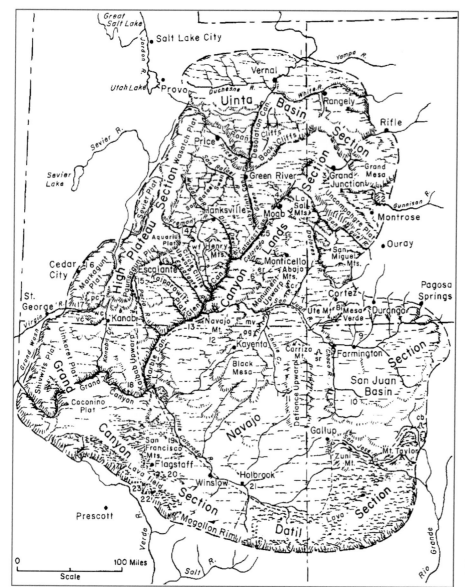

(From C. B. Hunt, *Natural Regions of the U.S. and Canada,* © 1974 W.H. Freeman & Co., used with permission.)

The Colorado Plateau

Notable Landforms, National Monuments (N.M.), National Parks (N.P.), and Population Centers

1 Dinosaur N.M.
2 Black Canyon of Gunnison N.M.
3 Colorado N.M.
4 Arches N.M.
5 Canyonlands N.P.
6 Natural Bridges N.M.
7 Hovenweep N.M.
8 Mesa Verde N.P.
9 Aztec Ruins N.M.
10 Chaco Canyon N.M.
11 Canyon de Chelly N.M.
12 Navajo N.M.
13 Rainbow Bridge N.M.
14 Capitol Reef N.P.
15 Bryce Canyon N.P.
16 Cedar Breaks N.M.
17 Zion N.P.
18 Wupatki and Sunset Crater N.M.
21 Petrified Forest and Painted Desert N.P.
22 Montezuma Castle N.M.
23 Tuzigoot N.M.
pc Pink Cliffs
wc White Cliffs
vc Vermilion Cliffs
wf Waterpocket Fold
er Elk Ridge
cr Comb Ridge
mv Monument Valley
ag Agathla Peak
sr Ship Rock
cb Cabezon Peak

sculptured by erosion they qualify as mountains in form and bulk though not in origin.

Upwarps such as the Zuni and San Rafael, monoclines like the Echo Cliffs and Waterpocket features, and basins like the Uinta and San Juan provide diversity on the Colorado Plateau. The scarp of the Grand Wash Fault, some 4,000 feet high at maximum, forms the border with the Basin and Range Province in the southwest, and the Hurricane Fault scarp, with up to 1,400 feet of relief, forms part of this border farther north. Other major faults, such as the Toroweap and Sevier, also control relief. Fault scarps, with wineglass valleys and other characteristic features, are seen where faulting has been relatively recent.

Here and there on the plateaus, volcanic activity has built numerous cones, as in the San Francisco Mountain and Mt. Taylor areas, in Arizona and New Mexico, respectively. These sectors and some others are covered partly by fissure flows. Numerous stocks have been emplaced, including those that fed the laccoliths of Utah's Henry Mountains. Other well-known laccoliths on the Colorado Plateau include Navajo Mountain, in Arizona, and the cluster of laccoliths in Colorado's La Plata Mountains. Many mesas on the plateau, including Grand Mesa, are kept high and steep-sided by basalt caps which are more resistant than the underlying sandstone or shale.

The Colorado River with its tributaries, including such

A Plateau on a Plateau: Extensive plateaus are dissected by streams to form what are essentially subsidiary plateaus; hence the term Colorado "Plateaus" is occasionally used instead of "Plateau." From Sunrise Point on one plateau at Bryce Canyon, Utah, another plateau, Table Cliffs, is seen in the distance. Note at right the so-called Sinking Ship, a tilt block well defined by inclined strata.

Anatomy of the Grand Canyon Area: This simplified diagram shows how the Colorado River has cut through rock layers of the Kanab Plateau near the edge of the Colorado Plateau. The lowest rocks shown are Precambrian granites and metamorphics. Above are sedimentary rocks, growing younger from the bottom up. The plateaus named here are subsections of the Colorado Plateau.

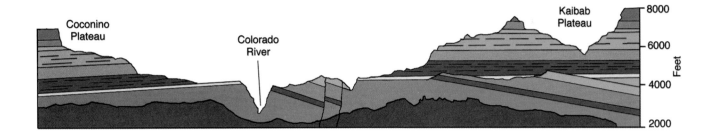

At the Edge of a Plateau: Gems of scenery are common along edges of highlands within the Colorado Plateau. Colorado National Monument, west of Grand Junction, Colorado, offers a view north over cathedral-like pillars to the Colorado River Valley (miles wide here) and the Book Cliffs.

Interior Plateau Topography: Along their higher edges, and along major rivers, plateaus often have sharp relief, but interior sedimentary plateau sections are more likely to have broad areas of gently rolling terrain, as we see here on the Appalachian Plateau near Glyde, Pennsylvania.

major streams as the San Juan and Green rivers, drains some 90 per cent of the Colorado Plateau. The erosional topography is characterized by multitudes of mesas and buttes, often with steplike sides shaped on strata of uneven resistance. The streams, many of them rejuvenated by uplift, occupy incised meandering valleys with structural benches and rock terraces. Relief forms tend to be angular and stairlike, because of the horizontality of the strata and also because of the climate, which is arid to semiarid except at relatively high elevations. These conditions tend to subdue weathering processes which would otherwise round off rock edges. Vertical sheeting of cliff faces helps to keep them steep and produces thick talus below. Because of the prevailing topography of vertical and level surfaces, sheetwash is the dominant erosional process, removing talus from the foot of cliffs and helping to shape pediments. Relief is spectacular where major rivers have cut deep, as in Canyonlands and the Grand Canyon vicinity. Mountainous sectors, notably the Aquarius Plateau, underwent Pleistocene glaciation.

The Appalachian Plateau

The Appalachian Plateau, extending from northwestern New York to northwestern Alabama, is bordered by the Ridge and Valley Province on the east and the Interior Low Plateaus province on the west. The plateau as a whole is on what is essentially a broad syncline, which began rising with the Folded Appalachians during the upper Paleozoic.

A Finger Lake: Valleys occupied by New York's Finger Lakes, deep in the edge of the Allegheny Plateau, are oriented in the direction of Pleistocene glacier ice movement. They were deeply eroded during the glaciation. Shown here is Lake Canandaigua, one of the smaller lakes; it is 16 miles long, with an average water level at 989 feet above sea level and a maximum average depth of 274 feet.

**The Appalachian Plateau
and Neighboring Provinces:**

On the west side of the "Appalachi-
ans" is the Appalachian Plateau.
East on the plateau are the other
Appalachian provinces and the
Atlantic Coastal Plain. Hatching
suggests the nature of plateau
topography compared to others.
Numbers and letters indicate high-
est elevations and other features.
National parks are: Acadia (Maine),
Great Falls (Potomac River at
Washington), Shenandoah (in Blue
Ridge south of the Potomac), and
Great Smoky Mountains (at 11).
(From C. B. Hunt, *Natural Regions
of the U.S. and Canada*, © 1974
W.H. Freeman & Co., used with
permission.)

1   Mt. Katahdin, ME – 5,266 ft.
2   Mt. Washington, NH – 6,280 ft.
3   Mt. Mansfield, VT – 4,393 ft.
4   Mt. Greylock, MA – 3,491 ft.
5a Mt. Marcy, NY – 5,344 ft.
5b Slide Mt., NY – 4,204 ft.
6   Mt. Davis, PA – 3,213 ft.
7   Backbone Mt., MD – 3,340 ft.
8   Spruce Knob, WV – 4,860 ft.
9   Big Black Mt., KY – 4,150 ft.
10 Mt. Rogers, VA – 5,719 ft.
11 Mt. Mitchell, NC – 6,642 ft.
12 Clingmans Dome, TN – 6,642 ft.
13 Brasstown Bald, GA – 4,766 ft.
14 Cheaha Mt., AL – 2,407 ft.

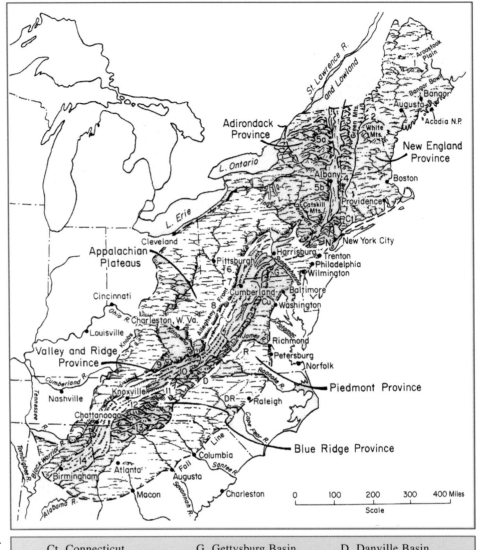

Ct Connecticut River Basin	G Gettysburg Basin	D Danville Basin
N Newark Basin	Cu Culpepper Basin	DR Deep River basin
	R Richmond Basin	

Upturned edges of the syncline form deeply eroded outfac-
ing escarpments of varying heights along most plateau
edges. Rocks are mainly Devonian, Mississippian, and Penn-
sylvanian sandstones, conglomerates, and shales, with occasional beds of limestone and coal. The strata, reaching thicknesses as great as 30,000 feet, are generally horizontal or
gently undulating except for broad, open flexures adjoining
the Ridge and Valley Province.

Drainage on the Appalachian Plateau, mainly westward, is
mostly dendritic, with exceptions such as the parallel pattern
in the Finger Lakes area. Major rivers include the Allegheny
and Susquehanna in the north and the Cumberland and Tennessee in the south. Relief is commonly 1,000 to 2,000 feet,
with landforms mostly rounded from humid-climate weathering and, in the north, with some streamlining and enlargement of valleys by glaciation.

For clearer understanding, the plateau has been divided
into two major sections, the Allegheny Plateau in the north
and the Cumberland Plateau in the south. These have been
divided further according to differing major characteristics.

"Mountains" on a Plateau: The highest part of the Allegheny Plateau has been dissected to form what are called the Catskill Mountains. These are in east-central New York state bordering the Hudson Valley. This winter view is northeast from Slide Mountain, highest of the Catskills. Flattish or slightly rounded summits, steep slopes, and mostly accordant (equal) summit elevations are typical of terranes near the edges of sedimentary plateaus.

One section of much scenic interest on the Allegheny Plateau comprises the Catskill Mountains (traditionally called mountains though they are plateau features). Shaped by deep erosion in the edge of the plateau, and standing as much as 3,000 feet above the Hudson Valley to the east, these highlands reach an elevation of 4,204 feet in Slide Mountain. Further west are the popular Pocono Mountains, carved from a domal uplift in the plateau.

To the northwest, in the northern edge of the Allegheny Plateau, are the eleven famous Finger Lakes, occupying basins in north-south valleys. These were gouged by Pleistocene glacier ice to great depths – in the case of Lakes Seneca and Cayuga, to below sea level. At the plateau's eastern edge in West Virginia, where elevations reach over 4,000 feet, the New River has cut a gorge 2,000 feet deep – the most impressive gorge east of the Rockies.

On the Cumberland Plateau, the mountain section is less dissected but nonetheless spectacular, with very rugged topography near the heads of some rivers, such as the Cumberland and the Kentucky. Cumberland Gap, through which westward-bound settlers streamed long ago, is a 600-foot-deep wind gap in Cumberland Mountain where Virginia, Tennessee, and Kentucky meet. Cumberland Mountain's great bulk rises commandingly 1,500 feet above Powell Valley, to the east.

Reaching into the Cumberland Mountains is a part of the Cumberland thrust block, some 125 miles long and 25 miles wide. This great slice of sedimentary strata, in the Appa-

A Plateau of Volcanics: For millions of years, cones and fissures in the Andes Mountains sector of South America have been producing lava flows and showers of pyroclastics. These materials have accumulated to great depths near the vents and have been uplifted by tectonic forces along with the main mountain masses to form plateaus. Here, north of Quito, Ecuador, valleys have been cut into a plateau edge by swift streams.

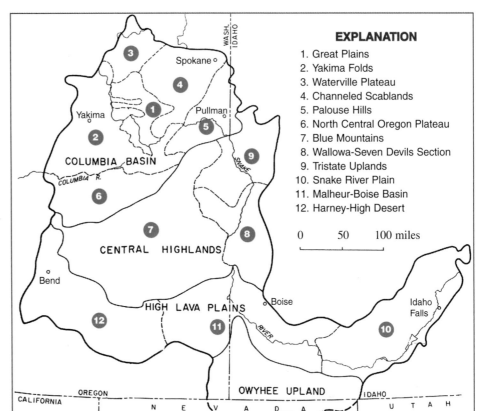

The Columbia Plateau

This physiographic region, which is known also as the Columbia Intermontane Province, has been divided into 12 parts, each characterized by a distinctive topography. Mountains, plateaus, plains, a basin, and other topographic types are represented.

EXPLANATION

1. Great Plains
2. Yakima Folds
3. Waterville Plateau
4. Channeled Scablands
5. Palouse Hills
6. North Central Oregon Plateau
7. Blue Mountains
8. Wallowa-Seven Devils Section
9. Tristate Uplands
10. Snake River Plain
11. Malheur-Boise Basin
12. Harney-High Desert

0   50   100 miles

lachian miogeosyncline sector, includes portions of Virginia, Tennessee, and Kentucky. It is believed to have moved northwest about 2 miles at its northeast corner and 10 miles northeast in the southwest sector.

Lava (Volcanic) Plateaus

Some lava plateaus are created by uplift of accumulated materials from volcanoes that develop in association with plate collisions and mountain building nearby; others result from the build-up of what are known as flood basalts. The latter come not from volcanoes but from long fissures in the crust, opened up where plates are separating or where zones within plates are being subjected to powerful tensional forces. Basalt floods could result also from breaks in the crust due to impacts of comets or asteroids. Being very fluid and voluminous, basalt flows can spread over hundreds of thousands of square miles. Individual flows generally are not more than 30 feet thick, but some reach 100 feet, and the longest can travel 100 miles. Over millions of years flows can accumulate to thicknesses up to 2 miles, deranging streams, covering lowlands, even engulfing mountains. Interbedded

with flows there are commonly layers of pyroclastics and breccia from cone eruptions, and deposits of sediments by streams.

In early stages of plateau buildup the topography may be essentially that of a high plain, with scattered peaks protruding above it. As time goes on, faulting may disrupt the land and raise it higher. Streams do their dissecting work as usual. Highlands and valleys take form with features typical of volcanic landscapes: more or less eroded cones, mud and debris flows, lava plains, zigzagging streams, waterfalls, and layers, columns, and pillows of basalt, along with beds of pyroclastics, exposed in cliffs.

The Columbia Plateau

The Columbia Plateau, in the northwestern coterminous United States, displays features which with some exceptions are much like those of lava plateaus elsewhere in the world. Columbia Plateau features are the higher portions of what is called the Columbia Intermontane Province, bounded by the Cascade Mountains on the west, the Rockies on the north and east, and the Basin and Range province on the south.

Volcanic activity in this province was originally associated with the rise of the Cascades and Rockies during Miocene and later time. It almost completely buried an existing terrane of old volcanic, sedimentary, and metamorphic rocks, and was followed by strong uplift. Even after 25 million years of erosion the plateau rises a mile above sea level.

Scenery on this plateau is dramatic, and challenging to curiosity. Major streams, notably the Snake, Columbia, and Spokane rivers, wind across the land. Lava flows as much as

Volcanics Displayed in a Lava Plateau: In the Columbia Plateau's central highlands, the John Day River has cut Picture Gorge. Basalt columns form buttresses on the valley sides. In the middle background the river cuts through pyroclastics. In the mountains beyond are seen edges of thick lava flows.

100 miles long fill huge valleys; in the east, a single flow covered about 28,000 square miles. The Blue Mountains, some rising more than 9,000 feet, are not really mountains; they are faulted, warped, and tilted basalt plateaus, or sharp-backed ridges shaped by streams that sliced swiftly through weak volcanic materials. In the Wallowa Mountains some peaks, predating the Miocene and later eruptions, rise as much as 9,600 feet above the flows. (Such "island" peaks on the plateaus are called steptoes, as if used by giants gingerly making their way across glowing lava.) In the uplands section of the province, which includes portions of Idaho, Oregon, and Washington, rivers such as the Snake and the Salmon have cut gorges 2,000 to 4,000 feet deep.

Exposed in valley walls of this plateau are towering basalt columns, layered lavas, and colorful beds of pyroclastics. Multitudes of springs – water that has percolated down through joints in the basalt rock and through loose materials – emerge from cliffs. Because of wide variations in resistance among the interbedded lavas and pyroclastics, the landscape where it has been well dissected shows a wide diversity of irregular and bizarre forms.

Quite as remarkable as the "museum" of volcanism on this plateau are landforms resulting from the Pleistocene glaciation. This region was at the southern edge of the huge Okanogan ice lobe, which reached down from Canada, covered lowlands, and filled mountain valleys. When the relatively rapid melting occurred, toward the end of the Pleistocene, flow from a huge meltwater lake overwhelmed the ice lobe, and torrents rushing over the surrounding terrain cut channels in the lava masses and deposited glacial debris thickly on lower areas.

The magnitude of these floods – and there were many of catastrophic proportions – is indicated by the dimensions of Grand Coulee, a channel cut by the Columbia River while it was temporarily displaced from its normal course. This

Plateau-forming Basalts: Stacked rock layers near Palouse Falls, Washington, represent basalt flows of the kind that, along with pyroclastics, built up the Columbia Plateau. Flows, mostly from fissures, covered some 200,000 square miles to a thickness as great as 2 miles.

Columns on a Plateau: Basalt columns rise majestically along a highway south of Grand Coulee, Washington. The columns are capped by a relatively formless mass of basalt called an entablature.

channel, now dry, is 50 miles long, nearly 1,000 feet deep at its origin, and up to 4 miles wide. Networks of smaller coulees cover much of the province. Today the majority of these are dry most of the time, but they still scar the land, and between them remain innumerable patches of glacial debris. Viewing this scene, a geologist once called it "Scablands," and the name has stuck. Incidentally, the Scablands are distinguished as the world's largest assemblage of rapidly created landforms of substantial size.

More Lava Plateaus

Among the most scenic lava plateaus are those of Ethiopia, in East Africa. Fissure eruptions in the Miocene and later, associated with the opening of the Rift Valleys, covered some 450,000 square miles with flood basalts 175 to 200 feet thick, and these accumulated to overall thicknesses of 6,000 feet or so. Plateau elevations average about 10,000 feet above sea level, a few peaks reaching about 13,000 feet. Many fault scarps rise 1,000 feet or more above surrounding terrain. A long north-south scarp in the east section towers 6,500 feet above the lowland to the east. The Blue Nile and other rivers, flowing north, have cut gorges a mile deep.

Also impressive are plateaus of the Brito-Arctic (Thulean) Province, in the North Atlantic region. Much of these plateaus, formed of flood basalts beginning in the Tertiary and continuing into the present, has sunk beneath the waves, but 60,000 square miles of the lavas remains above water, forming the Antrim Plateau of Northern Ireland, the Inner Hebrides, the Faroe Islands, Iceland (here eruptions continue), and East and West Greenland.

The greatest known basalt floods are the Siberian Traps, covering some 750,000 square miles, dating from the end of the Permian Period. The Deccan, in west-central India, dating from the end of the Cretaceous, covers some 300,000 square miles with flows up to 200 feet thick and accumulating to a thickness of 6,000 feet. Before plate drift created the Atlantic Ocean, the Central Atlantic Magmatic Province

"Scablands": Depressions resulting from erosion by glaciers and by torrents of meltwater, and masses of debris deposited by glaciers and streams, mark the Columbia Plateau. In this view, at Dry Falls, Washington, the Columbia River is in the foreground, Ancient Lake at left, and Dusty Lake at right.

Plateau Topography in Spain: The Old Castile area of north-central Spain is on a complex of rocks bounded by mountains but constituting a plateau, high enough for efficient dissection by streams. Mesas on sedimentary rocks south of Leon suggest those of the Colorado Plateau, but, in part because of central Spain's more humid climate, they are not as steep-sided.

extended from southern France to Brazil, covering over 420,000 square miles.

The Paraná Plateau, in southern Brazil and Paraguay, spreads over more than 600,000 square miles – two and a half times the size of Texas – with a thickness of 9,000 feet and more.

Compound Plateaus – Massifs

A relatively resistant, isostatically elevated mass of granite and metamorphic rocks 10 to 20 miles in diameter is called a massif. Some such masses are old mountain cores; some are crust that was subducted, recycled in the mantle, and uplifted – thus perhaps to be regarded as batholithic. Massifs are common in continental shield areas.

An extensive mountain mass eroded down to its core can be uplifted and shaped into a compound plateau, or "massif." The uplift may result from isostatic crustal movements or, as some geologists believe, from crustal adjustments due to

changes in Earth's axis of rotation – changes that would tend to cause shifting of mass in the crust to restore equilibrium. Whatever the cause, new uplifts can keep an old mountain mass high long after its "normal lifetime," with enough elevation and flattish areas to qualify as a plateau.

A prominent example in the United States is the Beartooth Plateau (also called the Beartooth "Mountains") of northern Wyoming and Montana, an uplifted mass of ancient mountain-core rocks flanked by sedimentary formations, and with a front rising 4,000 to 5,000 feet above the Great Plains to the east. Another, more complex example is the Massif Central in south-central France, which covers 32,819 square miles and, in the Dore Mountains, has elevations up to 6,186 feet. This massif exhibits basement rocks covered in some sectors by sedimentary strata, and is patched with volcanics in the Puy area. The Black Forest region, standing high east of the Rhine Graben, is a massif of nearly vertical uplift – a horst of complex structure. Loftiest of all massifs is the

A City on a Massif: Toledo, Spain, and its environs are on a massif of ancient rocks worn low by erosion, then uplifted in recent time and standing as a plateau. This historic city is on a granite knob, half-circled by the Ebro River.

Plateau Scenery in Australia: The so-called Blue Mountains, in eastern New South Wales, are actually a sedimentary portion of a plateau of varied rocks and rock structures in the Great Dividing Range. The sandstone terrane here is notably similar to some areas of the Colorado Plateau.

Tibet-Sikang-Tsinghai, consisting mostly of granite and metamorphic rocks, with most elevations at 12,000 to 15,000 feet above sea level and some exceeding 20,000 feet.

Certain plateaus are notable for their diversity of rock types, structures, and landforms. Among these, besides the Massif Central, is the Meseta of central Spain, bounded on all sides by moderately high mountains, which in the north-central section enclose a flattish terrane of sedimentary rocks with buttes and mesas suggesting Colorado Plateau scenery.

A full half of Australia is made up of the Great Western Plateau, with fold mountains such as the MacDonnell Range, which rises to 4,955 feet, and broad tablelands with scattered outliers of diverse rocks. The Great Dividing Range of eastern Australia, facing the sea, is the edge of another plateau, which, west of Sydney, includes the scenic Blue "Mountains." Table Mountain, rising above Cape Town, South Africa, is a remnant of an old, deeply eroded plateau surface which extended widely over southeast Africa.

A Volcano on a Plateau: In eastern Kenya the plateau rises gradually westward to the edge of the African Rift System. Plateau elevations have been increased by both crustal movements and eruptions of lava from volcanoes flanking the rift zone. Among the volcanoes is Mt. Kenya, a stratovolcano that once towered to 18,000 feet or so. In recent times erosion, including glaciation, has reduced the mountain's elevation to 17,058 feet. The volcano's highest peak, at center in this view, is a volcanic neck known as Batian.

On the Great Plains: The ability of erosional agents, given time, to transport sediments over long distances is demonstrated on the Great Plains of North America, which are covered with erosion debris mostly from the Rockies. Some 50 to 60 million years have been available for transportation of this material. The view is west toward Rapid City, South Dakota; the Black Hills are in the distance.

12
Plains: Wide and Mostly Flat

IN HUMAN history, plains have been known as places where horizons are far away, wild animals and hunters roam, crops grow, cattle graze, battles are fought, games are played, and jets take off and land. In geology, a plain is any land area that is flat or gently sloping and has very low relief. It can be as small as a floodplain a few feet square along a brook, or as large as the West Siberian Plain with its hundreds of thousands of square miles. Plains can exist on a great variety of crustal structures.

Plains and plateaus have some features in common, but sharp differences exist. A plain is created essentially by deposition or erosion and thus differs from plateaus, which result from uplift or the buildup of lava or, in cold lands, the buildup of ice. Plains are mostly flat, whereas plateaus are likely to have sharp relief from stream action, which is vigorous because of the elevation. Margins of plains usually have subdued relief, as do their interiors, but plateaus generally have sharp relief at their edges and in their interiors

where major streams cut deep valleys. Most plains are relatively low in elevation and thus are subject to less rapid erosion than occurs on plateaus.

A plain may be formed by the deposition of materials such as sediments or pyroclastics, or by lateral (sideward) erosion of extensive rock masses. It can be a meadow in mountains thousands of feet high, a stretch of sandy desert, or a low area along a coast. It may be forested, grassy, or bare. Rock structures beneath a plain may be horizontal, warped, basined, domed, or faulted. A plain may be almost perfectly level as far as the eye can see; or, it may have rising from it an occasional cuesta, or a scattering of highlands which are either fault blocks or erosional remnants of highlands.

High-standing plains undergo stream erosion that is relatively active; thus their horizontal surfaces become shaped into more or less sharp relief. As erosion proceeds and the plain is lowered, the pace of erosion diminishes, relief is gentled, and more and more sediments are deposited. When

On an Alluvial Plain: North of Luxor, Egypt, farmers eke out a living on a floodplain (one kind of alluvial plain) of the Nile River. Floodplain sediments are material for dwellings, and water pumped from the Nile or led to the fields by ditches makes the land fertile. Floodplains serve humankind well.

the plain is at or near base level, erosion virtually ceases and deposition reigns – until, perhaps, uplift starts the cycle all over again.

Plains of various kinds occupy some 55 per cent of the continental surfaces, and much of the sea bottoms also. Interestingly, on the continents the largest plain areas are paired on opposite sides of the equator, as are the great North American Plain in the north and the central plain of South America in the south. Most extensive plains are on ancient, worn-down basement rocks that form the continental shields, or cratons, and they are covered mostly with sediments from highlands being lowered by erosion. Such are the great central plains of the Americas, Asia, and Africa from the Sahara to the Congo and the Kalahari Desert. In Australia, lands between the central highlands and the east and west coasts are mostly plains, which are relatively little dissected because of their low elevation and the dry climate.

Plains with their nearly flat surfaces may look simple, but most originate by complex processes. Accordingly, geologists have classified them into about a dozen different types. Most of these can be recognized and more or less understood by non-professional observers who know a little about the locality where they are and what to expect there.

Built by Spreading Water: Alluvial Plains

Alluvial plains are built up of sediments deposited by streams and sheets of water. The plains may be formed on land or in water, and may have almost any kind of rock structure beneath them. Alluvial plains typically have low relief, with gentle slopes decreasing valleyward. Relatively small scattered highlands of solid rock not yet leveled by erosion may project like islands above the plain; for example, at Signal Butte and Scotts Bluff, Nebraska.

The basic types of alluvial plains are floodplains, delta plains, and alluvial fans. Floodplains are built by streams as

Part of a Delta Plain: Northwest of Vancouver, British Columbia, streams from the Coast Mountains have built up a broad delta on the coast. The land surface of these deposits is a delta plain. Subsidiary streams meandering seaward are distributaries, carrying sediments and building the delta outward.

A Plain on the Central Interior Lowland: The floor of the Mississippi Valley here is a broad floodplain – a kind of alluvial plain. The valley sides are seen in the distance, at left and right. The view is southwest from the valley side near Murphysboro, Illinois.

they flood over their channel banks, spread, lose velocity, and deposit that portion of their load which they can no longer transport. A delta plain is the surface above, or mostly above, water on the landward side of a large delta. Floodplains and deltas are described earlier in this book.

Alluvial fans are generally thought of as desert features, and indeed on arid terranes where they do exist they tend to be well developed and relatively easy to recognize. However, they are standard features on many humid lands also. All continents have alluvial fans spreading out below highlands, and in some regions these fans have merged and spread to become vast plains.

The size of a fan depends on the volume of sediments being transported and deposited by running water and on the rate at which the sediments are removed from the place of deposition. On arid lands lacking vegetation, sediments are easily collected by runoff on hillslopes and transported down to the lowland; but there, because rainfall is sparse, streams are not numerous or vigorous, and are slow to remove the deposits; thus on arid lands a large fan can accumulate. On humid lands with vegetation cover, usually the rate of sediment transport and deposition is relatively low and the rate of sediment removal by streams is – thanks to a climate with more rainfall – higher; thus fans, if any, are likely to be small. Where very large volumes of rock debris are available, as on a mountain range recently glaciated and with severe frost weathering, large amounts of debris will descend to the adjacent lowlands and spread out as fans merging with other fans. Depending on factors such as climate and uplift or regional subsidence, fan materials may spread to cover thousands or hundreds of thousands of square miles, filling valleys, engulfing low hills, and generally leveling the land.

A Plain on a Plateau: Plateaus that adjoin ranges of the Taurus Mountains of Turkey, in Anatolia, have broad, nearly flat surfaces on accumulated sediments brought down by streams from the mountains. These surfaces qualify as plains. The one shown here, little dissected, is west of Isparta.

The Great Plains

The Great Plains of the continental United States are the southern part of the great North American Plain. In many ways they parallel the history and characteristics of large alluvial plains elsewhere in the world. They lie on a geosynclinal area dating back to the Mesozoic. As the Rockies began rising from this trough 50 to 60 million years ago, erosion accelerated and fans started spreading eastward. For tens of millions of years this process continued, waning as the mountains were degraded by erosion, then accelerating as the mountains underwent successive uplifts. During the Pleistocene, glacial erosion and periglacial weathering in the Rockies produced huge amounts of rock debris. After the Pleistocene, even though the Rockies were condensing much moisture out of the prevailing westerly winds, there was enough precipitation in the rain shadow to enable streams to keep fans spreading eastward. Transport was mainly by streams but was aided by the westerly winds, which lifted lighter sediments, including loess, and carried them as far as the Mississippi Valley.

In recent time, fan materials from the Rockies have been undergoing removal, and their eastern "front" has been receding westward. However, they still cover areas as far east as central Canada, western North Dakota and western Kansas, and the Pecos and Rio Grande valleys in Texas. These vast deposits and the surfaces beneath them decline from about 6,000 feet of elevation just east of the Rockies to some 1,500 feet in the east, the average inclination being about 10 feet per mile. The area is drained by widely spaced rivers originating in the Rockies, such as the Missouri, Platte, Arkansas, Canadian, and Colorado.

On the higher Great Plains, just east of the Rockies, valleys are wide and shallow, and relief is subdued. Farther east,

A Landmark on the Plains: The Great Plains are not entirely flat. In some localities major rivers, such as the Missouri, have cut deep valleys; in others, mesas and buttes rise abruptly as erosional remnants. Scotts Bluff, Nebraska, shown here, was a landmark for pioneers on the Oregon Trail heading toward the promised land of the Northwest.

The Atlantic Coastal Plain – a Sample: At Newark, New Jersey, the coastal plain is low, with numerous tidal channels and brackish marshes. Uplift of the region followed the melting away of the burden of Pleistocene glacier ice. Recent coastal downwarping has lowered part of the plain.

erosion has spared a few outliers, such as the formations at Signal Butte and Scotts Bluff. Still farther east, near major rivers, tributaries have cut sharp and scenic relief, such as the badlands of North and South Dakota, and other strong relief in Missouri, Kansas, and western Oklahoma.

Basins are numerous in the Great Plains, major examples being the Denver Basin in Colorado and, in New Mexico, the Raton Basin with its volcanic scenery. Among significant domal uplifts are the Black Hills in South Dakota and the Llano uplifts in Texas. Plutonic and volcanic features associated with Rockies orogenies range from stocks and laccoliths in Montana and South Dakota to the Raton features and the Big Bend volcanics in Texas.

In the northern sector of the Great Plains the land is more or less covered with moraine (glacial deposits), in some areas to depths of hundreds of feet. Moraine fills or partly fills many preglacial valleys and generally levels the terrain. This, however, is laced with numerous shallow stream channels which are dry most if not all of the year; many of these were cut by streams from melting glacier ice, and are wholly dry now because of the disappearance of the ice and the drying of climate since the Pleistocene. The northern sector also has numerous "kettles" – depressions left where buried or partly buried, isolated blocks of glacier ice melted and earth settled over them. Some of these depressions reach below the water table (naturally or by human excavating work) and hold ponds. Other shallow depressions are thought to have been made by wallowing bison and other animals in the days when large herds roamed the plains.

Raised to Become Land: Coastal Plains

Any plain bordered by a large body of water (such as the ocean or a great lake) on one side and highlands on the other is a coastal plain. It may have been formed of sediments eroded from highlands or deposited in water, and the bordering body of water may be a lake or the ocean. Usually the term coastal plain is applied to a portion of sea bottom that has emerged to become land during recent geologic time. Extensive areas in both hemispheres which were depressed

Anatomy of a Coastal Plain: Many coastal plains, such as the Atlantic Coastal Plain from Maryland southward, exhibit a typical array of features. Successively, from the ocean edge to the inland edge of the plain, these features may include sedimentary deposits, a terrace, and cuestas with landward-facing scarps, beyond which are an inner lowland and the "old land" – land not part of the coastal plain.

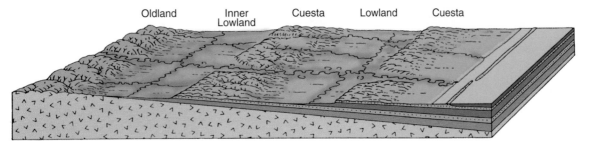

Oldland   Inner Lowland   Cuesta   Lowland   Cuesta

below sea level by the weight of Quaternary glacier ice have "rebounded" in recent time, rising again above the sea to become coastal plains; examples include much of Scandinavia and the Alaskan-Canadian arctic plain. Some coastal plains owe their rise above sea level to isostatic or other adjustments in the crust.

Underlying rock strata of a coastal plain, and its cover of sediments as well, incline very gently seaward. In a plain near shore there may be tidal channels and marshes in shallow basins made by warping or solution of underlying strata or by removal of sand by wind; a number of basins hold lakes. Some coastal plains are narrow strips along a coast; others extend scores or hundreds of miles inland, with terrestrial and marine sediments hundreds or thousands of feet thick. The world's coastal plains cover somewhat less than 1 per cent of the total land area.

Coastal-plain relief is generally low, but the plain may rise landward gradually in a succession of landward-facing cuesta scarps or hills, which can rise 300 or 350 feet and extend far inland. The scarps, which are on edges of tilted strata, have profiles reflecting variations in rock resistance and changes in sea level. Drainage is commonly by parallel streams trending seaward and joined along the way at wide angles by tributaries following the scarps.

Some coastal plains, such as the Florida terrane, are on marine sediments, deposited on the ocean bottom before the uplift. These are called marine plains.

The Atlantic and Gulf Coastal Plains

The Atlantic and Gulf coastal plains of the United States and Mexico are, considered together, the world's largest coastal plain, covering about 583,000 square miles. This plain has such a wide diversity of features that it represents a sort of catalog. It stretches from Cape Cod (which, like Long Island, consists mainly of Pleistocene glacial debris) to the southern tip of Florida and westward to Mexico, for a total length of 3,200 miles. The plain's width varies from a few miles at Cape Cod to a maximum of nearly 500 miles between New Orleans and the southeastern corner of Missouri, where the southern segment of the Mississippi Valley's alluvial plain begins.

From New Jersey to Maryland the Atlantic Plain is belted; that is, dissected by streams cutting landward-facing cuesta scarps on edges of seaward-dipping strata. The scarps more or less parallel the shoreline. In Georgia the plain has nearly horizontal wave-cut terraces, each representing a former sea level. From New Jersey to Cape Hatteras the plain is indented with numerous estuaries – river mouths flooded by the sea because of land subsidence and because of the rise of sea level during the world-wide melting of Quaternary glacier ice. Southward from Hatteras large volumes of sand have been shaped into bars and other forms by wave and current action. The renowned Sea Islands, a chain between Charleston, South Carolina, and the Georgia-Florida border, are above water because they are on rock strata that are relatively resistant and subsided less than those farther north. As mentioned, all of Florida is coastal – more specifically, marine – plain, on a broad, north-south trending anticline of marine limestone.

Beneath the Atlantic Plain, Mesozoic and Cenozoic sediments overlying Precambrian basement rocks are as thick as 17,000 feet. Beneath the Gulf Plain marine and continental sediments cover basement rocks at a depth of 30,000 feet.

A Marine Plain: All of Florida is coastal plain, and most of this is marine plain that is overlain with marine sediments. Uplift dates from the Pleistocene. The land is part of the top surface of the Floridian Plateau, most of which is still under the ocean. This view is westward from the Miami area to the Everglades, along Alligator Alley. Elevations here are less than 20 feet above sea level.

A Coastal Plain on Volcanics: Oahu Island, Hawaii, is ringed by a coastal plain as much as 6 miles wide and averaging about 8 feet above sea level. The emergence dates from the Pleistocene. The area seen here is west of Honolulu.

The eastern and western sections of this plain are belted; the central section is terraced and covered partly by the Mississippi's alluvial plain. The Mississippi sector has been of great interest especially because of the river's repeated changes of course and the enormous volumes of sediments it has dumped onto its delta, which some geologists say begins as far north as Cairo, Illinois.

From Mississippi westward on the Gulf Plain there are numerous salt domes and minor uplifts. Scattered along the plain's margin in Louisiana and Texas are hundreds of thousands of the strange "pimpled mounds," known also by such names as mud lumps and mud volcanoes. Up to 5 feet high and 100 feet wide, these features may be soft masses of fine sand squeezed upward here and there by the weight of new sediments being deposited in the vicinity. Along the western margin of the plain there have been igneous intrusions associated with a fault system.

Some Other Coastal Plains

In England a coastal plain extends westward from the English Channel to the Salisbury Plain, and north as far as Nottingham and The Wash. In the east it includes the high Dover Cliffs and in the north the Cotswold and Chiltern Hills, carved from cuestas. The Aquitaine region of France has about 15,500 square miles of coastal plain. Belgium, Holland, and northern Germany are coastal-plain lands. In South America a coastal plain of nearly 17,000 square miles spreads north, west, and south from Buenos Aires. The Nullarbor Plain of southern Australia covers 75,000 square miles. In north-central Asia, west of the Urals, is the Ob-Khatanga-Lena coastal plain, nearly 500,000 square miles in area. In south Asia, the coastal nation of Bangladesh is mostly a plain, often disastrously flooded because of deforestation of highlands to the north. In eastern China, a large

A High Coastal-plain Remnant: The famed White Cliffs on England's southeast coast are edges of a chalk formation that originated as ocean sediments and was uplifted during the Pleistocene to become dry land; now the chalk is being undermined and cut back by marine action. Here, at Beachy Head, west of Eastbourne in Kent, the coastal plain's elevation is about 600 feet. Rolling topography results from erosion since the ice age.

coastal plain spreads westward from the promontory on which Shanghai is built.

On Bedrock: Stratum Plains

In some highlands that are on horizontal strata, broad level areas occur between prominences such as buttes and mesas. These so-called stratum plains, or structural plains, are areas where erosion has removed regolith and weak rock strata, exposing strong strata which, in effect, become local base level. Small streams, sheetwash, and wind may keep stratum plains bare or nearly so. These plains tend to be short-lived, geologically, because if their elevation is high enough for erosion to strip them down to a resistant layer, it is high enough to make streams vigorous and thus able to dissect the plain relatively rapidly.

A Stratum Plain on a Plateau: In some sectors of the Colorado Plateau, broad and level expanses between mesas and buttes qualify as stratum plains. In this view near the Grand Canyon, Arizona, a broad stratum plain appears in the background.

Sand-dune Plains and Sand Plains

In areas of abundant dry sand and little vegetation, wind can build dunes over wide areas, forming a more or less rippling plain. Such features may be encountered on arid floodplains or along low-lying coasts as well as in deserts. Vast sand-dune plains exist in Egypt's Sahara and in the Arabian Desert, where they are called sand seas, or erg. Rows of dunes on these plains suggest immense ocean billows.

On terrains where the sand supply has become depleted, as by prolonged wind action or stream erosion, remaining sand may be inadequate for dune-building. Such terrains may be called simply sand plains – for example, those of Western Australia. Sand-depleted plains that are covered with lag deposits (material of calibers too heavy for the wind to transport) are called by the Arabs erg (already mentioned), and plains on bare rock are hammada.

Loess Plains

Loess (German *loess*, loose) is a yellow to reddish or brownish, homogeneous, loose material consisting mostly of silt and fine sand. It consists of rock ground fine by weathering and erosion, which often has included the grinding action of glaciers. Loess accumulates on alluvial plains and other flat surfaces. On arid terrains vegetation is sparse and is easily picked up and blown to a distance by winds; then it may be transported and sculptured by streams.

A Sand Plain in Egypt: West of Safaga, in the Western Desert, bedrock is near the surface and sand is insufficient for dune-building. Thus this area contrasts sharply with sand seas such as those of the Sahara and the Arabian Desert.

A Sand-dune Plain in a Lake Basin: At White Sands National Monument, New Mexico, a veritable sea of gypsum particles (a kind of sand), covering 230 square miles, has been shaped by wind into dunes. The gypsum was deposited as a Pleistocene lake here dried up, leaving a lake plain covered with the sand. Wind action then made the sand plain a sand-dune plain. On the sand-dune plain the sand is thick enough, though perhaps not extensive enough, to be called a sand sea.

In the United States much loess was produced by Quaternary glaciation in the Rockies. Loess covers broad areas in the Pacific Northwest and reaches east of the Rockies to the Mississippi Valley, where deposits up to 40 feet deep and deeper form plains in some localities. Interestingly, the 23,000 square miles of sand dunes in the Sand Hills sector of Nebraska may be mostly material that stayed in place while loess particles, which are lighter, blew away.

Loess deposits occur all the way from central and eastern Europe to eastern China. Loess plains of eastern Europe, between the Alps and the Carpathian Mountains, have been called the continent's breadbasket. The often-dusty air of eastern China, as at Beijing, is laden with loess wind-borne from the Gobi Desert and neighboring lands 500 miles and more to the west.

Tubular passages left in loess by decaying grass roots may become filled with natural limy cements deposited by percolating groundwater. These cements stiffen the loess, so that it can stand as cliffs tens of feet high. The flatness of clay particles in loess facilitates tight fitting and increases the material's stability. Thus loess deposits can be hollowed out to make dwellings, as was done in China during the past.

Loess Exposed in a Roadcut: Large amounts of loess transported by prevailing winds and streams from the Rockies were deposited as far east as the Mississippi Valley. Near Memphis, Tennessee, this 20-foot scarp of loess rises above limestone strata. As often happens with loess, masses of it are well enough consolidated to stand and even overhang without collapsing.

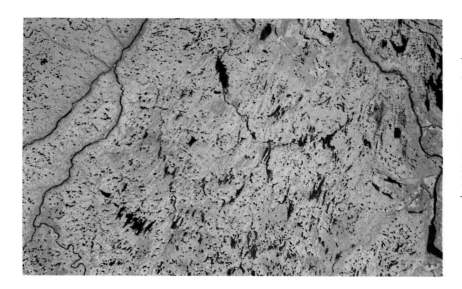

A Plain Leveled by Glacial Action: This Landsat view of terrain near Ungava Bay, Quebec, Canada, shows grooving by continental glacier ice, water-filled basins made by glacial grinding and plucking, and (in detail revealed by greater enlargement of the photograph) masses of glacial rock debris in the form of eskers, drumlins, and natural lake dams.

Leveled by Glaciers: Glacial Plains

Continental glaciers – those not restricted to valleys, but spread over wide areas – act somewhat like giant bulldozers, grinding bedrock, scraping up loose earth, and spreading this debris widely as they advance. Thus land can become more or less leveled, becoming a glacial plain, with low relief. Some such plains have been scraped almost bare, and today exhibit glacier-made grooves and scattered depressions, occupied by lakes and swamps. Examples abound in high-latitude regions such as northern Canada and Greenland. Other glacial plains, as in Illinois, Indiana, and neighboring Midwestern areas, are till plains, covered with rock debris compressed and plastered down on the land by miles-thick glacier ice.

Sites of Vanished Lakes: Lake Plains

Any nearly level terrain which was formerly the bottom of a lake is a lake plain. It is covered with sediments brought in by running water, gravity movements, and wind while the depression held the lake and, probably, later. Lakes can go dry if they fill with sediments, the climate becomes arid, the basin's side is cut through by stream action, a natural or artificial dam is removed, or tectonic activity tilts the basin or otherwise causes it to drain. Vulnerable to change, lakes are short-lived unless artificially maintained.

Lake plains are common on relatively low areas where meltwater from glaciers accumulated as the climate warmed toward the end of the Quaternary and perhaps later. More

A Lake Plain in the Alps: As a land with many basins formed by crustal deformation and by dams of glacial debris, the European Alps have a wealth of lakes and lake plains. At Malles Venosta, in the northern Italian Alps, flatness of the terrain indicates that a lake existed here. In right background is, apparently, a hanging delta, built out into the lake by a large stream when the lake level was higher.

A Land of Lake Plains: During the humid Pleistocene, the basin section of the Basin and Range Province held at least 140 lakes. Few of them – notably Great Salt Lake – remain today. The dry lake bottoms are now lake plains, covered with sands and clays deposited by Pleistocene streams from nearby highlands.

than 140 such plains have been counted in the Great Basin, the largest being the plain of ancient Lake Bonneville. To the east was glacial Lake Agassiz, which in drying up left a plain about 700 miles long and as much as 250 miles wide, covering an area including northern Minnesota, North Dakota, and an adjacent part of southern Canada. Lake Lahontan, mainly in northwestern Nevada but also reaching a short distance into California and Oregon, had a maximum length and width of 250 and 180 miles, respectively. Such plains are familiar sights throughout the world not only where glacier ice melted but in neighboring regions, such as the Great Basin, where climates were humid because of the wide spreading of glacial meltwaters.

Where Lava Spreads: Lava Plains

A lava plain is a surface that becomes mostly leveled as wide-spreading lava flows and wind-blown pyroclastics fill valleys and other depressions. From the plain may rise scattered cinder and spatter cones, pressure ridges, dike ridges, and flow features such as lava trees and cinder crags.

Streams may dissect lava plains into sharp relief. Irregular surfaces result from pronounced variations in the resistance of flow lavas and pyroclastics. In volcanic regions lava plains can cover hundreds or thousands of square miles. Relatively large examples in the United States include those in Idaho at Craters of the Moon, in Texas in the Big Bend area, in Arizona near San Francisco Mountain, and in New Mexico in the Raton Basin. All these are relatively young, little-dissected plains.

A Lava Plain: Basalt lava can be so fluid that flows a few feet deep can spread over many square miles before solidifying. Here, at Craters of the Moon, grasses cover an old flow but not a more recent one, which is about 500 years old. In the distance is Cinder Butte, a large cinder cone, with eroded smaller cones beyond.

Glacier Panorama: Yentna Glacier, in Alaska's Alaska Range, shows features of alpine glaciation: merging valley glaciers with medial moraines (on the glaciers) and lateral moraines (along the sides), tributary and trunk glaciers, ogives, horn peaks, truncated spurs, and cirques with arêtes and cols. The view is northeast, from west of Mt. Denali.

Tourist's Favorite: The original basin for St. Mary Lake, in Glacier National Park, was eroded out by a large glacier; later it was divided into two basins by an alluvial fan deposited by Divide Creek. This view is from Sun Point westward, with Little Chief Mountain at left and Reynolds Mountain, a horn peak, at center.

13
The Works of Glaciers

FEW manifestations of nature are as "unlikely," not to say spectacular, as glaciers. These ice masses, made from snows of yesteryear, are a kind of rock – rock that can flow. When Earth's climate is in a cool phase, glaciers lock up as much as 32 per cent of surface water, lowering the seas by hundreds of feet and making some sea bottoms dry land. Pushing forward, glaciers sculpture mountains, streamline minor highlands, scrape and groove plains, enlarge valleys, grind out basins for lakes, disorganize drainage systems, and erase or drive away nearly all living things in their path. Then, with but a slight warming of the climate (slight considering the range of temperatures naturally possible on this planet), the ice-rock starts melting, its ponderous movement ceases, and its great bulk gradually dissipates into water and vapor. In melting, it leaves broad masses of ground-up rock on Earth's surface as plains, hillocks, and ridges, and as dams for lakes, while meltwater streams spread rock debris over millions of square miles, incidentally enabling the revival of ecosystems all but destroyed by the glaciation.

In our present relatively cold epoch, glaciers and polar ice cover about 10 per cent of the world's landscapes; during the Pleistocene, the figure was something like 30 per cent. Yet, in the long stretch of Earth history, the existence of glaciers has been unusual. The sequence of rocks beginning about 3.9 billion years ago, when the planet's crust was relatively new, indicates that 12 major ice ages (long periods of continent-wide glaciation), possibly more, have occurred. Judging from the most recent one (which seems to be still going on), the length of a glacial age averages just a few million years; thus ice ages taken together have lasted no more than a thousandth of the time elapsed since the crust formed. Types of fossils in the rocks show that the average Earth temperature during the past 600 million years has been about 72° F, hardly low enough for the continuing existence of glaciers except on very high mountains or in polar regions. Clearly, glaciers are extraordinary, and living in an age of glaciers, as humanity has done throughout our short history, is a privilege, at least for those who enjoy superb scenery.

Glaciers as Earth Sculptors

Although running water is supreme as a long-term land sculptor, glaciers are far more powerful in the short term. They have shaped into spectacular scenery most of the

Thanks to Glaciation: The American Midwest, almost as far south as St. Louis, Missouri, is blanketed with ground-up rock deposited by melting Pleistocene glacier ice. Farmers are indebted to the glacier not only for arable soil but for multitudes of small depressions in the land that hold ponds.

world's high mountains, such as the Himalaya, Alps, and Rockies, and the highest of the tropical mountains also. Glaciers converted broad, gently contoured valleys in Norway, New Zealand, Alaska, Greenland, and other sea-bordered lands into profound, steep-walled, awesomely beautiful fiords. Tens of thousands of lakes, ranging from the great North American water bodies to the inland seas of Switzerland and Italy, from the lochs of Scotland to the scenic lakes

of the English Lake District and New Zealand, lie in basins made by glaciers. The Great Plains of the United States and similar lands of central Eurasia are on loess and other glacial deposits, which make not only distinctive topography but rich farmlands, and incidentally provide virtually unlimited sand and gravel for human uses. Advances of glacier ice shifted great rivers such as the Ohio and Mississippi into new courses. Deepening of valleys by glaciers made cliffs for thousands of waterfalls, such as those in Yosemite Valley and Norway's fiords. The locking up of great volumes of water in glaciers has lowered the ocean, causing changes in coastal and shore forms around the world. During periods of glaciation some lands have become deserts, and during periods of glacier melting some deserts have bloomed.

Advancing ice sheets of the late Pleistocene made our forebears retreat southward in Europe. Depletion of the oceans exposed the Bering Land Bridge, allowing primitive humans to migrate from Asia to North America. Some northern-hemisphere plants and animals did not survive the cold and the advancing ice; others migrated south into warmer climes. Although the southern hemisphere experienced much less glaciation than the north, it too participated in the ice age. The impact of the Pleistocene glaciations on Earth scenery and on ecosystems would be difficult to exaggerate.

Living with Glaciers

Glaciers form at high elevations and under climatic conditions which, incidentally, tend to discourage human presence. Like other imposing natural phenomena, such as mountains, rivers, and earthquakes, glaciers during the past were considered to be the embodiments or handiwork of spirits, friendly or hostile. Today on remote terrains, as in the European Alps, the Himalaya, and Norway's mountains, cattle and goats graze along divides between glaciers, but people live at lower elevations and rarely venture out on the ice. Glaciers do present hazards. They have overridden mountain villages, though usually allowing time for villagers to get out

Water Supply for the Lowland: A perennial meltwater stream from a cirque glacier is one of many that provide plenty of fresh, pure water to villages in Switzerland's Val Ferret.

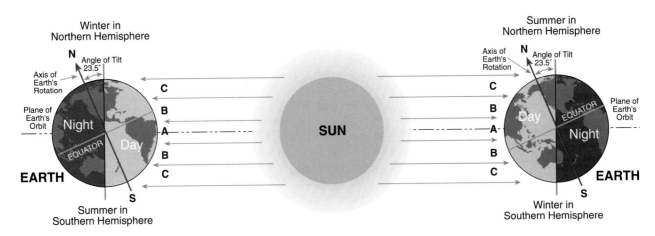

Insolation on Earth's Surface: Earth's axis of rotation tilts about 23.5 degrees from the plane of Earth's orbit; thus as Earth revolves around the Sun we have the seasons. Each hemisphere has winter when it leans away from the Sun, summer when it leans toward the Sun. Over millennia, the amount of tilt varies between 21.5 and 24.5 degrees in 26-thousand-year cycles, and the amount of insolation received by each hemisphere varies accordingly. In all seasons, intensity of insolation on any region diminishes from the equator toward the poles (A to C), because Earth's surface is curved. Insolation varies slightly also according to changes in the Earth-Sun distance.

of the way. Glacier ice teetering on a ledge can break off and fall, starting a destructive snow or rock avalanche. Occasionally ice that has accumulated on a volcano melts and breaks up during an eruption, and becomes part of a debris flow or lahar that brings disaster to the lowlands.

Remote and hazardous though they are, glaciers have been serviceable to man. They have spread ground-up rock over millions of square miles, creating the basic material for farming. They store snow and rain, then release water gradually to recharge aquifers and feed streams that afford water supply during dry as well as humid periods. Waterfalls they have made by deepening valleys provide water power. The scenery they make is a delight.

Glacier ice provides a record of recent and ancient volcanic eruptions that have spread dust around the world. It contains evidence of changes in the composition of the atmosphere, including alterations due to human activities, such as deforestation and use of fossil fuels. Changes in climate, and meteorite impacts as well, are recorded in accumulating glacier ice. Evidence of past glaciations on lands now temperate and even tropical has become useful in efforts to trace past movements of crustal plates.

Why Glaciations?

Much is known about glaciations, but they have been studied scientifically for only a brief period, and causes are not yet well understood. The major cause appears to be the one proposed early in the twentieth century by the Serbian astronomer Milutin Milankovitch: ice ages result from reductions in insolation — that is, radiation Earth receives from the Sun. The Sun's energy output does vary, being lower, for example, at times of high sunspot activity. Further, reductions in insolation occur during cycles of 100,000 and 400,000 years in which Earth's elliptical orbit around the Sun becomes more elongated and the Earth-Sun distance greater.

The amount of insolation on each hemisphere of Earth changes as the attitude of the planet as to the Sun changes. When a hemisphere is tilted away from the Sun, it receives less radiation. Attitude changes occur in three ways: First, as Earth orbits the Sun, the planet's axis of rotation remains tilted with respect to the plane of its orbit; this tilt accounts for our seasons. Second, Earth wobbles slowly as it rotates, completing one wobble in about 26,000 years. Third, the tilt of the axis varies between 21.5 and 24.5 degrees in cycles of about 41,000 years, adding another motion to the wobble. All these changes combine to cause, at times, an extraordinary reduction of insolation and an extraordinary growth of glaciers on one or both of the hemispheres. This explanation tends to be confirmed when the schedules of past glaciations are compared with the astronomical schedules.

Greater understanding of glaciations has come with studies of existing glacier ice, ocean-bottom sediments, glacial deposits on land, and plate movements. Studies suggest,

Evidence of Climate Change: Glacier Bay, Alaska, is a natural museum of glaciation and climate change. Here a glacier has melted back in its valley since the Pleistocene. Glacial erosion has produced the "hanging" valley and the U-shape of the cross section.

Cooling the Earth: In June 1991 a Philippine volcano, Pinatubo, began erupting huge quantities of ash and dust. These rose as clouds high in the atmosphere, spread around the globe, and caused noticeable worldwide cooling for several years. This photograph of the clouds was taken from eastern Maine in mid-July 1991.

among other things, that insolation can be significantly reduced by atmospheric dust and carbon dioxide from volcanic activity. Deep beds of volcanic materials spread widely over the ocean bottom in the North Pacific area indicate a period of high volcanic activity around 2.6 million years ago, the generally accepted date for the onset of the Pleistocene. This could be only a coincidence, but volcanic activity has apparently caused global cooling also during historic time. The eruption of Krakatoa Volcano in the Southwest Pacific, in 1883, blasted enough ash into the atmosphere to cause the world-wide Year Without a Summer. Ash from the 1991 eruptions of Mt. Pinatubo, in the Philippines, is reported to have caused as much as 1.5° F. of cooling around the world for years.

As the solar system moves through the Milky Way galaxy, relatively high concentrations of dust are encountered in certain zones. Several studies indicate a coincidence of these encounters with past glacial ages.

Since only one degree of temperature can mean the difference between freezing and thawing, and since the growth of glaciers is by small increments over millennia, substantial increases of dust and carbon dioxide in the atmosphere do seem likely as a cause of glaciations. On the other hand is the fact of global warming, which appears due to the greenhouse effect: trapping of both solar heat and Earth-produced heat beneath the world-wide layer of dust and carbon dioxide. Does this layer cause cooling or warming?

Tectonic theory offers another view. As plates migrate over the globe, land masses are subjected to climate changes. Fossils of tropical plants found in Antarctica and Spitsbergen, fossil coral reefs in the Canadian Rockies, and glacier-made grooves in the Sahara Desert bedrock – among many other evidences – suggest plate migrations as glaciation causes. Further, plate movements raise parts of the crust to elevations cold enough for glaciers to form, even at low latitudes, as witness the Himalaya in southern Asia and the Andes in northern South America. Still further, deformations of the ocean floor produced by tectonic activity change directions of ocean currents and thus cool some lands while warming others.

Even warming of the atmosphere can, in some localities, cause the growth of glaciers. The warmer the air, the more water it can evaporate from the ocean; and the wetter the air, the more snow it can deposit on high elevations. This scenario, however, seems not to apply world-wide. In this time of global warming, the great majority of glaciers are known to be shrinking, not growing.

Most if not all scenarios proposed to explain ice ages have some degree of plausibility. Most, if not all, may have a role.

A Snowline: In the French Alps northeast of Chamonix, the snowline – the lower boundary of the zone of perennial snow – is clearly visible. Glaciers can originate only above snowline.

Incidentally, the occurrence of ice ages such as the Pleistocene casts doubt on the saying that Earth's crust was shaped in the past much as it is being shaped today. There is a difference, to be sure. Conditions on the Sahara Desert today – blazing sun, intense dryness, blowing sand – differ radically from conditions 10,000 years ago, when the region was humid and forested, and from conditions during the Ordovician Period, when northern Africa was under glacier ice. Frigid Antarctica as we know it is not what it was during the Carboniferous, when warm subtropical swamps patched it and coal beds were being laid. True, the nature of geological processes remains about the same, but processes operate in different combinations at different times.

Biography of a Valley Glacier

Among high mountains where the climate is humid and cold, snow is frequent. Above a certain elevation, known as the snowline, annual snowfall exceeds annual melting and sublimation (sublimation is the conversion of snow into

A Snowline: In the French Alps northeast of Chamonix, the snowline – the lower boundary of the zone of perennial snow – is clearly visible. Glaciers can originate only above snowline.

water vapor without passage through a liquid phase). Snow accumulation is thickest in hollows at heads of valleys, because these locations are relatively high and cool, and have more or less protection from wind and sun. As snow accumulates in a hollow, there occurs some melting and refreezing because of compaction, earth heat, and temperature fluctuations from season to season; thus snow is converted into bits of ice collectively called firn, or névé, and

A Glacier Complete: The relatively small Pré de Bar Glacier, south of the Grand Col de Ferret, Switzerland, has a cirque with an arête separating it from a neighboring glacier; also, an icefall and a spreading terminus with crevasses that divide the ice into "toes."

Scenery on an Icefall: The Argentiére Glacier, in the French Alps, breaks up as it passes over the edge of a "stair" in its valley. The jagged blocks are seracs. This ice is approaching "rottenness."

these eventually combine and crystallize into solid ice. Usually a depth of about 120 feet (the exact figure varies with local conditions) is sufficient for completion of this process.

Gradually the hollow in the valley head fills with snow turning to ice. The hollow enlarges, becoming a cirque, as the rock walls weather and chunks of rock are plucked out by shifting ice. As new snow is added, the lower portion of the snow-and-ice mass bulges out, somewhat as in a mudpie. Outward movement of the mass, and melting caused by heat from the cirque walls, produce a narrow space called the bergschrund (German, "mountain crack") between the ice and the cirque's headwall. As the mass continues to bulge, part of the ice moves over the threshold, or edge, of the cirque and starts moving down valley.

A Valley Glacier at Work

The valley down which a glacier moves is not one of its own making; the valley existed before the glacier formed, and the glacier uses it and reshapes it for, one might say, the glacier's own purposes. On a broad, cold mountain terrain many glaciers following preexisting valleys form a glacier system just as a system is formed by streams of running water. There is, however, a great difference between the behavior of a stream in a valley and that of a glacier in a valley.

Glacier ice moves in different ways at different depths. From the surface to a depth of about 100 feet is the zone of fracturing. Here, stresses and strains due to movement break the ice into blocks and slabs of all sizes, which move more or less separately. At greater depths, in the zone of plastic flow, pressure is great enough to cause sliding within ice crystals, and melting and regelation (refreezing) occur as pressure and friction increase and decrease; meanwhile meltwater moves from zones of higher to lower pressure. In the zone of plastic flow the ice masses flow almost like tar, and continuously change form.

Under very cold climates glaciers are quite dry, because of the lack of melting. Lubrication by water being slight, the ice must deform in order to progress down its confining valley. Under warmer climates glaciers are better lubricated, but because of thaw-freeze, parts of the glacier intermit-

Longitudinal Cross Section of a Valley Glacier: The forms of a glacier and associated terrain features are here outlined from the cirque, where the glacier is born, to the terminal moraine, lake, or ocean, where it ends.

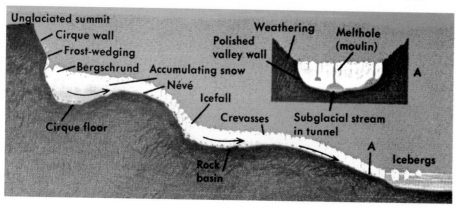

tently freeze to the bottom and sides of the valley and then, as the ice moves on, pluck out rock. Thus glaciers under moderately cold climates can be more erosive than those under very cold climates.

Proceeding downward, a glacier in its valley pushes aside loose material and breaks off minor projections from the valley floor and walls. Although the ice is less hard than bedrock, rock fragments embedded in the ice are very abrasive, especially at and near the glacier's bottom, where fragments are under high pressure. After a period of melting, refreezing of meltwater binds glacier ice to valley bedrock, and as the glacier resumes forward movement, plucking occurs. Thus as the ice moves downslope the channel tends to become straightened and deepened, and the valley walls and floor smoothed.

Glacier ice moves relatively slowly where it is in frictional contact with the valley walls and floor, but faster at midstream, where friction with the channel is least. Moving over uneven bedrock surfaces, the ice undergoes stresses and strains, and tends to fracture, forming deep clefts called crevasses. Ice deep in the glacier undergoes severe compression, which may cause melting or crushing. At a widening of the channel, the ice may spread, creating crevasses parallel to the direction of flow.

Going over a cliff, glacier ice forms an icefall, breaking up into a chaos of "seracs" – slabs, pinnacles, and blocks. Often an icefall displays alternating crescentic bands of clean and moraine-covered ice called ogives. These represent patterns of flow, and different colors represent different seasons.

Most valley glaciers move at a rate between a few inches and a few feet per day, large glaciers usually being faster than small ones. At icefalls, velocity may approach 20 feet per day. Outlet ice flowing from thick ice sheets into fiords, as on Greenland, may move 100 feet daily. Movement is usually faster in summer, when more meltwater is present beneath and around the ice mass to lubricate it and buoy it up.

Some glaciers, called surging or "galloping," have been known to move as much as 300 feet daily – hardly a "glacial pace." Alaska and Canada's Northwest Territories alone have at least 200 such glaciers. One prime example is Variegated Glacier, at the head of Yakatut Bay, Alaska, which surged four times between 1906 and 1983. The 125-miles-long Bering Glacier, North America's largest, originating in the Chugach-St.Elias mountains complex in southern Alaska, surges at about 20-year intervals; it moved more than 6 miles toward the Pacific Ocean between October 1993 and July 1994. Lengthening and stretching associated with surging cause thinning and increased crevassing in the ice mass.

Surging could be caused by various conditions, such as

Glaciation in Progress: In this view between Liskamm and the Breithorn, in the Swiss Alps near Zermatt, are seen several horn peaks, the edge of an icefield (*top*), junctions of small glaciers, icefalls, crevasses, a medial moraine, and a lateral moraine (the sharp-crested ridge at low left).

sudden adjustment to increased snow load or, more likely, an increase in production of meltwater. At least 10 different causes have been proposed.

A descending glacier picks up rock fragments from the channel sides and bottom, and these become incorporated in the moving mass. Meanwhile rock fragments loosened by weathering fall from the valley walls onto the glacier. All this material, in and on the ice, is called active moraine. It is this that forms the dark longitudinal streaks or bands on glaciers, showing directions of ice flow. Some moraine piles up along the glacier's sides, like snow alongside a plow; this is lateral moraine. Where a glacier from a tributary valley joins the main glacier, without merging with it, another streak or band of moraine is added. A glacier that has been joined by many others may, from the air, resemble a long white dress with wavy longitudinal stripes.

At higher elevations near its origin, a valley glacier is relatively narrow and twisting, and undulates as it passes

over humps and depressions. The glacier's top surface is likely to be rough with pressure ridges and crevasses. Some of the latter are veritable chasms, which one can cross only on a ladder or other makeshift bridge. Looking down into a crevasse, one sees sky-blue ice – blue because of absorption of red light rays by the ice and because of air trapped in it. If the glacier is melting, it may also be pocked with moulins, or meltholes, made by meltwater flowing down into the ice.

Crevassing in a glacier may create ice arches or ridges here and there. Where moraine has protected underlying ice from sun, ice mounds and ridges may rise, and also "table rocks" – rock slabs balanced on ice pedestals. Fresh snow may hide crevasses and meltholes, creating hazards.

Eventually most valley-glacier ice arrives at a level where the rate of melting and sublimation equals the rate at which ice arrives from above. Here the glacier is a mass of "rotten"

ice covered more or less by moraine, its surface patched with pools of meltwater, and rugged with crevasses and moulins. At the lower limit of this zone is the glacier's terminus, an irregular convex slope or wall, sometimes lobed like toes on a foot, and usually with an arched opening beneath it. This opening is the end of a subglacial tunnel, from which pours meltwater, usually milky and somewhat yellowish with its content of "rock flour" (rock ground fine by glacier action).

Glaciers in Many Forms

Glaciers occur in a multitude of sizes and forms. The smallest are diminutive masses of ice with barely enough bulk to deform and flow out of a mountain hollow. The largest are vast blankets of ice spread over half a continent.

An ice mass hardly bulky enough to move out of a cirque is known as a cirque glacier. It may be simply an accumulation where snows and prevailing temperatures are inadequate for growth to glacier status. Many cirque glaciers are remains of valley glaciers that have melted back to their birthplaces.

Here and there, ice builds up in a cirque or on a ledge above a cliff. As the mass grows, it spreads to the cliff edge, becoming a hanging glacier. If it continues growing, from time to time portions break off and fall down the mountainside. Often such falls trigger snow or rock avalanches. Masses of fallen ice at the foot of slopes may consolidate and become "reconstructed glaciers."

Valley glaciers made up of many ice streams that have joined are called compound glaciers. Broad, long, deep ice

Hanging Ice: Above the Valley of Ten Peaks, Banff National Park, Alberta, cirques feed ice onto a broad ledge. There it becomes a hanging glacier. Ice extending beyond the ledge forms a cornice, which at times collapses and falls into the valley.

streams nourished by many tributaries are called trunk glaciers. During the Pleistocene, major mountain ranges, including those in mid-latitudes, held trunk glaciers miles wide, thousands of feet thick, and scores of miles long. Today large trunk glaciers are seen only on polar lands, such as Greenland and Antarctica, and at very high elevations elsewhere, as in the highest European Alps and in the Himalaya.

On very cold lands, an ice stream may descend steeply and relatively rapidly onto a broad lowland and there spread widely. Spreading occurs mainly by crevassing, often in a radial pattern, causing the glacier edge to resemble a row of toes, as already mentioned. Thus is formed an expanded-foot glacier, one of the piedmont (French, "foot-of-the-mountain") group. A well-known example is seen near Dobbin Bay on Ellesmere Island, in the Canadian Arctic.

Another kind of piedmont glacier consists of two or more ice streams that have coalesced on a lowland. These masses can be enormous. During the Pleistocene in some regions they spread over thousands of square miles, hundreds of feet thick. Today they are generally smaller and limited mostly to the Alaska, Greenland, and Antarctic coasts, but they are still impressive. Malaspina Glacier, fed by the 3,000-foot-thick Seward Glacier and smaller ones, which descend from the St. Elias Mountains, covers 1,400 square miles (more than Rhode Island's area) of the coastal plain of south-central Alaska. Columbia Glacier, supplied with ice by Alaska's Chugach Mountains, is much smaller but still a respectable 40 miles long and 4 miles wide at its terminus, though shrinking fast. In Greenland and Antarctica, piedmont glaciers cover some of the coastal areas not under the ice sheets.

Ice Caps, Ice Fields and Ice Sheets

In a cold region with ample precipitation, a cluster of high mountain summits can be capped with a dome of snow and ice hundreds of feet thick and hundreds of square miles in area. Ice from these caps spreads, enters upper ends of valleys, and descends as "outlet glaciers." Ice caps cover mountain areas of Alaska and Norway, and polar islands such as Iceland, Baffin, and Spitsbergen. Relatively small ice caps

A Retreating Glacier: In a valley tributary to Ferret Valley, Switzerland, a glacier has melted back nearly to its cirque. Moraine covers remaining ice almost completely. Three lateral moraines, with typical sharp crests, are visible.

Terminus of a Piedmont Glacier: Alaska's Columbia Glacier descends from the Chugach Mountains into Prince William Sound. It has been receding at the rate of about ½ mile per year and discharging large numbers of icebergs into shipping lanes.

that cover a single summit, such as a volcanic cone like Mt. Rainier, are called carapaces, because in some instances they resemble the back of a turtle.

A thick accumulation of snow and ice on a high, nearly level area is an ice field. From its edges outlet glaciers flow to lower elevations, as on plateau areas of Iceland, Greenland, and Antarctica. In the Canadian Rockies the Columbia Ice Field, more than 100 square miles in area and up to 3,000 feet thick, is drained by three large outlet ice streams, among them the well-known Athabasca Glacier.

The supreme form of glacier is the continental ice sheet. It may build up either on a mountain range or on a low terrain – wherever snowfall is sufficient and low temperatures prevail. It can overwhelm mountains and, because of its own weight, spread out over regions of continental extent, grinding and plucking the land as it advances, and laying down huge volumes of rock debris as it melts.

An ice sheet differs markedly from a valley glacier in the conditions it meets and in the ways it adjusts. Coming to a valley, an ice sheet fills it and overrides it. Coming against a highland, ice low in the sheet may deform, dividing into streams that go around the highland while upper ice moves over it. Movement can continue as long as the ice can deform, spread by its own weight, and glide on its meltwater.

On Greenland today the continental glacier, with ice at its base about 125,000 years old, has a maximum thickness of at least 10,000 feet, depressing central Greenland below sea level. The ice covers all but a coastal strip and the summits of interior mountains. The summits, projecting above the ice, are known as "nunataks". Tongues of ice reach the sea through outlet glaciers, which contribute to the Atlantic Ocean about 10,000 icebergs per year.

The only other ice sheet surviving today is on Antarctica; it reaches a maximum thickness of 14,725 feet, with ice at its base about 200,000 years old. (Older ice has melted into the ocean.) Present ice covers all of Antarctica except a few mountain summits, and extends into the ocean as ice shelves averaging 600 feet in thickness, with top surfaces about 90 feet above the water. The Ross Shelf, the largest, extends into the Ross Sea with a front about 500 miles long.

"Only Yesterday"

The Pleistocene Epoch began about 2.6 million years ago as prevailing average temperatures in the northern hemisphere were sinking into the upper 40's (F). During tens of thousands of years, immense ice masses built up on northern highlands spread, completely covering millions of square

Polar Glaciers: Where ambient temperatures remain extremely low and thaw-freeze occurs rarely if at all, the viscosity of glacier ice is increased and the ice may move almost like molasses. Such are these glaciers flowing from a small ice-field on Axel Heiberg Island, in Canada's Northwest Territories.

An Expanded-foot Glacier: Some valley glaciers arriving on a flat lowland spread out into a bulbous form. This is done by crevassing, which may make the ice margin look like a row of toes. So with this glacier near Dobbin Bay on Ellesmere Island, in the Canadian Arctic.

miles before reaching the oceans or wasting away at lower, warmer elevations. In North America one ice sheet covered all but the loftiest peaks of the Cordillera from Alaska (not including Alaska's dry central area) to northern Montana, and valley glaciers existed as far south as New Mexico. The Laurentide sheet, centered on the Hudson Bay area, covered central and eastern Canada to depths of nearly 3 miles. The same sheet spread down into the present United States along a front extending from Montana southeastward almost to the junction of the Ohio and Mississippi rivers, and thence through Pennsylvania and northern New Jersey to Staten Island, New York. In the Midwest and parts of the Northeast the thickness of the ice approached a mile. At maximum, glacier ice covered some 5.2 million square miles of North America – a larger sheet than the one on Antarctica today.

While most of the northern half of North America was under ice, Greenland was completely covered, to a thickness of 11,000 feet or so. A Eurasian sheet about 2 miles thick at maximum spread over Scandinavia (which bore two thirds of the ice) and the Baltic region, invaded the British Isles as far as southern England, and reached eastward to the northern Siberian highlands (most of Siberia stayed ice-free). Central Asia, being far from oceans as a source of atmospheric water, lacked an ice sheet. In Europe, for a similar reason, the ice advanced only to the central plain, short of joining the ice cap on the Alps. As in North America, mountain regions in Eurasia to the south of the ice sheets acquired valley glaciers or lesser ice accumulations.

In the southern hemisphere temperatures were more moderate; only Antarctica, where glaciation had begun 20 million years earlier, had a broad ice sheet. However, the higher South American mountains had valley glaciers; so did Hawaii's Mauna Kea Volcano, Mts. Kenya and Kilimanjaro, in Tanzania, and the highland of New Guinea, in the southwest Pacific. Altogether, glacier ice covered about 30 per cent of Earth's lands, and the overall level of the ocean, which was the ultimate precipitation source for the ice, was down perhaps as much as 400 feet. Meanwhile, periglacial conditions prevailed near edges of the ice and scores of miles beyond; and in many regions the ground froze to depths of 1,000 feet or more – in Siberia, to 4,000 feet.

As the ice sheet melted, multitudes of meltwater streams spread rock debris over the land. Lakes and swamps occupied depressions made by glacial erosion during earlier ice advances; they formed also in valleys blocked by glacial debris, and in basins formed by settling of loose earth over and around melting ice blocks. Frequent thaw-freeze shattered much bedrock, and talus accumulated at the foot of cliffs. Forests of high-latitude types flourished, along with mastodon, mammoth, and other cold-climate fauna and flora. All the major continents had large lakes. Lands hundreds, even thousands, of miles from the ice margins were under a humid, cool climate. During this pluvial period (Latin *pluvia*, rain), lands such as the now-dry American Southwest bloomed, and, in now-arid regions such as Israel's Negev and Egypt's Sahara, streams laced the land and forests grew.

Outlet Glaciers: In Northern Victoria Land, Antarctica, ice from the Admiralty Mountains drains into ice-covered Robertson Bay. At high center is Dugdale Glacier, which has cut a deep trough. In addition to U-shaped valleys, erosional features here due to glaciation include cirques, arêtes, horn peaks, and truncated spurs.

Continental Glaciations During the Pleistocene: Ice cover was limited mostly to the northern hemisphere, as shown here. In the southern hemisphere there were mountain glaciers, even in the tropics, but no ice sheets except on Antarctica.

Judging from the distribution and layering of glacial rock debris and from radiometric dating of its organic contents, the Pleistocene Epoch saw perhaps four cycles (the exact number is controversial), in each of which ice sheets, fluctuating more or less, built up for 60,000 to 75,000 years and then melted away in 15,000 to 20,000 years, after which an interglacial (and pluvial) period of 200,000 to 300,000 years would ensue. In the coldest phases, average annual temperatures in the northern hemisphere reached lows around 48° F compared to today's 58° F – seemingly a minor difference, but sufficient to make an ice age.

In North America the traditionally recognized (but not fully verified) periods of continental glaciation are called, in order of increasing age, Wisconsinan, Illinoian, Kansan, and Nebraskan, according to where the basic studies were made. The Wisconsinan appears to have begun about 110,000 years B.P. (before the present), reached a maximum 18,000 years B.P., and ended some 6,000 years B.P. with the disappearance of all substantial ice sheets other than those on Greenland and Antarctica. Similar cycles occurred simultaneously in northern Eurasia; there they have different names.

Except on Greenland and Antarctica, the great continental glaciers are no more; only comparatively minor ice caps and valley glaciers remain. These are still in many ways impressive, however, and they can be seen not only in polar and subpolar regions but at higher elevations in the mid-latitudes and even in the tropics, notably on lofty African and South American volcanoes.

If the weight of glacier ice is taken as 56.78 pounds per cubic foot at maximum, Pleistocene glacier ice a mile thick must have borne down on some lands with a pressure averaging 145 tons per square foot. Mile-thick ice is believed to depress the crust about 1,320 feet, forcing it deeper into the mantle. Thus the crust under the ice sheets, perhaps nearly 3 miles thick at maximum, became strongly warped.

During thousands of years since the ice began melting away, the crust has been "rebounding." Along Scandinavian seacoasts, former beaches rising like stairs above present sea level testify to hundreds of feet of uplift. One authority estimates uplift in northern Finland at 1,500 feet during the past 10,000 years – a time span indicated by radiometric ages of beach sediments. In Canada's Hudson Bay region, which was under nearly 3 miles of ice, uplift has been at the rate of

Maximum extent during Pleistocene Epoch

3 feet per century, and if uplift continues it may some day empty the bay. Uplift in eastern Maine has been about 215 feet and goes on. Calculations of uplift include an allowance for the rise of sea level due to glacier melting.

Sculptures by Valley Glaciers

Glaciers exist today in relatively few regions; many people will never see one. However, many lands temperate today show numerous signs of a glacial past. Since the last maximum of the Pleistocene, a mere 18,000 years ago, erosion

Cirques, Arêtes, and Horns

Most valley glaciers originate in cirques (in England, "corries"; in Wales, "cwms"). Well-developed cirques are semicircular, amphitheater-like hollows. They are shaped mostly by intense frost weathering and erosion by shifting ice. On the mountain side of the hollow is a cliff called the headwall; below it is the basin. The basin floor rises toward a threshold over which glacier ice pushes up to make its exit. Beyond the threshold is the valley down which the ice moves, or the ledge on which the ice becomes a hanging glacier. After the ice melts away, cirques may hold lakes.

As a row of cirques enlarge during glaciation, divides between them narrow and may become the sharp, jagged ridges called arêtes. This is the term used also for narrow divides between valley glaciers, and for ridges that divide cirques which face in opposite directions.

Low parts of ridges formed where cirques intersect are "cols"; often these are used as mountain passes. Peaks sharpened into towerlike forms by the growth of cirques around them are known as horns or, if they are very high and thin, as needles. These various features are superbly developed in the Canadian Rockies, the Alps, and other mountains where strong valley glaciation has occurred.

has not yet had time to destroy all these signs. They invite recognition and understanding quite as much as a real glacier does. Lands glaciated during the Pleistocene are, so to speak, museums of glaciation, and they offer what many observers consider to be the most beautiful Earth scenery of all.

The most striking glacial sculptures are those made by valley glaciers. These have hollowed out cirques, widened and deepened valleys, narrowed divides between valleys, shaped mountain summits into horns, carved coasts to make fiords, and sharpened mountain forms generally.

Rock Basins and Stairs

As a valley glacier melts away, depressions it has made in weaker zones of the valley floor become exposed. Water may fill these, becoming scenic lakes. In recently glaciated areas, where streams are still carrying rock flour to the lakes, the white to yellowish color of the flour, added to the blue light rays scattered in the water, gives the lakes a startlingly beautiful turquoise hue. Such bodies of water in a chain are sometimes called "paternoster" lakes, because they suggest a string of beads.

Postglacial Rebound: The New England coast was depressed several hundred feet by the weight of Pleistocene glacier ice. Long after the ice melted, the coast is still rising. Islands in Frenchman Bay, at Bar Harbor, Maine, may some day become part of the mainland – as they were before the ice age.

Birthplaces of Glaciers: Along Icefields Parkway, Jasper National Park, Alberta, many cirques are seen. This pair once shed glacier ice into the great trunk glacier of the Athabasca River Valley. Today there is snow in these cirques in all seasons, but not enough for glaciers to form.

A Cirque Lake: On a trail above Lake Louise, which lies in a glacial trough in the Canadian Rockies, hikers pass Mirror Lake, in a cirque which once spawned a small glacier. In the background, on the other side of the lake, are horn peaks and avalanche chutes. Cirque lakes and other small lakes in basins made by glacier action are often called tarns.

At icefalls, and where tributaries join a trunk glacier, the valley floor may abruptly "step" toward a lower level, because at such locations erosion of the floor below the step has been more vigorous. The upper level of the floor there is called a stair.

Rock Knobs and Rock Drumlins

A rock mass projecting from a valley floor may be shaped by moving glacier ice into a rock knob. This shape develops as ice moves up and over the projection. On the upstream side, moving ice grinds the projecting rock, smoothing it,

A Roche Moutonnée - I: An entire hill of bedrock can be shaped into a roche moutonnée as glacier ice abrades the upstream side, making it smooth, and plucks the downstream side, making it rough. This is The Beehive, on Mt. Desert Island, Maine, shaped by ice moving north (*at right*) to south.

and because of the pressure, some of the ice melts; then, as the ice mass turns down on the downstream side, pressure is relieved and meltwater freezes to rock beneath it. As the ice moves on, it plucks out blocks of rock to which it has frozen, thus leaving the downstream side rugged. In this way a rock knob is formed, with an abraded, smoothed surface on the upstream side and the plucked, rugged surface on the downstream side. Rock knobs are often called sheepbacks or *roches moutonnées* (French, "fleece rocks") from their supposed resemblance to the rounded backs of sheep. Relatively uncommon in narrow glaciated valleys, they are numerous on terranes of continental glaciation.

Projections of rock streamlined from end to end, steep in front and trailing out toward the rear, are known as rock drumlins. They are not common in glaciated valleys but are often seen on land eroded by ice sheets. Being of solid rock, they differ from ordinary drumlins, which are of glacial drift.

Valleys of Trunk Glaciers

Trunk glaciers erode valleys to great depth and width. These valleys, called troughs, have been more or less straightened by glacial erosion, have very steep walls, and have a U-shaped cross section. The U-shape results from the fact that erosive action by ice low in the glacier is very powerful because of the tremendous weight of ice above it. The U-shape allows passage of ice with least expenditure of energy; in other words, this shape presents the least resistance to ice advance.

Along the sides of the trough there may be ends of tributary valleys which, having held less ice, were eroded less deeply than the trough and thus remain high above the trough's floor. These are known as hanging valleys. As the ice melts away, many become sites of waterfalls.

Hanging valleys closely spaced and cut deep into the side of a trough are separated by sharp-backed ridges. Ridge ends shaved off by the glacier have triangular cross sections, looking like facets on young fault scarps. Ridges thus cut off are known as truncated spurs.

A Roche Moutonnée - II: Relatively small roches moutonnées are made by both valley glaciers and ice sheets. This one is on a Ramapo Mountain summit in northern New Jersey. Note striations on the otherwise smooth upstream end (*at right*).

A Fiord Panorama: At an altitude of perhaps 35,000 feet a clean airliner window affords a view of the ria coast of southeast Greenland. An icefield, glacier-sculptured highlands, outlet glaciers entering fiords, and a scattering of icebergs are visible.

In some troughs an abrupt change in the steepness of the walls may be noticed; often this is about a third to two thirds of the distance from the top to the valley floor. The line along which the steepness changes is the "trim line"; it indicates the height reached by the glacier ice during the most recent glaciation. Often the slope above the trim line is relatively rough and more or less vegetated; below the line the valley walls are too smoothed and steepened for much vegetation to take hold.

Troughs are seen in all highlands that have undergone valley glaciation. Famous scenic troughs include (among many) California's Yosemite Valley, the Mistaya Valley in the Canadian Rockies, Switzerland's Rhône Valley, the Valley of Chamonix in the French Alps, and the fiords of Norway and New Zealand.

With their massive erosional powers trunk glaciers have created basins for lakes, some large enough to be called inland seas. A few of the most notable are Lake McDonald in Glacier National Park, Peyto Lake in Canada's Banff National Park, Lakes Como and Maggiore in Italy, Lake Windermere in the English Lake District, and Scotland's lochs.

Erosion by Valley Glaciers: Moving ice alters details of scenery, enlarging valleys and narrowing divides, but the overall pattern of mountains and valleys remains about the same. This kind of glaciation is also called "mountain" or "alpine" glaciation because it develops best among high mountains.

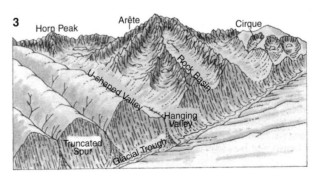

Trunk glaciers that reach the sea may erode the lower ends of their valleys to a depth below sea level. These lower ends, which become filled by ocean water as the ice melts back and sea level rises, are fiords. Numerous fiords were carved in high, cold coasts during the Pleistocene, when sea level was far down. Most famous among fiords are those of Norway's mountainous coast, but there are many also in coasts of Greenland, southern Alaska, southern South America, and New Zealand. Antarctica has fiords as deep as 6,900 feet. The eastern coast of the United States has two fiords: Somes Sound, in Maine, and the gorge of the Hudson River through the Hudson Highlands north of New York City. Montenegro's magnificent Gulf of Kotor, sometimes referred to as a fiord, looks like one but was produced by faulting.

Sculptures by Ice Caps and Ice Sheets

Because ice sheets and ice caps cover land completely, or almost completely, their sculpturing somewhat differs from that of valley glaciers. Valley glaciers deepen valleys and sharpen ridges between them, thus making overall relief more rugged. Ice sheets and ice caps while overriding the entire terrain shave summits and other projections, thus reducing elevations and smoothing contours; also they deposit much rock debris not only in mountain valleys but on lowlands, thus aggrading them and often leveling them. Hence the emphatic difference between rugged topography such as that of the European Alps and the much gentler relief of highlands such as the Laurentians in Canada, the Green and White mountains of the northeastern United States, and the highlands of the English Lake District. But an ice-sheet glaciation often is preceded or followed by valley glaciation, which may leave some of its own rugged signatures. Thus a landscape may reflect both kinds of glaciation.

Pleistocene ice sheets that built up on lowlands, such as the Hudson Bay area in Canada and north-central Europe, had no high mountains to reshape, but they wrought significant topographic changes. Travelers by air over Labrador, central Canada, and the northern coterminous United States – notably Minnesota, northernmost New York state, and Maine – see below them, in patches, wide expanses of bedrock scraped more or less clear and deeply grooved. Here rock knobs and rock drumlins are common, and low hills have been sculptured by ice following their contours. Multitudes of lakes lie in basins made by glacial abrasion and plucking; many are elongated in the direction of glacier movement. Among the largest of these water bodies are Canada's Great Slave Lake and Great Bear Lake.

Similar scoured landscapes exist in Scotland, Scandinavia, northern Russia, and Siberia. These terrains are to be distinguished from those shaped more by glacial deposition (areas such as the American Midwest and north-central Europe) than by glacial erosion.

Basins made in valleys by lobes of ice sheets compare with those made by valley glaciers. Examples of lobe-sculptured basins include the Great Lakes in the United States and the

Left High and Hanging: Usually, as the early geologist John Playfair observed, streams on erosional landscapes meet at about the same level. The Merced River and Bridalveil Creek probably did so long ago in Yosemite Valley. In time, uplift speeded the Merced's flow, and it started cutting below the junction. Then a Pleistocene glacier deepened the valley about 1,500 feet more, to a total of about 3,000 feet. Bridalveil Creek was left hanging, and so a lovely waterfall came into being.

Valley of a Trunk Glacier: Chamonix Valley, in the French Alps, is a glacial trough with classic lines. A huge glacier once filled or nearly filled it. Trim lines (*background*) show how high the ice reached during the glacier's last major advance.

Fiord Panorama: Among the most spectacular features in nature are fiords, those breathtakingly deep valley mouths eroded out by glaciers and later filled by the sea. This is Sør Fiord, near Kinsarvik, Norway. The fiord's walls are actually steeper than the wide-angle camera view suggests.

"Good Medicine": Medicine Lake, in Canada's Jasper National Park, has been called good medicine for the eyes. It lies in a glacial rock basin dammed by moraine. In the distance here are the sawtooth ridges of the Queen Elizabeth Range, in the Front Ranges of the Canadian Rockies.

Smoothed by Glaciation: New York's Hudson Highlands show smoothing and streamlining by half-mile-thick Pleistocene glaciers. Rounding of hills is due in part to weathering of nearly homogeneous rocks. View is south from Bear Mountain.

Glacial Polishing: On the shore at Peggys Cove, Nova Scotia, granite was spectacularly smoothed and grooved by advancing Pleistocene glacier ice. Smoothing rather than breakup of the granite was favored by wide spacing of the joints.

A Glaciated Terrane in Canada's Northwest Territories: Grooving and basin-making here by Pleistocene glaciers are spectacular. Grooves and elongation of lakes (dark patches) indicate direction of ice movement, upper right to lower left.

fabled lochs of Scotland. After such valleys were ground out they were dammed by glacial debris. New York's Finger Lakes, likewise, lie in troughs deepened by ice tongues; the rock below the bottoms of Lakes Seneca and Cayuga is below sea level and as much as 1,200 feet below the pre-Pleistocene valley bottoms.

Minor Glacial Sculptures

Minor glacial sculptures and other features associated with glaciation are common and often easy to identify. They can be convincing evidence that an area has been glaciated. The most familiar forms are grooves and scratches on bedrock, chatter marks, and fractures. Nivation hollows and thick talus accumulations are among indicators of periglacial weathering – that is, weathering of land near glaciers.

Glacial Striations and Grooves

Scratches made on bedrock by rock fragments embedded in moving glacier ice are called striations. These occur mostly but not entirely on bedrock surfaces that are relatively soft and widely jointed. Grooves, wider and deeper than striations, often are up to a foot deep and 2 to 4 feet wide when made by a large boulder or cluster of stones. The famous grooves in limestone on Kelly Island, Ohio, are even larger. Some grooves are imposing landforms; for example, those in the

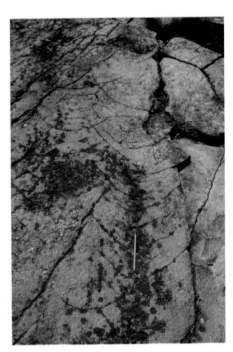

(left) Striations vs. Joints: Glacial striations on rock surfaces must be distinguished from joints and from furrows resulting from differential weathering. On Bearfort Mountain, in southeastern New York state, striations (parallel to hammer handle) are glacial; they are at an angle to the joints and differentially weathered edges of the upturned sandstone layers.

(right) Crescentic Fractures: These crescentic fractures, about a foot wide, in Schunemunk Mountain sandstone, in southeastern New York state, are as usual nested, and concave downstream. (Pencil points in direction of ice movement.) Though called crescentic, the fractures are paraboloidal; the arms if extended would never meet. Such fractures are mostly on upstream slopes, because ice pressure was greatest there.

MacKenzie Valley, in Canada's Northwest Territories, which are about 300 feet wide, 100 feet deep, and 7 miles long.

Striations and grooves parallel the direction of ice movement. Since this direction may be somewhat different during different episodes of glacial advance, striations and grooves in the same vicinity also may vary in direction. An abrupt end to a striation or groove is likely to be the downstream end, where motion of the rock fragment suddenly stopped or the rock was completely crushed. Striations and grooves on bedrock that has been covered by other bedrock can last for millions of years, striations covered by loose earth may last for thousands of years, but those exposed to weathering can be erased within centuries.

Chatter Marks and Fractures

Chatter marks are slight, irregular, sometimes crescentic depressions, usually no more than a few inches wide, made where rock fragments embedded in glacier ice moved over a bedrock surface under great pressure in an intermittent stick-slip (stop-start) fashion. Embedded boulders moving similarly may make curved fractures 1 to 2 inches deep and usually 6 to 10 inches wide, concave on the downstream side; these are called crescentic fractures. Also on these rock surfaces one may see "crescentic gouges," which are depressions left where chips were removed; these may be about the same size

Where Chips Were Broken Out: Stick-slip (stop-start) movements of rock fragments embedded in the bottom or side of a glacier often break out chips, perhaps especially where rock is highly resistant or the angle of pressure or impact is relatively high. These crescentic gouges, about a foot wide (unusually large), are in granite on Bear Mountain, in the Hudson Highlands. They are, as is typical, convex downstream.

as crescentic fractures, but the upstream side is concave. Chattermarks, crescentic fractures, and crescentic gouges often occur in nested series, sometimes in broad grooves, with which their axes are parallel.

Nivation Hollows

Nivation hollows develop in valley heads where snow is deep and thaw-freeze frequent but snow accumulation is not sufficient for development of valley glaciers. A nivation hollow somewhat resembles a cirque, but the bowl form is less pronounced; there is no basin; the floor inclines down valley without a threshold. These hollows, numerous on glaciated lands, are seen also on lands that have not been glaciated but

A Nivation Hollow: The great hollow in the west side of Giant Mountain, in the Adirondacks, is a feature resulting probably from periglacial weathering, not valley glaciation.

have been under much snow. In the eastern United States nivation hollows are well developed in highlands as far south as the Great Smokies.

Features of Glacial Deposition
Glacial Drift

That glaciers loosen, break up, and transport great amounts of rock is attested by the quantities of debris deposited on landscapes by melting Pleistocene glaciers. Much of this debris still lies on highland summits, hillslopes, and valley floors; even more has been spread widely by streams over lowlands extending far beyond the maximum reach of the glaciers. This material is called glacial drift, the term used by early geologists. They recognized it as glacial in origin but thought it had drifted south on pack ice from polar regions and had been deposited on warmer lands during some past era when these lands were submerged beneath the ocean. Previously, such material had been regarded as deposits from ancient floods, especially the one that brought fame to Noah.

Drift from Pleistocene glaciers covers or patches nearly all of Canada, the upper half of the coterminous United States, most of the British Isles, and northern Eurasia as well, not to mention the many glaciated regions of the southern hemisphere. Deposits on North America's Great Plains generally resemble those on other glaciated lands. In the Midwest, drift transported from as far north as central Canada (as indicated by the mineral content of the drift) covers the terrain, filling valleys to depths of several hundred feet. Lowland areas covered by drift have numerous depressions, some now occupied by ponds or swamps, where more or less buried blocks of glacier ice melted away. Drift deposits diverted rivers of the Midwest and Northeast from their courses, dammed valleys to make lake basins, and overloaded rivers. Such deposits make up landforms as diverse as Boston's Bunker Hill, portions of Nebraska's Sand Hills, and much of Long Island and Cape Cod. The vast Mississippi Delta, with its estimated volume of 25 cubic miles, consists mainly of drift brought

Bulldozed Drift: Masses of rock fragments transported by a glacier and then deposited, without stratification, make up glacial till. This chaotic mass of till, in New Jersey's Ramapos, consists mostly of angular, little-weathered rock fragments covered by a foot of topsoil.

Stream-deposited Drift: Stratified drift is exposed in the wall of a sand quarry near Greenwood Lake, New York. The sand, with a few cobbles, was deposited by meltwater streams running off the edge of glacier ice. The layering is irregular because of changes in conditions during periods of deposition.

down by the river from its drainage basin during and following the Pleistocene.

Drift ranges from house-size boulders to the finest silt and clay particles. Some drift has been more or less sorted as to caliber and has been deposited in layers, by water or wind; this is called stratified drift. Unsorted and unlayered glacial material, including material spread and compacted beneath the glacier (such material is called ground moraine), is known as unstratified drift, or till.

Landforms consisting of drift often are far more complex and more difficult to identify than those consisting of non-glacial alluvial or windblown sediments. Drift features can form in contact with glacier ice or at a distance. Some result from the melting back of a glacier. Features may form at any angle to the horizontal, in any position, in valleys or on hill-sides or summits. Deposits may form regardless of rock structures next to them. They may involve either till or unstratified drift, or both, in orderly or chaotic arrangements. Any particular drift feature may be subject to the principle of equifinality; that is, its form could have been produced by processes other than glacial.

Precise identification of drift features is a job for experts. However, these features are interesting and, for ordinary observers, identifiable in a general way if the area is known to have been glaciated.

Lateral Moraines

As a valley glacier grinds downward, it picks up and transports rock debris that has gravitated from the valley sides or been broken off by the glacier. Much of this material is pushed up to form lateral ridges along the sides of the glacier, just as snow piles up on both sides of a plow. These ridges, called lateral moraines, often have sharp crests and steep sides, at the angle of repose. Rainwash may sculpture them into "demoiselles" – pillars of till, each protected by a

Demoiselles: Lateral moraines are easily sculptured by rain-wash and gravity effects. When a boulder protects underlying material from erosion, a pillar may develop with the boulder on top. Some pillars do resemble ladies with big, fashionable hats; others, more like hoodoos, are named correspondingly. Features here are at Evolene, in Val d'Herens, Switzerland.

boulder (the demoiselle's hat) on top. Being of loose material, lateral moraines may be eroded away relatively soon after the glacier melts.

Terminal Moraines

Just beyond the front of a valley glacier or ice sheet there is commonly a ridge of till brought forward during the most recent ice advance or accumulated during a long stationary

A Dam for a Trough Lake: Lakes on glaciated lands often are in valleys dammed by a terminal moraine. Lake Louise, in the Canadian Rockies, is dammed by a terminal moraine at its east end (*low center*). On the dam a hotel has been built.

period. Such features are known as end moraines. If the feature represents the limit of any advance by a glacier, it is a terminal moraine. In the United States, terminal moraines from Pleistocene ice sheets have been identified over most of the line of farthest glacier advance from Maine to New Jersey, southern Ohio, and Montana. Cape Cod was created as a terminal moraine deposited by three lobes of melting glacier ice. The original deposits have been reshaped by marine and wind action.

Another kind of till ridge is the recessional moraine, built up at the front of the glacier during its retreat. Still another consists of material that filled crevasses near the terminus and remained after the ice melted; these crevasse fillings may be stratified or unstratified.

Kames, Kame Terraces, and Kame Moraines

Where a thick ice mass containing or covered by rock debris is melting, streams run off its edges, carrying more or less rock debris and depositing it on the terrain below, against the ice wall. The deposit from a single stream is likely to be a pointed cone with relatively steep sides; with time, gravity movements and erosion round off the hill. Such features, stratified and commonly up to 100 feet high, are called kames (from Scottish "comb," perhaps because kames sometimes occur in even rows). Kames are common on plains that were covered by a waning ice sheet. Rarely, they occur well out on floors of glacial troughs.

Useful Glacial Drift: Kames are convenient sources for sand and gravel. This one in Bergen County, New Jersey, when partly excavated showed the typical overall kame profile and stratification. Like many others, this kame now exists only in pictures.

A Drumlin's Profile: Seen from the east side, this drumlin shows the typically steep upstream end (*at right*) and the downstream end with its gentler slope. Drumlins like this, which is near Monroe, New York, are common in glaciated areas with much fine-grained material, occasionally with sand and gravel.

Along the edge of a valley, kames may coalesce to make a kame terrace. Along a glacier's front sometimes there is a row of kames which are parts of the end moraine; these are called kame moraines. They often occur along the front of receding glacier ice.

Drumlins and Fluting

Somewhat resembling rock drumlins in form is the drumlin made of till. It is a low, elongated hill composed of varying sand or gravel units. From the upstream end, which is relatively steep, the drumlin trails back with a gentle, streamlined form. Its length-to-width ratio is usually between 1.5 and 4 to 1, the maximum height being near 175 feet. Often occurring in swarms, drumlins have been compared to schools of dolphins.

Drumlins are relatively common in areas of ice-sheet glaciation where fine-grained material is abundant. Even in the eastern coterminous United States, where the most recent ice sheet had melted away by 11,000 to 10,000 years ago, numerous drumlins survive; for example, in the Saranac Lake sector of New York state, in the Boston area (where drumlins include historic Bunker Hill), and as far south as northern New Jersey.

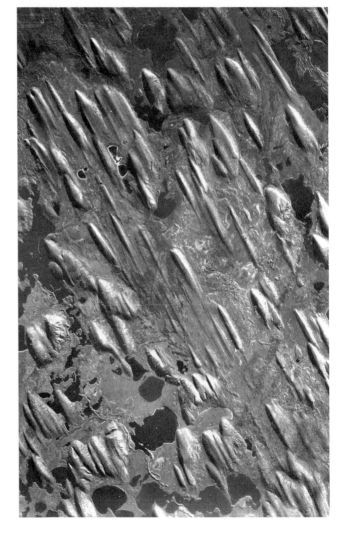

A Swarm of "Dolphins": This high-altitude aerial view of drumlins of drift in Ontario, Canada, shows characteristic steep-fronted noses and streamlined bodies narrowing toward the rear. These drumlins are narrower than most. Ice movement was from upper left (north or northwest) to lower right. Small, winding, threadlike features, notably those at lower left, are eskers; dark patches are lakes.

Some drumlins are built as glacier ice comes up against an obstacle and rides over it. The obstacle may be a rock outcrop, a mass of very sticky clay, or any other especially resistant body. For other drumlins the obstacle may be nothing more than the resistance of the drift. According to a recently proposed theory, some drumlins have been created in drift by floods of meltwater beneath a waning ice sheet.

Arrays of drift deposits that resemble drumlins but are more elongated are called fluting.

"Snakes" on the Land: Eskers

An especially odd form of drift deposit is the esker. This is a ridge of stratified sand and gravel laid down by a meltwater stream flowing on, within, or beneath receding glacier ice. As the ice melts back, the ridge is left lying on the land, undulating across it like a huge snake. Its course is not graded, but conforms to unevennesses of the underlying ground surface. It may have branches or it may be discontinuous, consisting of unconnected segments.

Eskers can be 125 feet or more high and hundreds of feet wide at the base; some extend tens of miles, but most are much smaller. They are seen extending out from the foot of shrinking valley glaciers and also on lowlands once covered by ice sheets. Eskers are numerous on terrains glaciated relatively recently, as in Alaska, Finland, or New Zealand, and they have survived fairly intact in some localities where the glacier ice melted away 11,000 to 10,000 years ago, as in interior Maine and the Saranac Lake vicinity of New York state.

"Pavements" of Drift: Ground Moraine

As glacier ice advances, some rock debris may be deposited and compacted beneath it, and as the ice melts, still more material is deposited. This unsorted, unstratified, often hard-packed material is known as ground moraine. Extensive areas of it are called till plains. Sandy, clayey tills form relatively smooth, undulating plains; stony, bouldery till usually has rough surfaces. Till plains may grade into terminal moraines.

Glacier-transported Drift

Some rock debris is transported by the glacier over distances of scores and even hundreds of miles, then deposited more or less randomly and formlessly. Drift transported from Canada by mile- or 2-mile-thick ice sheets lies several hun-

A Pentagonal Stone: A stone embedded in a glacier may be shaped into pentagonal form by abrasion as it is moved against bedrock or through loose rock. The stone's upstream end becomes pointed because that shape facilitates penetration. Usually the stone has at least one flat side (the side that moved mostly against bedrock) and shows scratches and nicks.

A "Lonely" Erratic: A solitary boulder graces a farmer's field in central Iowa. There is no nearby bedrock from which it could have become detached in recent time by ordinary weathering and erosion. Assuming it was not placed there by man (an unlikely event), it must have been brought by a glacier, perhaps over a distance of hundreds of miles.

dred feet deep on the Midwestern states. In the East, drift to lesser depths lies patchily on areas from eastern Canada southward to New York, northern Pennsylvania, and New Jersey. This glacial debris was deposited on highland summits, hillslopes, plains, and valley bottoms. Much has since been redistributed by runoff and gravity movements, mixed with nonglacial materials, and shaped into forms unrelated to glaciation. Today the greatest concentrations of drift are on valley bottoms and broad lowlands to which valleys lead.

The glacial origin of this material can be inferred in certain situations. Ground moraine, for example, is likely to have been strongly compacted by glacier action. It may consist of material not indigenous to the region where it is deposited; for example, limy ground moraine deposited on granite bedrock probably has been transported from a region where limestone is the native bedrock. Stones present in ground moraine or lying on the ground surface may be scratched, chipped, pointed, or somewhat rounded, as is usual with transported drift. If these stones, called erratics (Latin *errare*, to wander), are of a rock type different from the local bedrock, this is further evidence of transport. Transport is suggested (though not proved) also where boulders lie on a terrain lacking visible rock outcrops from which they could have been weathered.

On slopes below rock outcrops, erratics often mingle with fragments weathered off the outcrops. Many such fragments were weathered off the outcrops during the periglacial period, when weathering was severe. Fragments from weathering are likely to be angular, sharp-edged, because there has been little time for weathering to round off edges. Erratics are often somewhat rounded from abrasion during transport and from washing in meltwater streams.

Erratics range from pebbles to blocks with diameters of 20 or 30 feet; many have been scratched; some have been planed smooth on one or more sides by scraping against bedrock during transport. Stones flattened, polished, and scratched by grinding against bedrock beneath moving ice are called

A Glacial "Shoe": A boulder embedded in the bottom of a glacier may move over bedrock while held against it, like the "shoe" which rides an electrified rail. Such boulders can become flattened, striated, and polished on the contact side.

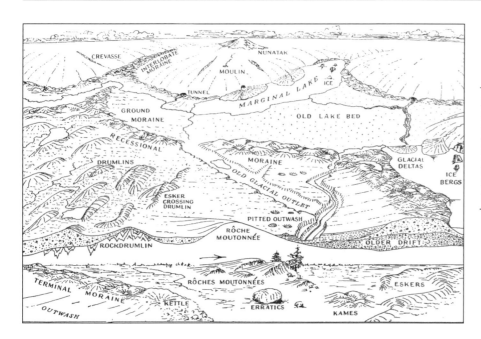

Some Features of Glaciers and Glaciation: Glaciers and related features occur with many variations. Those shown here appear in common forms and in their usual settings. The upper three fourths of the drawing represents a high-angle view; the bottom portion a low-angle view.

"shoes," the flattened side being the "sole." Stones held in one position for a long time during their journey may be pointed on the upstream end (the point allows easier progress) and may have an overall pentagonal (five-sided) shape, like that of some desert ventifacts.

Talus often lies thick at the foot of hillslopes on periglacial

terranes, because thaw-freeze in such areas is frequent. Talus accumulations are impressive even in the southern Appalachians and the lower southern Rockies, testifying to highly active weathering toward the close of the Pleistocene.

Ponds and Lakes in Drift

Among multitudes of basins for lakes made by Pleistocene glacial deposition are those in valleys dammed by a glacier and by rock debris pushed up along the ice front. As the ice melted back, water accumulated against the dam. Water bodies thus formed are called moraine lakes. Some survive today not only in mountains but in glaciated highlands of modest elevation, as in New York state and New England.

On plains, many basins exist where drift was deposited unevenly. As the ice age waned, these basins filled with meltwater, but since then most of them have dried up. In the Midwest these basins are often called potholes (a term better used for holes in bedrock made by stream action). Many have been dug deeper, below the water table, to make farm ponds.

During the Pleistocene, "ice-marginal" lakes dammed by ice were formed by hundreds, and some were truly inland

Clustered Erratics: Many streambeds in New York's Adirondack Mountains, as in other glaciated areas, are half-choked with boulders. Most were rounded by Pleistocene glacier action, then deposited as the ice melted, and since then have gravitated to lower levels. Smaller-caliber material that was deposited along with the boulders – cobbles, gravel, sand – has been mostly removed by stream action.

seas. North America's largest, Lake Agassiz, occupied parts of North Dakota, Manitoba, Saskatchewan, and Ontario, covering at least 100,000 square miles at the glacial maximum. Lake Dakota, mostly in what is now South Dakota but reaching into North Dakota, was about 100 miles long. In New Jersey, Lake Passaic reached a length of 30 miles, a width of 8 to 10 miles, and a depth of 240 feet. But the lifetimes of these ice-dammed lakes were limited. As the ice sheet melted back, outlets for most of the lakes opened and the waters – even those of mighty Agassiz – drained away. Among the survivors is Lake Winnipeg, in south-central Manitoba, 9,465 square miles in area, 266 miles long, and up to 60 feet deep.

With the passage of centuries, erosion of the edges of basins and filling by sediments have eliminated most drift-dammed lakes, and all that remains of them today is swamps or lake plains. Beneath the now-dry lake bottoms are scientifically valuable legacies in the form of varves: layer upon layer of sediments in which histories of the lakes, in terms of both events and time, as well as flora and fauna, were precisely recorded.

As an ice sheet melted, it separated into blocks. Where drift was thick, many blocks were more or less buried in it. As blocks melted, sediments settled around them, and when the ice was gone, depressions remained. Such depressions, called kettles because of their usually circular or elliptical shape, were relatively shallow, but those with bottoms below the water table held ponds. Since the Pleistocene, most kettles have filled with sediments and gone dry, or have been emptied by outlets opened in their edges by overflow. Some kettles still hold ponds or lakes.

Kettles are numerous on broad valley bottoms and on

The Great Lakes of the United States in the Pleistocene: Drawings represent changes in the Great Lakes as the ice sheet melted back during the Wisconsinan, beginning with (1) the Lake Maumee stage and ending with (4) the Lake Algonquin stage. Arrows show courses of rivers that drained the lakes. (From C. B. Hunt, *Natural Regions of the U.S. and Canada,* © 1974 W.H. Freeman & Co., used with permission.)

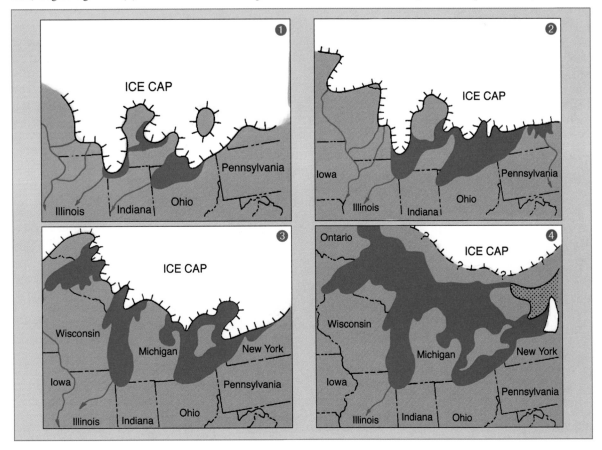

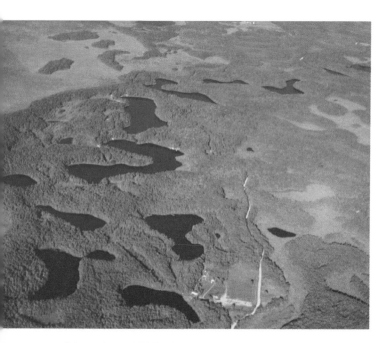

Lakes in Glacial Drift: As the last Pleistocene ice sheet waned, drift settled around and over melting blocks. When the blocks were gone, depressions remained. Called kettles, many of these today are basins of lakes and swamps. On this north-central Wisconsin landscape, rounded basins appear to be kettles; some of the curved ones are oxbow lakes. Pale areas are depressions that have filled with sediments.

Spread Out by Water: Outwash

Drift distributed by streams beyond the terminus of a valley glacier or the margin of an ice sheet is known as outwash. This is sorted and stratified. When confined in a valley and extended longitudinally along the valley floor, it is a valley train. When spread out over a broad area, it is an outwash plain, which may be pitted where ice blocks have melted. Outwash may occur also as an apron along the base of an ice wall, or as a fan built by a meltwater stream flowing down the side of a terminal moraine.

Young outwash deposits are familiar on lowlands near mountains where valley glaciers have existed during recent centuries. On lowlands covered by Pleistocene ice sheets, broad blankets of outwash still exist; in humid regions they are now usually well covered with vegetation.

Since the Pleistocene glaciers melted, streams on all the continents have been busily carrying the outwash to lower and lower levels. The amount of outwash already transported over distances of hundreds, even thousands, of miles is aptly suggested by the Mississippi Delta, whose volume of some 25 cubic miles includes much glacial debris, along with other sediments, from the river's vast drainage basin.

plains where drift is deep. On the Great Plains and the Interior Lowlands they can be counted by thousands. They are seen in patches of drift in the Northwest and the Northeast. The drift areas of Cape Cod and Long Island are pocked with kettles and freshwater kettle lakes.

A terrain with numerous kames and kettles is said to have kame-and-kettle topography. Somewhat similar is knob-and-kettle topography, where moraine takes the forms of irregular knobs and mounds with a scattering of kettles.

An Outwash Plain: Glaciers produce huge amounts of debris, and meltwater streams spread it over lowlands. In this valley of the Sunwapta River, in the Canadian Rockies, drift from nearby mountains is thick and coarse, and the stream braids to get through it. In the background here is Mt. Athabaska.

Looking Ahead

In the future, will glaciers now and then grow and spread over much of our planet, incidentally causing a bit of human inconvenience? Glacial ages are uncommon but do recur; in fact, during the past 2 million years they have been more common than not. Morainal materials, along with glacial grooves and striations, have been found at many locations on oldlands of the continents. Such evidence has survived from all the geologic eras.

Today most mountain glaciers are shrinking; a few, because of unusual local conditions, are growing. Portions of the Antarctic ice cover are growing; other portions are shrinking. The Greenland sheet seems to be holding its own. Sea level during the past century has been rising because of world-wide glacier melting and expansion of seawater due to rising water temperatures. Progressive warming of the atmosphere – "global warming" – now recognized as fact, is causing concern. Over all, we appear to be in a warming trend.

A return to Earth's former average temperature of 72° F. may lie ahead. This would mean accelerated glacier melting, a rise of around 200 feet in sea level, and the flooding of most of the world's port cities. Parts of the United States bordering the Gulf of Mexico, and also Europe's Low Countries, much of southern Asia, and other extensive regions, would be inundated. True, a rise of but 5 feet in sea level could take a century or two at the present rate, but even 5 feet would mean disaster for cities on very low coasts.

According to some glaciologists, a different view is in order. To them the time scales and patterns of glaciations in the past strongly suggest a return of ice-age conditions within 10,000 years. With present average annual temperatures at 58° F., far short of the historical average of 72° F., Earth may be regarded as still in an ice age, perhaps in an interglacial period to be followed by new glacial maxima. That would be consistent with Earth history, and would involve sea-level lowering and world-wide chilling.

An optimistic view can be taken. If global warming does make most glaciers melt away, it will also make many now cold and forbidding lands more habitable. If, on the other hand, the current ice age intensifies, covering high-latitude areas and tropical highlands with ice, areas now tropical can become more habitable, and many a desert will bloom. In either case, the change will take thousands of years – long enough for an intelligent species to adapt.

Always to Be on Dry Land? If the present worldwide melting of glaciers continues for centuries, sea level will rise enough to inundate many coastal cities. Then buildings on Lower Manhattan Island, in New York City, will be islands.

Desert Panorama: In the view east from Stovepipe Wells, Death Valley, are many features typical of deserts – barren, angular mountains behind fault scarps; alluvial fans; sand dunes; hardy but scattered desert vegetation, lag deposits; and an aspect of aridity and starkness over all.

14
Realms of Dryness and Wind

THE word "desert" (Latin *desertis*, "barren") evokes many images – fierce sun, billowing seas of sand, sidewinders, palms and pools dancing on the horizon, the dying prospector, riders in the purple sage, irrigation wheels, a saguaro at sunset, and off-road vehicles in a cloud of dust racing to Las Vegas. Traditionally, deserts have been for most people (except Bedouins and their kind) forbidding wilderness, where few plants can grow, water is rare, and man does not belong. Today, with aqueducts, air conditioners, and the family car, deserts are less threatening, even friendly, and what was once stark and ugly is now beautiful. Whatever our impressions of deserts, or our attitudes toward them, they are unique and remarkable places, not to be blindly conquered but rather to be better understood and, therefore, more wisely used and even enjoyed. The present trend in many parts of the world, notably Africa and Asia, toward desertification, with the loss of wildlife habitats and arable soils, and consequent human hardship, can be checked only as conditions that make deserts become better understood.

Why Deserts?

What is a desert? By the simplest definition, it is a relatively barren terrain on which annual rainfall is less than 10 inches per year, and vegetation is therefore very sparse if it exists at all. Under this broad definition even barren polar lands are deserts. In a narrower concept, deserts are places not only of low rainfall but of high temperatures and high evaporation levels, which may be 15 to 20 times those on humid lands. Some terrains, such as parts of the Colorado Plateau, can become extremely hot and dry in summer, but receive too much precipitation to qualify as true deserts; they are called semiarid. About a third of all continental surfaces do qualify as deserts. About 5 per cent of North America is desert. The only continent without a desert is Europe.

The origins of deserts can be traced, directly or indirectly, to movements of Earth's plates and other tectonic activity. Under these influences land masses are subject to changes in location with respect to ocean currents, and there are changes also in wind directions and velocities, prevailing temperatures, relative humidity, and elevations. It is a combination of such factors that makes some lands desert and some humid. The distribution of deserts today around the world is temporary; with time the distribution will change as tectonic and climatic influences change.

The shortage of water for deserts may result from any of several conditions. One involves wind patterns. Prevailing winds that travel great distances from the ocean lose much of their moisture along the way; thus they provide little rain for areas far inland. This situation occurs commonly in the middle latitudes, 30 to 50 degrees north and south of the equator; for example, in Mongolia's Gobi Desert and in the Desert of Turkestan, east of the Caspian Sea.

Some deserts, though nearer the ocean, are in the rain shadow of mountains. Prevailing winds cool as they rise against the windward side of the mountains, and moisture

In a Rain Shadow: Southern California's Owens Valley is on the lee side of the snow-capped Sierra Nevada. Rain here is rare. Fine, salty sediments deposited in former Owens Lake are blown across the parched flats; heavier material (lag deposits) is left behind. Note the relatively young fault scarps on this segment of the Sierra Nevada.

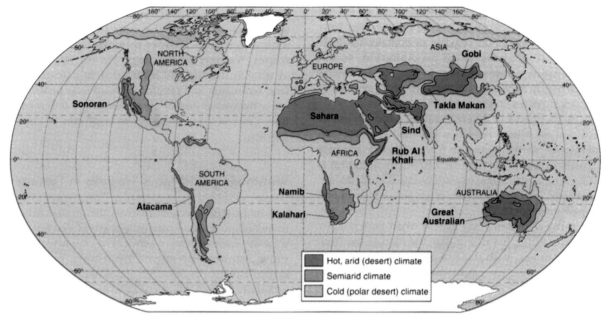

Desert Areas of the World: Deserts commonly are areas where, for various reasons, rainfall is minimal and temperatures and evaporation rates are high. Such areas often are far from oceans or to leeward of mountain ranges which precipitate moisture from prevailing winds. (From Skinner and Porter, *The Dynamic Earth*, © 1995 John Wiley & Sons, Inc., used by permission.)

condenses out of the air and falls as rain; then the air, descending the lee side of the mountains, becomes warmer and dryer, and thus has little moisture left for the lowlands. Such is the main reason for the aridity of the Mohave Desert, in the southwestern United States, which lies east (to leeward) of mountains. In South America, some lands east of the Andes also are subject to rain-shadow effects. So is the Gobi Desert. Rain-shadow deserts, as well as deserts remote from the oceans, characterize mid-latitudes. Warm-season temperatures there may reach 110 degrees F. or higher.

Lands between about 15 and 35 degrees latitude north and south of the equator are swept by trade winds. Blowing toward the equator, these winds become warmer and therefore dryer, and thus produce little rain. Also in these latitudes high atmospheric pressure is common, with air descending and thus becoming warmer and dryer. So-called tropical, or low-altitude, deserts produced by these conditions include among others the Sonora of the southwestern United States and Mexico, the Sahara in Africa, the Arabian Desert, the Thar of southwestern Asia, and the Great Australian Desert.

In tropical deserts temperatures can run extremely high; for example, the 136.4 degrees F. once registered in Tripoli, Algeria. Correspondingly, rainfall can be almost non-existent; thus in the Atacama Desert of Peru and Chile annual rainfall may be under 0.5 inch. Tropical deserts near the western edges of Africa and South America – for example, the Namib and Atacama deserts – are sometimes more or less covered by fog, produced by contact of warm tropical air with cool ocean currents. Still, rainfall there is very sparse.

Most deserts are in basins produced by downwarping or downfaulting. Here sediments may accumulate to thicknesses of hundreds or thousands of feet. Air entering these basins sinks and is thus compressed and warmed – the more so if the basin is bordered by mountains. These basins are, then, in what might be called extreme rain shadows. In the deep graben of Death Valley, where air descends from high mountains to elevations below sea level, temperatures up to 134 degrees F. have been recorded.

The Look of Deserts

Entering a desert, we soon perceive that we are on a unique landscape. The initial impression is one of bareness and starkness. Most deserts are rocky, not sand-covered. Highlands, if any, are likely to rise abruptly from valley bottoms and plains, with rough, angular profiles. Hillslopes are rocky and, except at high elevations (which get the most rain), are bare or nearly bare of vegetation. At the foot of slopes and gullies, talus aprons and cones may exist, and at lower valley ends there are likely to be alluvial cones and fans. Sand and gravel are spread patchily or thickly over valley floors. Where sand is thick on desert plains, it may be blown into dunes; where it is thin, bedrock or masses of large rock fragments polished by windblown sand may be

The Look of a Desert: Egypt's Eastern Desert, west of Safaga, appears as bleak as a land could be. No vegetation is evident; watercourses are dry. Naked mountains, seemingly chaotic in structure and partly buried in their own erosion debris, rise abruptly and steeply above the lowlands.

exposed. Watercourses may be abundant, showing evidence of erosion by torrents, but – unless rain has occurred very recently – they will be dry. Vegetation will be low-growing and scattered, if it exists at all.

From locality to locality, desert surfaces vary widely. The great Arabian and Sahara deserts feature the "sand sea" – a vast, deep sand deposit accumulated during the ages in a basin or on a plain where wind has been more or less moderated by a highland barrier. On these deserts dunes are built to great sizes. Some other deserts are covered by shallower sand deposits with dunes of modest size. Terrains covered with thin sand sheets lack dunes; such areas are called by the Arabs "erg." A stony terrain with much loose rock is "reg"; for example, at California's Joshua Tree National Park. Terrains of wind-scoured bare rock, as in parts of Egypt's Western Desert, are "hammada."

Desert Forms in the Making
The Works of Weather and Gravity

Despite being subdued for lack of water, weathering in deserts makes a strong contribution to landform sculpturing. Rain that does come, especially at higher elevations, provides moisture for hydration, carbonation, and oxidation of granitic, basaltic, and other rocks. Chemical and physical changes in massive or loose rock due to wetting and drying can be significant over long periods. At higher elevations, in mid-latitudes, freezing may occur in crevices, slowly breaking up large rock masses. Expansion and contraction of rock surfaces due to the sharp temperature changes characteristic of deserts cause exfoliation and granular disintegration, the rate being very slow but substantial over geologic time. As elsewhere, rock masses – especially granite – formed at depth will tend to exfoliate when unloaded.

Thick talus at the foot of many steep desert slopes may suggest that, in deserts generally, gravity movements occur at a relatively high rate. Actually, in warmer deserts such movements tend to be subdued because weathering is sub-

dued. If talus at the foot of slopes is thick, this is likely to be due mainly to the paucity of running water, which if present would tend to carry talus away. On polar deserts, talus tends to accumulate thickly because of the frequency of frost-prying and because here, too, running water is scarce.

Especially where bedrock is stratified and about horizontal, desert valley walls tend to be steep, even cliffy. The reason is that stream downcutting, even though subdued by lack of rainfall, still is dominant over weathering and gravity movements. Desert rock profiles may be angular in part because of faulting or because of the lack of chemical weathering, which if present would help to round off prominences.

Testimonials to Desert Weathering: These boulders, remnants of a once-solid mass of granite, are up to 20 feet in diameter. Known as "The Devil's Marbles," they are seen in the Tennant Creek area of Australia's Northern Territory.

Sculpturing by Rain Spatter and Streams

Erosion by spattering raindrops is more significant in the desert than elsewhere. There is little vegetation to intercept the drops before they reach the ground. Relatively few roots are present to hold soil in position. Although in some areas desert regolith may have become somewhat consolidated by wetting, natural cementation, and drying, in other areas

Erosion by Desert Streams – I: In one sector of Death Valley, rainfall on steeply sloping soft materials shapes a typical dense drainage system with nearly straight watercourses and sharp-backed divides. Compare with the photograph below.

material remains loose and easy to scatter. Rainfall on a slope of loose material can dislodge much of it, preparing it for removal by gravity, sheetwash, or streams.

In dry areas, sculpturing of rock masses by rainwash is favored by the lack of vegetation. Earth pillars, pinnacles, demoiselles, and mushroom rocks are produced especially where rock masses are vertically jointed or consist of layers of varying resistance. Goblin State Park, in Utah, and the Bisti Badlands, in northwestern New Mexico, are among many localities that offer dramatic examples.

Running water in the form of sheets, or "sheet floods," is familiar in deserts, especially on steep slopes. Rain falling on broad, relatively flat areas is channeled little if at all by vegetation, and it can run down for some distance as sheets. Then, finding and following natural depressions, it gradually divides into rivulets which, in turn, converge into stream channels small and large. Much rock debris is picked up by sheets and moved over broad surfaces, eroding and smoothing them, before collecting in valleys and basins. Erosion by sheetwash probably helps to create pediments.

Desert streams that are permanent are rare, being limited to a few that originate in relatively humid highlands and have such volume that they can survive a long desert journey. Examples include the Nile, the Indus, the Tigris and Euphrates, and – until irrigation projects all but dried it up – the Colorado. Desert streams are generally intermittent, but often torrential because of the nature of desert surfaces. On desert slopes lacking vegetation, rainwater runs relatively fast, does not readily sink into dry earth (it tends to form dust-coated balls of water instead), and soon arrives in streambeds. Torrents can be produced even when rain is moderate, particularly if it is temporarily heavy, as during thunderstorms. Runoff collecting in a canyon can be so loaded with sediments that it becomes a mudflow or a debris flow.

Rainwater falling on an unvegetated slope picks up loose material readily and quickly transports much of it into streambeds and basins. Streams thus produced are, with their velocity and heavy sediment loads, highly erosive. In many

Erosion by Desert Streams – II: In another sector of Death Valley, rainfall on gently sloping soft materials forms a less dense drainage system with meandering watercourses and round-backed divides.

A Steep-walled Desert Valley: In the Siq, at Petra, Jordan, stream downcutting has been dominant over weathering and gravity movements. Downcutting was done mostly during the Pleistocene and before. The valley floor is covered with rock fragments of cobble size, indicating thousands of years of rockfalls and infrequent stream action.

A Desert Wash: In Arizona's Tucson Mountains, a typical streambed extends down a long slope. The streambed is dry most of the year. Rainfall is frequent enough to support considerable vegetation, but not frequent enough to clear the terrain of rock fragments that have weathered off the mountains and masswasted downslope for countless thousands of years.

deserts, especially the warmer ones, very steep hillslopes demonstrate the dominance of erosion by running water over weathering and gravity movements. So it is in southern Arizona, where, though erosion has been dominant for many millions of years, most mountains still have steep slopes.

Some lands which are now deserts were under a more humid climate during the past, notably during the ice age. Much erosion by running water occurred then; in fact, all or nearly all major rock sculptures in deserts date back to humid periods. In some sectors, such as Egypt's Western Desert and the Arabian Desert in Jordan, the terrain is laced with watercourses that rarely if ever convey water today. Infrared satellite photographs of the Sahara show, below the sands, networks of ancient stream channels that could have been cut only under a relatively wet climate. Since then, under a dryer climate, channels have been covered by blowing sand.

Blowing Sand at Work

Thanks to the paucity of vegetation in deserts, dust particles (that is, particles less than 0.01 inch in diameter) are selectively picked up by desert winds, carried high in the atmosphere by air currents, and spread near and far – even around the world. Thus in the United States dust from the Sahara is found in Florida and Arkansas, dust from the Gobi Desert reaches California, and Europe has frequent gifts of dust from Texas – to cite just a few examples. All this is to be expected when we recall that ash from major volcanic eruptions commonly rings the planet. Each year an estimated 500 million tons of dust is transported by winds.

Blowing dust can do substantial rock sculpturing over long periods; blowing sand naturally is much more effective. Although material of sand size (0.01 to 0.25 inch in diameter) cannot be lifted except by very strong winds, sand is

Sandblasted Terrain: Near Barstow, California, the Mohave Desert is well exposed to prevailing winds from the west, and sandblasting of rock surfaces is frequent. In the foreground here are lava-flow remnants grooved, pitted, and streamlined by blowing sand.

moved along the ground by creep and saltation (Latin *saltare*, to leap), leaving depressions where it has been removed and piling up as ripples or dunes elsewhere. Even relatively fine sand grains are too heavy for all but very strong winds to lift more than a few feet. Thus the belief that mesas, natural bridges, and other sizable landforms have been created by natural "sandblasting" is faulty. Experiments conducted in deserts indicate that maximum abrasion by blowing sand on level ground occurs at a height of between 6 and 10 inches, and the highest reach of abrasion is 4 to 5 feet. High-standing rock can be significantly abraded only if sand particles have a gentle slope, or ramp, up which they can move in stages by leaping. For the most part, natural sandblasting is limited to the base of cliffs, low outcrops, and scattered stones. Over lengthy periods, however, wind can move enormous amounts of sand over long distances, leaving depressions behind and piling up sand against highlands ahead.

Wind action in deserts can do remarkable things, as witness the well-known Moving Rocks, in the southwestern part of Death Valley. These 200- to 600-pound limestone fragments have been moved over the level playa surface, leaving tracks. Tracks of the smaller boulders are up to 800 feet long and have right-angle jogs and loops. The public loves a mystery, but geologists see a natural cause: very powerful

Bared by Wind: The terrain of Joshua Tree National Park, in the Mohave Desert, stands high above the Salton Trough, to the west, and is often windswept. Weathering of the very resistant bedrock (quartz monzonite) is slow, producing relatively little sand. Wind has been able to sweep parts of the area bare, creating a hammada terrane.

A Basin and Range Profile: East of Yuma, Arizona, fault-block highlands after millions of years of erosion have rough, steep slopes suggesting dominance of stream erosion. Facets were eroded away long ago. Lying against the slopes are thick aprons of talus – evidence that removal of erosion debris by streams is relatively slow.

wind gusts, perhaps up to 100 miles per hour, moving the rocks during wet periods when the playa is muddy and slick.

Stark and Beautiful: Desert Highlands

As mentioned earlier, most large-scale rock sculpturing in deserts dates from humid eras when there was more stream activity than there is now. Sculptured forms reflect variations in rock types and structures. Variations are the more evident because weathering and gravity movements, which tend to round off edges, are subdued and because vegetation, which tends to conceal rock surfaces, is sparse or absent. Deserts show the bare bones of geology.

The roughness and angularity of most desert mountains are accentuated if the topography has been produced directly by faulting rather than erosion. Among such landscapes are those in the Basin and Range and in the Gobi Desert. The mountain mass usually is bounded by major faults and more or less divided into separate blocks by minor faults, which, as with step faults, are expressed in angular mountain profiles. Fault scarps tend to be steep, especially if faulting has been relatively recent; thus they provide high gradients for streams and are sculptured into sharp relief. Scarps on bare or thinly mantled bedrock end rather abruptly at valley floors, where deposits of alluvium begin. Some mountain profiles are extended by pediments reaching far out from the base of the scarps. Where gravity movements are limited and little talus accumulates, scarps may retain their steep slopes as they retreat, so that even where the topography is "old," highlands are steep-sided.

Desert valleys in relatively strong rocks tend to be steep-walled, even more so than in humid regions, because of the dominance of stream erosion. Mesa forms are common where rocks are sedimentary and strata are horizontal; at times the resistant cap is part of a lava flow or a sill, as in the Mohave Desert and on the Colorado Plateau.

In desert areas of soft rock and little or no vegetation, temporary streams easily cut small, closely spaced, V-shaped gullies, thus creating badlands. Death Valley has good examples in the Zabriskie Point sector. Similar topography is seen on some semiarid lands, such as South Dakota's White River badlands.

A desert terrane in late stages of erosion may, like some humid lands, display the numerous scattered, small highlands called inselbergs (German, "island mountains"). Some are steep-sided in part because they are capped by relatively resistant strata and are subject to undercutting. The American Southwest displays many examples; so does the Arabian Desert. The Australian Desert has fine inselbergs in Ayers Rock and the Mt. Olga group.

A Lava-capped Mesa: Mesas with protective resistant caps are common on terranes of horizontal sedimentary strata. They are especially common in the Southwest. Lava caps here are the more notable because of the sparseness of vegetation. This is Black Mesa, south of Espanola, New Mexico.

A Desert Dome? Cima Dome, in California's Mohave Desert, has long been a subject of geological discussion. Is it a dome of uplift or simply an erosional remnant in dome form? A recent study concludes that it is a pediment, its scattered hills inselbergs. This view is from the west.

A "Little Head": Tanzania's Serengeti Plains feature numerous koppies. A koppie (from Dutch *kopje*, little head) is a highland remnant on a surface which apparently was previously worn low, then uplifted. This one, of gneiss with a layered structure, has weathered to rounded and slabby forms. Koppies of massive homogeneous granite usually weather to roundish forms. Koppies can be regarded as miniature inselbergs.

Inselbergs of massive homogeneous rock such as granite usually are somewhat rounded. Called bornhardts, they are familiar especially in the Sahara and Kalahari deserts. They occur in some humid regions also (such as Sugarloaf in Rio de Janeiro, Brazil), raising the question as to whether such regions were formerly arid. Closely jointed granitic masses that were subject to strong subsurface weathering before becoming exposed by erosion may occur as tors.

On deserts of low relief, prolonged erosion by sheetwash, small streams, and wind action can shape a highland of homogeneous rock, such as massive granite, into a broad, low, nearly flat feature called a desert dome. If this is elongated it is called a desert arch, because of its profile. Cima Dome, in the Mohave near Cima, California, was once generally regarded as a desert dome but is now said to be a pediment.

Highlands of other plutonic rocks, which like granite tend to be dense and homogeneous, may preserve much of their original shape as they are lowered by erosion; for example,

Limestone, But Still High-standing: The San Andres Mountains, in New Mexico, are a block raised and tilted millions of years ago. Since the block contains much limestone, under a humid climate it probably could not have endured so long as a lofty highland. In this view of the eastern side, north-facing scarps produced by step faulting are evident.

the rounded profiles of the Navajo and Packsaddle mountain laccoliths. Volcanic necks, which usually include masses of varying resistance and thus erode non-uniformly, may dot a desert landscape as rugged pillars, "cathedrals," or stumps.

Where highland rock has closely spaced vertical fractures, a topography of fins and pinnacles may be seen, as in Arches and Bryce Canyon National Parks. Where strata are horizontal, with widely spaced joints, sheetwash and streams may shape "pancake rocks." Such features occur also on humid terrains, but in deserts they are more noticeable because of the lack of covering vegetation.

A desert feature that seems mostly wind-made is the yardang. This is a minor elongated highland that parallels the dominant wind direction; it can be up to 150 feet high and a mile or more long. Seen from the air, yardangs look like hulls of inverted ships. Usually the rock of a yardang is soft and easily undercut by sandblasting. Yardangs often occur in groups separated by valleys with U-shaped cross sections enlarged by wind action. Relatively large yardangs occur in Peru, Iran, and Libya. Small ones, as in the Mohave Desert, may be only a few feet long.

In humid regions limestone masses usually are valley-forming, because of solution, but in deserts solution is minimal and limestone often has equal standing, so to speak, with other rock types. Even after prolonged erosion, mechanically strong limestone may stand relatively high, as on upper portions of New Mexico's San Andres and Sandia mountains. In fault-block topography, limestone may remain high-standing not only because solution is subdued but because an erosional topography, shaped on weak-strong patterns, has not yet developed.

Pediments, those broad, low-angle, bedrock surfaces extending out from highland margins, are associated especially with arid and semiarid conditions. They may develop by lateral planation. This involves sheetwash and small streams spreading from valley ends at the base of steep scarps and eroding bedrock with their heavy loads of sediments. A pediment lengthens as the mountain front retreats. The bedrock of a pediment may be exposed near the base of the highland, but lower down it is commonly covered by alluvium. Pediments can merge to form what are called pediplains.

A Mohave Pediment: Near Baker, California, the wide, nearly flat surface sweeping down from a highland demonstrates a pediment, the bedrock surface of which lies beneath a cover of sediments.

Sculptured by Sandblasting: Here in the Mohave Desert, blocks of basalt have been pointed, fluted, and pitted by blowing sand. This ventifact lay in several different positions during thousands of years and thus acquired several points.

A Desert-pavement Closeup: These rock fragments, an inch or two wide, are embedded in duricrust, a layer of naturally cemented sediments. Although long-weathered, the stones have sharp edges, due probably to wind action.

Sculptures on Valley Floors

On portions of a desert floor which winds have swept clean of sand and dust, bare bedrock – "hammada" – may be exposed. It may be patched with loose rock fragments too heavy for the wind to remove; these are known as lag deposits, called reg (in the western Sahara), serir (in the eastern Sahara), or gibber (in Australia). Lag deposits may become close-packed or cemented together by hardened mineral materials such as salt or silica, and smoothed by wind abrasion; then the deposits are called desert pavement or

pebble armor. Crusts of naturally cemented soil particles are called duricrust, various types of which are known as caliche (lime-rich material), silcrete (silica-rich), and ferricrete (iron-rich).

Individual stones polished and more or less shaped by blowing sand are known as ventifacts (Latin *ventus*, wind, and *factus*, wrought). Ventifacts that have been in one position a long time tend to be triangular, with a point on the windward side. These are "dreikanter" (from German, "three-edged"). Those that have changed position may have more than one point.

Isolated erosional rock features in deserts may be shaped by natural sandblasting into what are called mushroom rocks, or pedestal rocks. The upper portions are wider than the base, because they are not reached by sandblasting; hence the mushroom form. If the rock at the base is less resistant to rainwash than the rock above, the "stem" of the mushroom can be very narrow. Mushroom rocks look somewhat like demoiselles, but the latter differ in origin: they are shaped by rainwash on glacial moraine.

A rock surface pitted, fluted, or otherwise shaped into designs by differential sandblasting is called etched. In sand-

Mushroom Rock: Among basalt boulders on the east side of Death Valley stands Mushroom Rock. Rainwash must be credited for overall shaping of this feature, but greater smoothness near the base suggests the work of blowing sand also.

blasted limestone, veins of quartz, being relatively resistant, stand out as ridges. Sandstone and conglomerate may be honeycombed with pockets where relatively soft material has been eroded faster than the matrix. Homogeneous rock may be polished by sandblasting rather than pitted or grooved.

In some desert and semiarid localities, as at Arches National Park, finlike rock masses have been eroded through, becoming arches and "windows." Finlike forms are favored by close vertical jointing. Openings in them may be produced by exfoliation and sheeting due to unloading or stress relief, and by chemical weathering. (Usually the openings are too high to have been made by sandblasting.) Desert windows may resemble natural bridges, but the latter differ in origin: they are remnants of stream tunnels.

Alcoves and niches, made at the base of cliffs by spring sapping or by the erosive action of a swinging stream, also are noticeable in arid and semiarid regions. Their lower parts may be smoothed by sandblasting.

Blowouts, also called deflation hollows, are depressions made in loose earth as material is removed by wind action. Some are thousands of feet in extent. They are likely to be seen where wind often is strong and there is little vegetation to hold soil. They form not only in deserts but on semiarid plains, along sandy shores, and in other localities where vegetation is sparse. On the Great Plains small blowouts were once thought to be bison wallows; today most are considered wind-made.

Blowouts may enclose small mesas, pedestal rocks, and other wind-sculptured minor erosional remnants. In some blowouts, roots of hardy plants such as yucca may hold sand tightly, preventing it from blowing away; thus as surrounding sand is removed, the plant is left standing on a sand pillar. Sand removed from blowouts is deposited somewhere to leeward, often forming "blowout dunes."

Deep blowouts, such as some in the Sahara, may reach to the water table or near enough so that vegetation can flourish and water can be obtained by digging. These sites are oases. Oases occur also in low areas where water seeps from beneath rock strata or from faults.

Deposits by Running Water
Alluvial Cones and Fans

Alluvial cones and fans are prominent in deserts at the foot of steep, narrow ravines. The more noticeable because of the lack of covering vegetation, they are favored by desert conditions. One condition is steepness of slope, due to dominance of stream action and a low rate of gravity movements or, in some cases, to the fact that the slope is a fault scarp. Another condition is the often torrential nature of temporary streams in ravines and their corresponding ability to transport much abrasive rock debris. Again, prolonged deposition of sediments and the lack of sufficient stream action to carry them away may have aggraded the valley floor so much that streams reaching the floor cannot travel far from the mouth of the ravine before depositing their sediment loads. In any case, a stream descending a ravine is torrential and meets the valley floor abruptly; therefore the sediment load is deposited abruptly and over a small area. An accumulation of such deposits is a cone or a fan.

Facets and Fans: Many fans at the base of this fault-block mountain range in southwestern Arizona have coalesced to form a broad bahada. Some major channels divide to form distributaries. Minor channels join to form major ones. Successive rows of facets indicate many episodes of uplift by faulting.

An Alluvial Fan Par Excellence: The fan spreading from the mouth of Copper Canyon, in the Black Mountains fringing Death Valley, has classic form. About 2 miles wide at the base, it consists of relatively fine materials. Being fine, these can be carried far out over the valley floor; thus the fan's gradient is relatively gentle. A fault scarp separates the bedrock from the fan. The wavy line near the fan's edge is a highway.

Cones are made by streams that originate near the edge of the highland and thus have high gradient. Fans are built by streams that originate in the highland's interior and have cut to a lower level. Both cones and fans are steepest at the mouth of the ravine. Cones may incline almost as steeply as the vertical; fans incline from about 7 degrees at the top to ½ degree at the edge. Alluvium in cones, having been deposited by swifter streams, is coarser than alluvium in fans, and in both features the caliber of the material decreases toward the outer edges. Cones are built up close to the base of the highland; fans spread out onto the valley floor, in some cases to distances of thousands of feet.

The continuum from talus cones and sheets to alluvial cones and fans may be hard to trace. Occasionally the material in a cone is "contaminated" with talus; thus determining whether the feature is an alluvial cone or a talus cone may be difficult.

A temporary stream running out onto a fan encounters thick sediments and, like a stream on a delta, divides into a network of smaller streams – "distributaries" – to get through. These streams cut channels, known as washes, or wadis (WAH-deez, the Arabic name), which are dry except in rainy periods; they can be from a few feet to a hundred feet deep and wide. Except when rains are extremely heavy, distributary streams soon die in the sand and deposit their loads before reaching fan edges. Streams that do reach the edges build the fan outward. Water that has sunk into the fan may emerge as a spring at the fan's base.

As fans along a mountain front grow, they may merge to form a bahada (Spanish *bajada*, "slope"). This is a vast, undulating, gently sloping surface on sediments. A pediment may be exposed along a narrow strip between the base of the mountain range and the head of the bahada. Bahadas are especially well developed in the Basin and Range Province.

Basin Fillings

On a fault-block terrane, as in the Basin and Range Province, mountains may enclose a basin with interior drainage, called a bolson (Spanish, "large purse," or "basin"). Runoff from heavy rain may flow down adjacent bahadas, finally collecting as a shallow lake in the bolson. The lake usually vanishes within hours or days (possibly not for months) because of absorption and evaporation. Absorption can be rapid where, as in southern Arizona, sediment deposits between the mountains are hundreds or thousands of feet thick and thus afford plenty of space for groundwater.

Sediments deposited in a bolson have a broad, level surface called a playa (Spanish, "beach"). Sediments are usually clayey, so that the playa may be called a clay pan. In dry weather the water table is far below clay pans, and these become mud-cracked. Where the deposits are of salt rather than clay, the surfaces are called salt pans, or salinas. The water table is usually within 10 feet of the surface, and the salt may remain wet during much of the year.

Salinas, common in the Basin and Range, are spectacular in Death Valley. On one sector of the valley floor, salt deposited by streams from adjacent mountains has built up in rugged masses several feet high, forming a sort of badland, called Devils Golf Course. In February 1973 Death Valley pans were converted into veritable lakes by the heaviest rains in decades. Salt deposits form spectacular towers in shrinking Mono Lake, east of Yosemite National Park.

Devils Golf Course: During the Pleistocene streams from nearby mountains, carrying sediments, flowed into Death Valley, creating Lake Manly. As the climate grew dryer, the lake shrank, finally going almost totally dry about 20,000 years ago. Masses of crystallized minerals, mostly salt, were left on the lake bed. Periodically these masses are washed, partly dissolved, and lowered by rain, but when dry conditions return they grow higher again as salt in solution rises by capillary action and recrystallizes. A tourist's pamphlet explains that the salt formations are a fairway for the Devil's midsummer golfing.

Deserts that are areas of fault-block topography and interior drainage would have numerous permanent lakes but for the aridity. During the Pleistocene, when humid conditions prevailed, many basins in areas that are now deserts held large, deep lakes with lifetimes of thousands of years. Shorelines in Death Valley show that it was occupied by a lake or successive lakes up to 116 miles long, 10 to 12 miles wide, and 600 feet deep. The last of these water bodies, Lake Manly, had dried up by about 20,000 years ago.

Desert Varnish

"Desert varnish" is a familiar aspect of rock scenery. This shiny, usually black or reddish coating on rock, especially sandstone and basalt, consists of clay (up to about 70 per cent) and iron or manganese oxide (up to about 30 per cent), often with some silica. Sandblasting may increase the shine. Formerly the "varnish" was thought to have exuded in solution from the rock's interior, then precipitated on the surface; now it is believed to derive from desert dust or, perhaps, the activities of microorganisms that live on the rock.

"Desert varnish" occurs not only in desert and semiarid regions, such as the Mohave and the Colorado Plateau, but in humid regions too; so a better term for it is "rock varnish."

Deposits by Wind
Sand Ripples

Long, low ridges of sand built up by wind action are known as sand ripples. Commonly the ridge crests are somewhat sinuous and about transverse to the wind direction. Rarely, ripples are made by wind simply pushing the sand en masse; these ripples are called aerodynamic. All other ripples result from impacts as blown particles saltate (leap), strike other particles as they fall, and drive these forward, rolling, sliding, and hopping.

Impact ripples develop as moving sand encounters a small barrier or irregularity on the ground surface. The wind moves the sand upward as far as gravity permits; then the sand drops. Thus a backslope (the windward side of the ripple)

Site of a Pleistocene Lake: Soda Lake, west of Baker, California, is one of a chain of dry lakes occupying grabens that trend southeastward from Death Valley to the Colorado River on the Arizona border. Soda Lake and its neighbor, Silver Lake, together are about 60 miles long and 5 to 6 miles wide. Here in the Basin and Range, worn fault-block mountains rise from the salt-covered lake floor.

A Desert Dust Storm: Strong winds can occur even in the deep Death Valley graben. Here, east of Stovepipe Wells, sand dunes are seen under blowing sand and dust. The dunes are mostly barchans.

with an inclination of about 8 to 10 degrees, and a concave slipface (the leeward side) at 20 to 30 degrees, take form. Ripples range in height from less than a half inch to about 4 inches. Height increases as wind velocity increases and the caliber of the sand particles decreases. The space between ripples increases with their size, up to about 65 feet.

Sand Dune Areas

Sand dunes, popularly associated with deserts, are relatively rare on deserts. Dune formation requires plentiful supplies of sand as well as certain other conditions. Most deserts are rocky or only patched with thin veneers of sand, not enough for dune formation.

Sand can accumulate thickly and become duned under certain conditions. It can be trapped and build up thickly in a depression, as in the Basin and Range Province. It can accumulate where the prevailing wind blows against any highland barrier, as at Great Sand Dunes National Monument, in southeastern Colorado. Finally, sand has accumulated thickly where it was produced plentifully in the past but where, because of climatic and topographic conditions, as in parts of the Arabian and Sahara deserts, winds even if strong have lacked the power to transport much of the sand elsewhere.

Dunes cover only about 1 percent of American deserts, less than 20 percent of the Sahara, and no more than 30 percent of the great Arabian Desert. Spectacular dune areas are seen on the Arabian Desert's 400,000 square miles of sand sea, in the Sahara sand sea, and on the deserts of central Asia.

Dunes can form only on extensive flat lands where winds blow without topographic obstacles, sand is thick, and vegetation is absent or sparse. Desert dunes are most abundant on the lower parts of large basins, understandably so because much sand has been deposited there by running water and because the basin's sides limit removal of sand by wind. Such depressions include those of the Great Basin in the Southwest, the Gobi in China, and the Qatara in Egypt. Dunes generally tend to build up on very gentle slopes of fans leading down to playas. Dry lake plains, such as the one covered by scenic white gypsum dunes at White Sands, New Mexico, are common sites for dunes.

Sand movement and dune-building increase with aridity; however, dunes are not limited to deserts. They are familiar on certain kinds of humid terrains also, notably sandy coasts such as those of the Pacific Northwest, the Northeast from Cape Cod to Florida, and the Texas Gulf Coast. In Europe, shores of the Baltic Sea have multitudes of dunes. Dunes build up on floodplains and deltas where vegetation is sparse, as along some segments of the Mississippi.

Areas that were periglacial during the Pleistocene, including many parts of the Great Plains (notably the Nebraska Sand Hills), southern shores of Lake Michigan (including Indiana Dunes), and north of Albany, New York, have well-developed dunes. In the Moses Lake area of Washington state an impressive barchan field has been built of glacial outwash. On the Colorado Plateau much of the sandstone, such as the crossbedded Navajo and Toroweap formations, consists of "fossil dunes" – rock masses formed from dunes that built up on Mesozoic deserts.

Building Dunes

Dunes generally originate where movement of wind-blown sand or dust is restricted or guided by an irregularity in the

A Bit of Desert in Texas: East of El Paso, on a broad plain at Salt Flats, wind work is well displayed. The dune here is a sharp-edged barchan, with horns pointing east, according to the prevailing westerly winds.

sand mass or some obstacle, such as vegetation or a rock outcrop. Backs of dunes usually are convex upward, but very large dunes may lose this convexity. Generally dune shapes are determined by sand supply, caliber of the sand particles, presence of vegetation (if any), variations in wind direction and strength, ground contours, damp patches, wind eddies, or air currents caused by differential heating. That long list suggests the complexity of dune formations.

Most dunes are between 10 and 300 feet high; rarely they attain 1,500 feet. The windward side commonly has a slope of 10 to 15 degrees; the leeward side (slip face), 30 to 34 degrees. Dune width averages about 10 times the height.

Dune types grade into one another. One or several types mingled can form assemblages. Some dunes develop on top of others. Most migrate, at rates up to 60 feet per year; smaller dunes travel faster. Dunes may change in character en route.

Holding On: In a blowout, on New Mexico's White Sands National Monument, a yucca's roots have held sand tightly while surrounding sand has been blown away. Thus the plant now stands on a sand pillar.

Dunes and Blowouts: On the White Sands lake plain are rows and rows of well-developed dunes and blowouts. In the foreground here, a blowout has been formed in the lee of a parabolic dune. Plants have established themselves in the hollow, taking advantage of shelter from the wind.

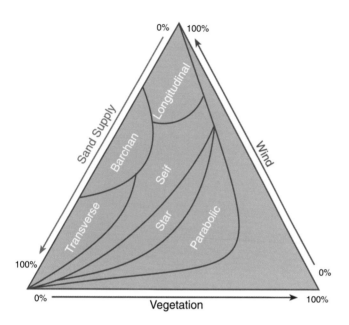

Dunes as natural phenomena tend to be so dynamic, so complex, so sensitive to changing conditions, that classifying them and tracing their evolution from one form to another can be baffling. Dune formation remains a subject of frequent argument among experts. However, most dunes do represent variations or combinations of a few basic types. These can be described in order of increasing sand supply.

How Dune Forms Are Determined: The kind of dune formed by blowing sand seems determined mainly by three variables: wind direction and strength, amount of vegetation cover, and sand supply. Particle size also is an influence. The complex interplay of factors is not fully understood.

The simplest of dunes is a sand dome, seen where sand is very thin over bedrock or some other hard surface. Circular or elliptical in plan, with a convex profile and uniformly smooth slopes, it is very low but can cover square miles.

The barchan (from Turkish, "sandhill") is a crescentic dune, as high as 100 feet and with a length and breadth as great as a half mile. Barchan "swarms" build up where sand supply is sparse-to-modest and wind is almost always from the same direction. Where sand is sufficient, blowing particles meet an obstacle or irregularity and, by saltation and creep, pile up against it at an angle of 10 to 15 degrees, creating a mound. Reaching a height determined mainly by wind strength and particle size, particles lose velocity and fall or slide down the lee side of the mound. This is a steep slope, the slip face. Meanwhile particles near the sides of the mound, where sand is thinner, advance farther, forming "horns" to leeward. Hence the barchan form: horns curving down and leeward from the crest, and a hollow with low wind velocity between them. As winds blow and sand continues to move to the dune's lee side, the dune advances.

Like other dunes, barchans may have sand ripples on their

Land of Giant Dunes: Colorado's San Luis Valley, in the rain shadow between the San Juan and Sangre de Cristo mountains, is dry and sandy. In a sheltered corner just north of Blanca Peak is Great Sand Dunes National Monument. Here prevailing southwesterly winds transport sand up to the mountain barrier, deposit it, and shape it into dunes rising as much as 700 feet above the valley floor. In this view, visitors are barely visible against the giant barchan in the background.

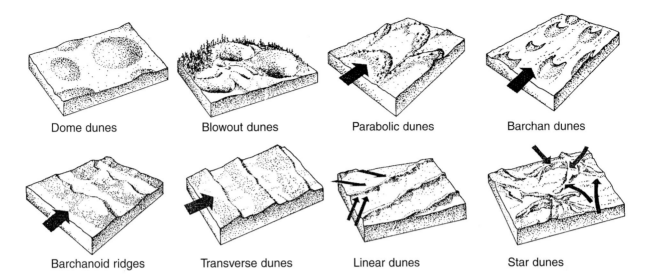

Dome dunes   Blowout dunes   Parabolic dunes   Barchan dunes

Barchanoid ridges   Transverse dunes   Linear dunes   Star dunes

Major Kinds of Sand Dunes: Dune types are determined by sand supply, caliber of particles, variability in wind strength and direction, and existing vegetation. Some dunes travel; others don't. One type may merge into another as conditions change. In these drawings, arrows show wind directions.

backs. As sand supply increases, barchans may connect to form rows, called barchanoid dunes.

With a larger sand supply, larger sand particles, and the wind still unidirectional, transverse dunes may form. Up to ⅔ mile long and ⅓ mile wide, these are more or less irregular ridges perpendicular to the prevailing wind direction, with steep slip faces on the leeward side. These ridges, sometimes with scattered vegetation, may acquire barchanoid forms, with slip faces either concave or convex downstream, due to eddies formed by wind action or heating of the sand. With a still larger sand supply, an assemblage of transverse dunes may evolve into a complex pattern of sinuous ridges called an akle (French *écaille*, fish scale) field. If, on the contrary, sand supply decreases downwind, transverse dunes may convert to barchans.

Another basic type is the longitudinal dune, which appears in several varieties. The relatively common seif (Arabic, "sword"), averaging 10 to 12 miles long, 800 feet wide, and 300 to 400 feet high, has a sharp, sinuous crest and two slip faces. It appears to result from winds blowing in obliquely different directions in summer and winter. It is seen best in the Sahara and Arabian deserts.

The linear, nearly straight longitudinal dune, up to 100 feet high and 15 miles long, is seen in Africa's Kalahari Desert and in Australia's Simpson Desert. It forms parallel to the resultant ("net result") of several wind directions.

The mightiest of longitudinal dunes are the whalebacks, or draas, seen mostly in the Sahara and Namib deserts. As large as 150 feet high, 2 miles wide, and 200 miles long, these round-backed, slow-moving monsters may take as long as a

million years to construct. Whalebacks appear to form by the joining or coalescence of other dunes, notably star dunes.

Star dunes, also called pyramidal dunes, are most remarkable when seen from the air. They may begin as relatively isolated sand masses with a central peak up to 800 feet high, from which three or more arms radiate, starlike. These dunes are shaped by winds blowing alternately from all sides. The dunes do not migrate, but tend to stay in place for centuries. They may appear as interruptions on whalebacks and may, with or without other dunes, combine to form whalebacks.

Intermediate between transverse and star dunes is the reversing dune, a transverse ridge with a slip face on each side. It is shaped by opposing winds of similar strength, alternately dominating as seasons change. The dune migrates a short distance in one direction, then goes into reverse, over and over, and thus tends to remain in about the same place.

The presence of vegetation, which holds sand, deflects wind and reduces wind velocity. It tends to deform dunes and to guide the shaping of distinct types. One of these is the blowout dune, built up on the lee side of a blowout where sand is limited and vegetation cover is increasing. On the windward side is the circular deflation hollow. These dunes are common in the Southwest.

Another vegetation-influenced dune is the U- or V-shaped parabolic type. The hollow is often a blowout between vegetated areas, which restrict wind. The nose of the dune faces windward. These dunes are seen often in the Southwest. There are fine arrays on Cape Cod east of Provincetown, Massachusetts, where they are shaped by onshore winds.

Like deltas, dunes may be crossbedded; that is, made up of

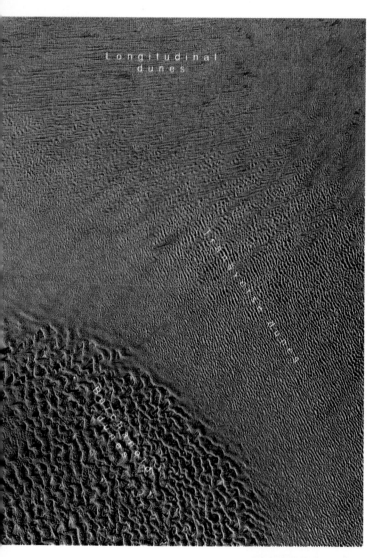

Longitudinal dunes

Transverse dunes

Barchanoid dunes

Star Dunes: In this Landsat photograph of an area in the Grand Erg Oriental of Algeria, Tunisia, and Libya, dunes are a half mile to one mile in diameter and rise 500 to 800 feet above their surroundings.

Over the Arabian Desert: A single high-altitude photograph taken by an earth satellite displays barchanoid, transverse, and longitudinal dunes. North is at left. These are in the Arabian sand sea, where building material for dunes is abundant.

Linear Dunes in the Namib: In a view from a passenger jet, seifs extend to the far horizon. They develop parallel to the prevailing wind, becoming more or less wavy because of slight variations in wind direction. The Namib is desert because air cooled by the south equatorial ocean current is warmed and dried as it comes in over the warm land.

Barchanoid Ridges: Near the airport at Cairo, Egypt, barchanoid dunes appear below the wing of an airliner. These dunes have developed on unirrigated terrain. In the distance, terrain irrigated by water from the Nile is covered with vegetation and lacks dunes.

successive layers of sand or clay deposited at different angles with respect to the horizontal and the vertical. Crossbedding is due to changes in wind direction and wind strength. It is best seen where dunes have been eroded enough for cross sections to be exposed, as in the Zion Canyon "fossil" dunes and elsewhere on the Colorado Plateau.

Most dunes migrate ceaselessly, climbing hillslopes, filling valleys, burying forests. Some degree of dune control is practicable by fencing and by plantings, but transport of sand by wind is almost unstoppable; it continues until sand supplies are reduced to amounts inadequate for dune-building. Sands may eventually arrive in basins too deep for excavation by wind, or against mountains too high to climb, or in the ocean, where they become distributed along the shore and outward on the continental slope.

Some dunes emit singing or booming sounds from rhythmic vibrations when sand particles avalanche down the slip face. Such dunes commonly have unstable slipfaces and sands of medium caliber; their sands have been transported a long distance, are very dry, and have been etched by weathering and erosion. They are rare but were known more than 1,500 years ago in the Middle East and 9th-century China. In the United States they occur at Kelso Dunes, California; Sand Mountain, Nevada; and Mana, Kauai (Hawaii).

Deserts and the Future

With time, as temperatures, sand supply, prevailing wind directions, vegetation cover, land elevations, ocean currents, and other conditions change, humid lands may become deserts, and deserts may become humid lands. Deserts may become sandier or rockier; highlands may wear down or be uplifted again; dunes may disappear or multiply; land will be lowered by erosion. A new continental glaciation would profoundly modify deserts as well as humid lands. Change is in store for all deserts – and for all lands now humid.

Earth in the future will probably have more, not less, areas of desert. Although in any era there are conditions militating for and against desert development, humankind, with its increasing populations, is tipping the balance toward desertification. Deforestation, overgrazing, high-intensity agriculture, off-road vehicle traffic, heavy demands for water, and global warming due to burning of fossil fuels – all these activities seem significant for the future.

Desert in the Making: In the Amboseli Game Preserve, Kenya, an elephant wanders in search of forage. This terrain, once green savanna, is being converted into desert because of overgrazing by cattle and intensified foraging by wild animals as humankind takes over their natural habitats.

Remnants of an Ocean Bottom of Long Ago: Monument Rocks, in Gove County, Kansas, have been carved from the edge of an escarpment on chalk. Caps of hard limestone, relatively insoluble, have helped to preserve pillar forms. This chalk, deposited on a Cretaceous sea bottom 60 to 130 million years ago, is similar to the chalk, of similar age, in the famous White Cliffs of Dover, England.

15
Scenery of Soluble Rocks

AMONG all rock types, the soluble ones – especially limestone – yield the most distinctive and perhaps the most interesting landforms. Far more than other rocks they are shaped by solution on and in the crust. Topography so shaped is called karst, from the Karst Plateau, in the former Yugoslavia, famous for solution forms.

As described in earlier pages, limestone consists entirely or mostly of calcium carbonate ($CaCO_3$). Its close relative, dolostone, is similar except that magnesium has replaced some of the calcium. These rocks form either from materials chemically precipitated from water or from limy organic remains, such as coral and shell fragments, deposited in water. Sites of deposition are bodies of seawater or, less commonly, fresh water. Rock structures may include crossbedding, stratification, lineation from current action, or graded bedding (bedding with materials sorted as to size).

Limestone and dolostone masses exist on all continents, in many islands, and on some of the sedimentary platforms that fringe continents. These rocks contain detailed records of the past, indicating former locations of oceans and continents as well as former climates and ocean-current patterns, varying compositions of seawater, impacts of extraterrestrial objects, and evolution of aquatic organisms. Incidentally, limestone and dolostone contain Earth's greatest store of the element carbon.

Because limestone and dolostone look much the same and yield similar scenery, in these pages a distinction is made between them only where necessary.

A Rock of Many Meanings

Karst terranes offer an astonishingly variegated, sometimes bizarre, often beautiful array of features, including needle- or tower-like mountain summits; closed valleys with streams entering and exiting through tunnels; cavernous, conelike hills; clusters of sinkholes like bomb craters; broad and bare expanses of rock flowingly smooth, or divided by joints into neat flat-topped blocks or hosts of ragged "fins." Beneath each limestone surface, because water so easily penetrates and dissolves this rock, there is often a dreamlike realm of caverns, shafts, and tunnels; giant rooms; stalactites and stalagmites; "frozen waterfalls" and "curtains" of stone; pools with strange flora and fauna; and streams that emerge from, and disappear into, dark tunnels.

Influences of karst landscapes on humankind have been powerful. Because rainfall on a limestone terrane quickly finds its way underground, little water is available for flora and fauna at the surface, chemical and mechanical weathering there are subdued, and soil formation is limited; thus these terranes are more or less barren. With few exceptions, such as vineyard areas, limestone highlands are not friendly toward crops, and tillers of the fields are less likely to be seen there than shepherds with goats and sheep. Limestone topographies often are of the badland type – deeply gullied, rugged – discouraging travel and intercommunication. From the age of grass huts to the age of skyscrapers, parts of these terrains have been prone to collapse on occasion, engulfing livestock, buildings, even people. Historically, these lands

Hardly for Farming: On an elevated limestone terrane, storm water quickly percolates down through the bedrock and thus supports little vegetation at the surface. Little topsoil is formed. The result is a relatively barren landscape with scattered residual boulders. This example is near New Braunfels, Texas.

Demonstrating Solution and Erosion: At The Grottoes, in Bryce Canyon, Utah, rain-wash and subsurface water working downward through weak siltstone and limestone have cre- ated a maze of scenic features - tunnels, grot-toes, arches, towers, needles. Erosion is swift because of the solubility of the rocks, close vertical jointing, and steepness of gradient.

have been unable to support relatively large populations, and population centers have been few and scattered. Life on limestone has been hard and discouraging.

The nether worlds of limestone have had extraordinary meanings for humankind. Caverns, with dripping rooms, weird natural ornamentation, mysterious pools, mazes of tunnels, and enveloping darkness have stimulated the human imagination and become important sites for religious and artistic activities, as Cro-Magnon cave paintings and Greek myths testify. In war, limestone caverns and tunnels have been exploited for attack and defense, as in ancient Israel and modern Vietnam. On the peaceful side, underground cavities in limestone have long served as reservoirs for precious water, today easily reached by well-drilling. And, too, lime-stone caverns rank high among the marvels of nature to visit – to wander through with guides who point out stone "gob-lins" and "pipe organs," or to explore adventurously with scuba gear and flashlight, and perhaps to understand.

Into Solution and Out

Basic in limestone sculpturing is the solution process. Cal-cium and magnesium carbonates, the minerals that constitute pure limestone and dolostone, respectively, are readily solu-ble in carbonic acid (calcium carbonate being more soluble than magnesium carbonate). The acid is formed as water from precipitation combines with carbon dioxide (CO_2) from the atmosphere and from decaying organic materials in soil. (Soil can contain 15 times as much carbon dioxide, by vol-ume, as air.) Solution is favored where the rock is pure car-bonate, water and carbon dioxide are ample, temperatures are cool, the water and gas are under pressure (as when con-fined in rock pores and fissures), circulation is active (removing saturated solution rapidly and replacing it with new water), and time is long.

Important for scenery is the fact that the solution process is reversible. As conditions favoring solution are removed – for example, as pressure is relieved or the air becomes warmer – some carbonate will precipitate out of solution and become stone again. Thus in the making of limestone sce-nery there is not only decomposition and disintegration of rock but the creation of new rock with new forms.

Not all limestone is pure carbonate; often it contains in-soluble or only slightly soluble foreign materials, such as iron oxides, sulfides, and silicates. As the carbonate is dis-solved and carried away, the insoluble materials remain,

Solution and Collapse: Near Cave City, Kentucky, in a limestone area, a roadcut exposes the cross section of a sinkhole. Water working down through joints has dissolved much of the rock and carried it away in solution, leaving reddish insoluble material ("terra rossa") in cavities. The large cavity was enlarged by collapse of its sides.

accumulating on the land surface and in rock joints and cavities, more or less protecting carbonate surfaces around them, and slowing circulation. If the proportion of magnesium carbonate in dolostone is as high as 2 or 3 per cent, the solubility of the rock mass as a whole is radically reduced. Minerals in limestone give it various colors.

Often interbedded with limestone is gypsum: hydrous calcium sulfate ($CaSO_4 + 2\ H_2O$). This rock consists of precipitates from water; when dried it is anhydrite. Relatively soft, gypsum is soluble and yields certain unique features of its own. Some of the most remarkable depositional forms in limestone caverns are of gypsum, as in Carlsbad Caverns.

Solution of limestone, with the sculpturing and building up of its special landforms, is favored where the climate is humid, affording ample precipitation, and where the terrain is well covered by vegetation and soil that is rich in decaying organic materials. The rocks of arid and semiarid terrains undergo little solution, but many if not most such terrains exhibit solution features formed during a humid period in the past, as is true of Carlsbad Caverns. Although cool temperatures favor solution, in very cold lands vegetation is lacking and so is the needed carbon dioxide; hence solution and limestone sculptures there may be minimal. In humid tropics, solution can be relatively active, despite warm temperatures,

Cavernous Erosion and Weathering: This rock fragment, fallen from a limestone cliff at New Braunfels, Texas, is from a stratum riddled by percolating groundwater. The same process of erosion and weathering creates limestone caverns.

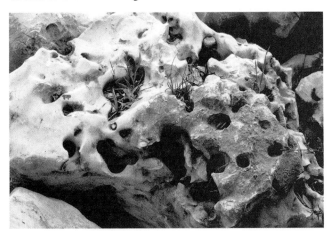

Smoothed by Solution: This limestone boulder in northern New Jersey has the smooth surface seen on many limestone outcrops after long exposure to a humid climate. The ridges are edges of strata of superior resistance.

Furrowed by Solution and Erosion: The steep side of a dolostone mountain near the Icefields Highway in Alberta, Canada, shows deep furrowing due to stream erosion and solution. These furrows are large examples of lapiés.

A Mountain of Massive Limestone: In Mt. Inglismaldie, rising above Lake Minnewanka, in Canada's Banff National Park, limestone rock layers are massive and strongly folded, with little vertical jointing. Accordingly, erosion has left the mountain with a massive aspect, without needles and pillars.

because of the abundance of rain, thickness of soil, and plenitude of decaying organic materials.

Solution rates are influenced by rock structures. The more joints there are, the greater is the total area of rock exposed to solution. If strata are tilted, circulation is speeded by gravitation, and so is solution. In chalk, which is permeable and little jointed, circulation is very slow and solution minimal; but in oolitic limestone, which is porous, solution can be relatively rapid.

Another significant factor in the solution process is mechanical resistance. Rock of low compression strength tends to crush under overlying rock. Crushing closes joints and other spaces through which water could circulate. In stronger rock, cavities remain open and, with prolonged solution activity, can grow to substantial size, becoming caves.

Relatively rapid solution depends ultimately on good drainage – that is, prompt removal of saturated solution and replacement by fresh water. For the creation of sizable solution features, such as tower forms, sinkholes, and caves, the limestone mass must drain toward lower surroundings. Generally its thickness should be at least 400 feet.

At the Surface: Karst Highlands

Limestone highlands take numerous forms. Most spectacular are the very tall pinnacles, needles, or towers occurring where the rock has joints closely spaced. Joints enlarge relatively rapidly by solution and thus can accommodate large, vigorous streams, which cut down rapidly because corrasion is accompanied by solution. Needles and towers can be hundreds of feet high if the rock is resistant and structurally strong, for then gravity movements are subdued and high vertical slopes can be maintained. Many fine examples are seen in the deformed but strong limestones of the European Alps, especially in the Italian Dolomites. Limestone needles make spectacular landscapes in the Kunming sector of mainland China and in the Bemaraha Tsingy National Reserve of Madagascar, to mention only two impressive sites among many. Where vertically jointed highland limestone is soft, needles are smaller and more fragile, yet still impressive, as at Bryce Canyon, Utah, where limestone layers alternate with siltstone, which is weaker, and sandstone.

Fold mountains on tilted, hard limestone strata with widely spaced joints do not erode into towers and pinnacles; contours more commonly resemble those of similarly jointed sandstone. Notable examples include the huge limestone homoclines of the Canadian Rockies, such as those of the Queen Elizabeth Range, and similar ridges in the French and Swiss Alps. In some instances the broad, sweeping dip

Not for Caverns and Pillars: In this outcrop of flaky, clayey, soft limestone near Union, Kentucky, groundwater circulation is limited. Vertical joints are few. Settling of the weak rock layers prevents formation of sizable cavities by solution. Such is not a rock type in which needle, tower, or cavern forms could be sculptured.

slopes exhibit solution furrows called lapiés, and scarp slopes may be pocked with entrances of caves and tunnels. In arid regions, because of the lack of solution and stream erosion, limestone weathers and erodes like other rocks, such as sandstone, and may lack distinguishing solution features. So it is in New Mexico's San Andres and Sandia mountains. But in the same state are the grand Carlsbad Caverns, most of which formed when the climate was humid.

In Croatia the Karst Plateau, which rises more than 8,200 feet above sea level and extends more than 300 miles, from Istria to Kotor, exemplifies karst topography par excellence. Uplifted only recently, geologically speaking, its nearly horizontal layers of pure limestone are as much as 4,900 feet thick. The plateau is still not deeply dissected, except at the edges and where major streams have been at work. On the plateau, surface streams are rare, being confined mostly to floors of major valleys, where they flow on poorly soluble strata or accumulated insoluble rock debris. The countryside has very high areas but is largely rolling, with numerous dry valleys and with gentle slopes and valley floors pocked with sinkholes, especially in the north. In the north are hills, as at Plitvice National Park, with mazes of caves and tunnels and now-you-see-it-now-you-don't streams. One of these, the

Timavo River, flows underground 30 miles, emerging on the coast of Italy. Also in the north is Postonja Cave, claimed locally to be the world's most extensive cavern (it is not), in which visitors tour by railway.

Another, smaller picturesque karst landscape is The Burren, in County Clare, Eire. Here rolling hills and limestone platforms are corduroyed with clints and grikes (alternating ridges and furrows that follow joints or bedding planes). Another such landscape has been developed above Malham

Pillars and Needles a-Making: Vigorous erosion in closely spaced vertical joints in strong highland limestone favors needle and pillar forms. Vertical joints and solubility of the rock allow rapid downcutting by streams, while the hardness of the rock limits mass movements. This scene is in the eastern Italian Dolomites at Pordoi Pass, west of Cortina d'Ampezzo.

On the Burren: In County Clare, Eire, barren hilltops feature spectacular clints and grikes. Even under the humid climate, vegetation can hardly take hold; drainage down into vertical joints is too rapid. Sharp-pointed rocks in the background here are *chevaux de frise*, planted centuries ago for defense against attack.

Cove in Yorkshire, England. In Virginia, areas of higher elevation on the floor of the Shenandoah Valley have both sinkhole topography and caverns.

At the Surface: Valleys

Valleys in karst terranes depart in many ways from the standard valley forms developed in non-soluble rocks. Their walls are likely to be more or less barren, in part because

storm water is soon absorbed and is not available to support much vegetation. More commonly, walls are cavernous from sapping, and from the openings many springs, called rises, may issue, some forming sizable waterfalls. Walls may be corduroyed with solution furrows. Edges of relatively resistant strata may form ledges along the walls. If strata are tilted, the valley cross section may be asymmetrical.

Minor valleys in limestone are likely to have few if any surface streams. After cutting for a time, such streams reach an opening in the valley floor and go underground. Minor valleys, like youthful valleys generally, are likely to be rock-floored and rough, steep-walled, twisting, of relatively high gradient, and with few sinkholes. Floors of major valleys have less gradient and are likely to be on an insoluble layer or more or less covered with accumulated insoluble erosional debris; thus the stream is held at the surface. On these valley floors, too, sinkholes may be lacking. Any valley may grade into a soluble zone at one or both ends, and drainage may be in one direction in one segment of the valley and opposite in the rest of the valley. Three-dimensional drainage often creates an impression of chaos.

Karst valleys, known also as solution valleys or valley sinks (in former Yugoslavia, "uvalas"), are closed depres-

A "Stone Forest":   At Bryce Point in Bryce Canyon National Park this almost vertical valley wall is being rapidly dissected by swift temporary streams. Relatively weak limestone strata, even weaker siltstone layers, and close vertical jointing account in part for pinnacle forms and rapid erosion.

Panorama of a Dry Valley: On the limestone terranes of Provence, in France, dry valleys are common. No stream is visible in this scene near Manerbe. The land is made fertile by water from deep wells.

sions formed by coalescence of sinkholes. So-called dry valleys are those cut during the past by a stream that has since gone underground. (These valleys differ from valleys left dry by stream piracy, climate change, or some other such event.) A valley with a head enclosed by steep walls and a stream emerging from the base of the headwall is known as a pocket valley, or steephead. A blind valley is one that ends at a steep wall at the lower end; at the base of the wall the stream, if any, vanishes into a tunnel.

A valley that passes through a karst region without losing its river to underground drainage is termed a through valley. Often these valleys, such as the Yangtse in mainland China, the Tarn Gorge in southwestern France, and the Dove Valley in Derbyshire, England, originate outside the karst region. The stream may survive either because it has plenty of water or because it flows on insoluble materials, or both.

The term gorge is applicable to vertical-walled valleys everywhere. In karst regions it is applied to a valley created by collapse of roofs of a chain of caves. These valleys, such as Cheddar Gorge in Somerset, England, tend to have steep, rugged, cavernous walls (sometimes including natural bridges) and rough, ungraded floors.

The interior valley (Yugoslav *polje*) is a depression 2 to 3 miles in diameter, steep-walled, with a flat floor usually

covered with insoluble alluvium. It occurs in a structural depression such as a syncline or a fault trough, and commonly is closed, with no surface outlet. On the Karst Plateau such features reach lengths of 30 miles and widths of 3 to 4 miles. They are drained by swallow holes (Yugoslav *ponors*), which connect with subterranean passages. When swallow holes become blocked by sediments, flooding of the valley floor results.

A Small Dry Valley: On the Karst Plateau in Bosnia, a valley made by a stream of yesteryear is now essentially dry. The day after a heavy rain, no water is noticeable in this valley.

"Cockpit Country": North of Jajce, Bosnia, the karst highland is densely pocked with sinkholes. Most are smooth-sided and at the center of broad depressions that collect rainfall; hence they are swallow holes rather than collapse sinks.

Three-dimensional Erosion: The park at Plitvice in Bosnia's karst country is a wonderland of streams vanishing into sinkholes and pouring from openings in cliffs. These openings are entrances to tunnels made by underground streams in relatively pure limestone.

Routes to the Underground: Sinks

It is on elevated, horizontal or gently sloping limestone strata that the sinkhole, or sink (Yugoslav *doline*, or *dolina*; in Jamaica, banana hole) is best developed. One variety is the swallow hole, a funnel-shaped opening through which a stream dives underground. Other sinkholes are made by collapse of the roof of a tunnel or cave; these are known as collapse sinks or, in Central America, cenotes if they form natural wells. (The dark well at Chichen Itza, into which Mayan sacrificial maidens were cast, is a cenote.) Any sinkhole that leads down to a cave is known as a jama in (former) Yugoslavia, and elsewhere is sometimes called a karst pipe or a pothole (the latter term is better reserved for a hole made in bedrock by an eddying stream).

Sinkholes range in size from those that can be stepped over to basins 3,000 feet in diameter and 900 feet deep. The depth is usually about a third of the diameter, and the sides commonly slope about 20 to 30 degrees toward the center. The overall shape is likely to be asymmetrical if the underlying strata are tilted. Whereas swallow holes are smooth-sided, from shaping by solution, collapse sinks are irregular, with sides left rugged by rockfalls. Sinkholes tend to be more common in humid tropics than elsewhere, because solution is strongly favored by abundant rainfall.

Sinkhole formation is relatively rapid under snow because of the cool temperature and the steady supply of water from melting; it is rapid also under soil and vegetation, because these contain carbon dioxide and hold moisture longer than bare ground does. Sinkholes can become partly filled with accumulations of insoluble material, including glacial drift; thus many sinkholes can hold ponds, which may be long-lasting if sediment-filled and if the water table is high.

Although more common on high-standing terrains such as plateaus, sinkholes of various kinds can occur in valley bottoms if these are above the water table. They are seen in gypsum, anhydrite, and salt deposits as well as in limestone. Also they may occur in non-soluble strata where underlying soluble strata have collapsed over a cave. They do not develop on arid lands, but they may exist there as relics of a more humid past.

In the coterminous United States, the most notable sinkhole displays are in Kentucky (especially in the Mammoth Cave vicinity), southern Indiana, western Virginia, and Florida. In England, Yorkshire and Somerset have spectacular sinkhole terranes. Similar areas are seen on the vast limestone formations of Mediterranean lands such as Greece and the south of France and Italy, as well as former Yugoslavia.

The so-called blue holes, or ocean holes, in reefs are sinkholes formed when the reefs were above sea level. They be-

A Pond in a Collapse Sink: Though less soluble than limestone, gypsum can include remarkable solution features. At Bottomless Lakes State Park, New Mexico, ponds are in collapse sinks (note the rough sides of this one). Surrounding terrain, on deformed gypsum strata, is rugged, pitted, and barren.

Land of Sinks and Solution Pans: Much of Florida is patched with lakes and ponds lying in sinkholes or solution pans. Pans usually are floored with insoluble clayey materials which check solution. This view is southwest from north of Orlando, with the northern end of Lake George at upper left.

came covered by the ocean during the rise of sea level as the Pleistocene glaciers melted.

Sinkholes are to be distinguished from solution pans, which are broad, shallow depressions that may not connect with underground cavities. Pans are more or less random in occurrence, resulting from solution where water has collected in pre-existing depressions. Some pans are floored with insoluble materials and hold ponds. Numerous examples are seen in Florida, where, in many sectors, elevations are not high enough for the degree of drainage needed for sinkhole and cave development. In Florida, some pans were made by differential solution of the limestone when it was sea bottom, before it was uplifted to become land.

Sinkholes formed naturally can be substantial environmental hazards, and those formed as a result of man's activities can create even greater hazards. Among these activities are drilling for water and oil, and mining for coal and other resources. Removal of these materials (even of water and oil) deprives cave roofs of support, and collapse results. Where collapse, which usually is sudden, occurs beneath buildings, highways, bridges, and other structures, consequences can be grave. So it has been in Virginia, Alabama, Florida, and Texas, among other areas. In Paris's Montmartre vicinity,

ground has been collapsing over subsurface gypsum quarries, roofs of which have been weakened by solution. Archaeologists excavating the fabled city of Uban, in southern Oman, have found evidence that the city's catastrophic destruction, toward the end of the Roman period, occurred when its major buildings collapsed into a huge limestone cavern, which had been used as an underground reservoir.

A Chain of Caves Uncovered: Cheddar Gorge, in Somersetshire, England, was created by collapse of roofs of a chain of caves. Cave or tunnel entrances are seen in the gorge walls, which rise as much as 400 feet. The original boulder-covered floor of the gorge was cleared for the highway.

Karst Windows and Limestone Residuals

Depressions formed by collapse of a cave roof over a stream are called karst windows. Looking down into the hollow, one may see the stream emerging from a tunnel at one end and disappearing into a tunnel at the other end.

Prolonged erosion with deep dissection may reduce a limestone highland of regional extent to a scattering of low, conical or pyramidal hills. Under temperate climates these residuals generally are limited to elevations of a few hundred feet, as in Kentucky, where they are known as haystack hills, and on the Karst Plateau, where they are "hums." These hills

"Cone Karst": Rising behind Puerto Rico's coastal plain are residuals of reef limestone. These, locally called "pepino hills," correspond to the "hums" and "haystack hills" of other lands.

usually are developed on horizontal or gently dipping strata and may be capped with an insoluble layer. Also, with rapid drainage, they may get riddled with caves; thus surfaces can remain rough and rugged as erosion cuts farther and farther into the rock mass.

Under wet tropical conditions, residuals may form "cone karst": closely clustered hills rising 300 to 500 feet, with sides as steep as 40 degrees. The hills are separated by concave, star-shaped depressions called cockpits (actually, they are sinkholes). Such terranes are seen in Puerto Rico (where they are called "pepino hills"), Cuba and Central America ("mogotes"), and the Philippines "tit-hills"). The greater height relative to width of these residuals over those in temperate lands seems due to the high rate of stream downcutting and lateral erosion under rainy conditions.

Among the world's finest displays of residual limestone forms are the "tower karst" features in China's warm, humid Kwangsi sector. These truly marvelous forms, which grace traditional Chinese painting, rise with nearly vertical walls as high as 1,000 feet above the surrounding alluvial plain. Their great height relative to thickness is attributed to a rare combination of vertical jointing, rock resistance, spring sapping, and undercutting at the base by frequent floods.

Tower Karst: From a boat on the River Li near Guilin, China, tourists see perhaps the finest of the world's limestone tower forms. These features, developed under subtropical humid conditions, have inspired Chinese painting and other art forms for centuries.

Entrance to (or Exit from) an Underworld: East of New Middleton, Tennessee, a roadcut has been made through a segment of a tunnel or a cave. The limestone is well jointed, but a slab successfully holds up the roof.

Tunnels and Caverns

Rainwater and meltwater on a limestone terrane collect in depressions. There the water, absorbing carbon dioxide from air and soil, finds its way into rock joints, which have already been widened by solution and which branch in various directions below the surface. Following joints, water eventually encounters an obstacle, perhaps an insoluble joint filling or a poorly soluble stratum, and there spreads sideward until it again finds an unobstructed route downward. Sooner or later, perhaps a few inches or hundreds of feet below ground, the water reaches the water table, the boundary between the vadose (unsaturated) zone and the phreatic (saturated) zone.

Arriving at this boundary, water spreads through branching passages. En route it may follow zigzag and up-and-down paths, here and there siphoning under hydrostatic pressure (that is, the weight of water above it). Eventually the spreading waters reach an outlet at the ground surface, at a lower level, usually in the side of a slope or at its foot; and there water emerges as what is called a rise. Then, after flowing over the surface for a time, the water may find an opening into which it dives for a new journey underground.

In early stages, when routes for circulation are being established and passages are narrow, weathering and solution are the dominant sculpturing processes. When joints have been widened to ¼ inch or more, a fairly steady flow is established, with turbulence that spurs solution. As passages widen and trickles become vigorous streams, corrasion becomes dominant and shafts, tunnels, and caverns open up.

Erosional activity in caves appears to be at maximum near the water table, where water is in continuing contact with rock and circulation is vigorous. Here and there, insoluble portions of rock, such as iron carbonate, bauxite, and quartz, block passages, or poorly soluble strata are encountered, so that spreading water must find other routes, however roundabout. It does find routes, and in time reaches the surface.

If rock structures and other factors are favorable, spreading underground water creates a regional system of tunnels and caverns. Extensive systems, which involve mazes of passages large and small, many of them water-filled, are impossible for spelunkers to explore fully, but the strategic use of dyes may reveal the sources of water and the routes by which it has traveled. Thus it has been possible to map, at least approximately, systems with hundreds of miles of passages.

The popularity of caverns is well deserved. Regrettably, some guides offer the public little more information about cave features than fanciful names like Wedding Cake and Lion's Cage, which convey no understanding of the features' origins. Other guides are well informed and are glad to explain when questions are asked.

The formation of a cavern begins with water moving from the ground surface down through the vadose zone. This zone may reach only a few feet or hundreds of feet down. Its upper

Inviting the Imagination: Like other limestone caverns, the huge Postonja Cave in Slovenia offers countless forms suggesting animate beings. This scene could suggest religious figures standing among various fixtures in a cathedral. Such suggestions help to interest sightseers, but learning how such forms take shape can be more interesting still.

A Cavern Spectacle: Luray Caverns, at Luray, Virginia, is known for fine displays of speleothems. Giant's Hall, shown here, offers unusually spectacular views of stalactites, stalagmites, and columns with capitals suggesting the Corinthian style. Human figures near the bottom of the picture provide scale.

portions may be all but dry, especially if the regional climate has become less humid, as in the case of Carlsbad Caverns. In the vadose zone, rock is eroded and dissolved by descending water, and new rock is created as calcium carbonate is precipitated from solution. Reaching the saturated zone, water comes into contact with all exposed surfaces, and solution and corrasion continue, most rapidly just above, at, or below the water table – wherever circulation itself is most active. It is in this region, mainly, that water flowing through joints and filtering through rock makes tunnels and caves.

Cave formation and enlargement cannot occur much below the cave's water table, which usually is the same as the regional table. Solutional activity can be halted by subsidence of the region if this causes a rise in the water table, and it may resume later if the region is uplifted. The process of cave formation will stop also where solutional activity, progressing downward, encounters an insoluble layer.

Caves are predominantly vertical or horizontal in plan.

Vertical caves develop at the intersection of vertical joints or faults; they are commonly in the vadose zone. Horizontal caves form along bedding planes, mainly in the saturated zone. Among folded rock layers, caves can become highly complex, because of variations in attitudes of bedding and joint planes.

As vertical joints widen, they become shafts. As horizontal joints are opened up, tunnels form. Large cavities called chambers, or rooms, develop where tunnels intersect with caves or other tunnels. As the water table falls, more and more tunnels and caves become exposed to the air, to the processes of deposition, and, incidentally, to the wide eyes of visitors.

Nature's Fantasies:
Deposit Features in Caves

No less remarkable than tunnels and caves are the stone features created in them as calcium carbonate and other min-

Tubes of Travertine: A stalactite fragment *(left)*, recovered from a Kentucky roadcut, shows the interior passage through which solution moved down to the tip. A broomlike cluster *(right)* of stalactites hangs from a cave ceiling in Luray Caverns. Tips of some stalactites have been broken off by thoughtless sightseers.

erals precipitate out of solution. Precipitation occurs rapidly as solution emerges from rock pores, joints, and other openings; precipitation results less from evaporation than from pressure relief, agitation, and warming. The resulting rock, travertine, is finely crystalline, usually dense and massive, often with a concentric fibrous or splintery structure; it may occur also as a spongy mass called calcareous tufa. Colors are usually white to tan, with banding due to oxidation of iron and manganese minerals during pauses in growth.

Deposit features are known as speleothems (from Greek *spelaion*, cave, and *thematikos*, relating to), among which stalactites and stalagmites are most familiar. Deposit features

form not only in caves but above ground, notable examples including the colorful deposits at Mammoth Hot Springs, in Yellowstone, and those at Thermopolis, Wyoming, the largest hot-spring area in the world.

Caves vary widely in their displays of deposit features, for these depend on varying factors such as temperature, wetness of the cave, jointing of the rock, impurities in the rock, and rock structures.

Dripstone

Formation of the kind of travertine called dripstone starts as mineral-bearing solution emerges slowly from an opening

Sites of Limestone and Gypsum Caverns in the United States: Caverns are most numerous in the Folded Appalachians and in relatively high-standing areas of the Midwest. Although Florida is a limestone state, its caves are few, because the terrain is low and water tables are mostly at or near the surface. Hawaii and Alaska have comparatively little limestone.

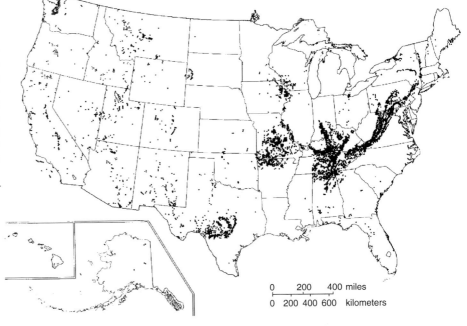

| 0 | 200 | 400 miles |

| 0 | 200 400 600 | kilometers |

"Bacon": Travertine strips in Shenandoah Caverns form as solutions emerge slowly from long fractures in cave walls. Banding results from interruptions of growth during which precipitate changes in composition.

Flowstone and Stalagmites: A flow of solution, being continuous, may produce forms that are wider and smoother than stalactites and stalagmites. Drip from flow forms can build stalagmites and columns, as here in Shenandoah Caverns.

Flowstone Draperies: Flow of solution from long cracks in a cave wall can form what are understandably called draperies. As mineral precipitates out, the flow forms narrow into "spearheads," some of which reach the cave floor to form columns.

in a ceiling, dripping rather than streaming. Pressure on the solution is relieved as the solution emerges, and some carbon dioxide escapes. Precipitation starts because of the lack of carbon dioxide to hold mineral in solution. Evaporation of water in the solution causes further precipitation.

Mineral precipitating around the ceiling opening often forms a ring. As more solution emerges, more mineral is precipitated and the ring elongates downward, becoming a tube. Thus a stalactite is formed and grows. In time, the tube may become blocked; then the solution creeps down the sides of the stalactite, continuing to broaden and lengthen it; or, drip will cease and the stalactite will stop growing.

On occasion, the rate of drip from a stalactite exceeds the rate of precipitation and evaporation, and the excess of drip falls to the floor of the cave, building up a stalagmite. If the drip is very rapid, it may form a pool, and around the pool precipitation and evaporation can create terraces.

Often a stalactite and a stalagmite below it grow until they join, forming a column, or pillar. With prolonged dripping from a cluster of ceiling openings, columns can grow to heights of 50 feet or more and diameters exceeding 25 feet, involving hundreds of tons of rock. Often the form is that of a pillar or pillars ringed by "organ pipes," which emit a deep musical sound when tapped by the cave guide.

The dripstone forms called curtains, draperies, or sheets hang from cave ceilings and on walls. Some start as a row of stalactites growing from a long fracture. As the stalactites grow larger, they join to form the continuous sheet.

Thin, ribbonlike features may take shape where solution emerges gradually from a short, thin slit in a ceiling or wall. These features, "bacon strips," may grow many feet long. With their longitudinal banding they do resemble bacon.

Helictites, usually less than a foot long, originate like stalactites but twist and turn erratically, forming branches extending in all directions. Deformation may be due to any or all of several causes, including blocking of tubes, capillary action (with water moving between rock grains from the inside to the outside of the feature), surface tension (which holds drops on the surface until they evaporate), or drafts of air which turn the evaporating water drops one way or another. An interesting possibility is that where drip is very slow, precipitation occurs at the helictite's tips before drops can form, and the tendency of the mineral to crystallize deforms it. This explanation is plausible because crystals accumulate to form nested cones, and because helictites occur always in dryer parts of caves.

A most delicate and exquisite type of dripstone is "popcorn." This consists of layered nodules, not more than a half

"Parachutes of Stone": The "parachute" is a flowstone feature of travertine projecting from a cave wall. Long, hanging cordlike parts strongly suggest the name. These specimens are in Lehman Caves National Monument, Nevada. Somewhat similar forms, not so strongly suggesting parachutes, are known as "shields" and "palettes."

inch in diameter, on very short, thin stalks. From these nodules crystals of aragonite (a form of calcite) may sprout, making "frostwork." Popcorn may result from splash caused by drip into a pool, or may grow where water seeps from the wall at the edge of a pool. It may also grow out from pores in cavern walls.

Flowstone, Ponds, and Other Cave Features

Flowstone forms from solutions that flow over limestone surfaces. It often builds up on walls as curtains or as "petrified waterfalls." Surfaces may have lapiés. On cave floors, flowstone spreads out to form series of terraces.

Water flowing as a sheet over an inclined limestone surface may dissolve portions of it, leaving depressions known as scallops. These can be a half inch to several feet long. The slower the flow, the larger the scallops. As in wave ripple marks, the steep face of the form is on the upstream side. Scallops develop similarly on some limestone surfaces above ground.

Among the most curious deposit forms is so-called box-

"In All Directions": Helictites may grow in all directions. These fine samples, growing from stalactites in Timpanogos Cave National Monument, in Utah's Wasatch Range, are pure calcite.

Cave "Popcorn," "Snowballs," and "Frostwork": Timpanogos has fine exhibits of travertine with sprouts of aragonite crystals. Such features, remarkably beautiful, occur in many other caves, also.

"Scallops": Water flowing over an inclined cave ceiling in Lehman Caves National Monument, Nevada, dissolved portions of the surface, forming depressions resembling scallop shells.

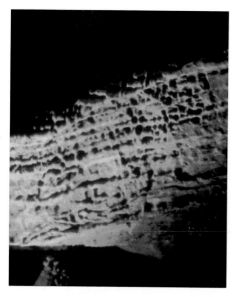

Boxwork: One kind of boxwork *(left)* consists of a "grid" of precipitates from solutions that emerge from crisscrossing fractures. This specimen is in Shenandoah Caverns. Another kind of boxwork *(right)* consists of intersecting plates. These are of material deposited in fractures and remaining after removal of surrounding rock by solution. This specimen is from an industrial excavation in West Virginia.

work. One variety consists of a cluster of thin, intersecting plates spaced inches or fractions of an inch apart. These are calcite fillings deposited by groundwater in fractures in limestone, and later exposed as enclosing, less soluble rock was removed by solution. Boxwork of this kind is considered rare but in the United States is known at various sites, including Wind Cave National Park, South Dakota; Shenandoah Caverns, Virginia; and a site in West Virginia. Another kind of boxwork consists of a checkerboard pattern of small ridges formed of precipitates from solutions emerging very slowly from intersecting fractures.

In dryer parts of caves, air currents crossing a rock surface may draw out tiny drops of solution from small rock pores, in a kind of capillary action. If evaporation occurs as fast as drops emerge, calcite may be deposited in grapelike forms. These can grow on other speleothems, such as stalactites and stalagmites, on sides that face air currents.

In many limestone formations there are deposits of gypsum, and this may appear in caves. It occurs only in the dryer sectors and is never associated with dripstone, the reason being that it is readily soluble in water and will soon disappear under wet conditions.

Calcite "Grapes": Solution that emerges very slowly from pores in a ceiling or wall, or in a speleothem, may evaporate as fast as it emerges. Resulting deposits may have the form of grapes. These are in Shenandoah Caverns.

An Anthodite: Among rare flower forms assumed by gypsum and aragonite crystals in caves is the anthodite (Greek *anthos,* flower). This one, in Skyline Caverns, Virginia, developed around an opening in a cave ceiling.

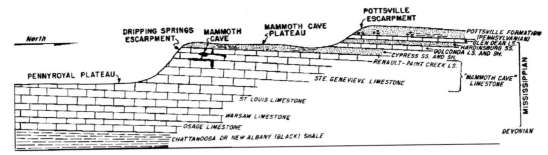

The Mammoth Cave Environment: A north-south cross section shows rock formations in the Mammoth Cave vicinity. The Pennyroyal Plateau, on soft, soluble limestone, is an area of caves and sinkholes; the highlands to the north are protected by sandstone caps. Mammoth Cave has developed behind the border, at the Dripping Springs Escarpment.

As a gypsum solution begins to emerge from a porous ceiling or wall, precipitation starts and crystals form, pushing other crystals ahead of them. In this way successive layers of gypsum grow out from a pore, becoming "blisters." These in time expand and break open to form exquisite white "flowers," "rosettes," or "snowballs," or long, slender, transparent crystals of the mineral selenite. Another gypsum form (one taken by aragonite also) is the anthodite, a cluster of delicate needlelike crystals radiating from a pore.

Among other minerals encountered in caves is manganese, which accumulates as thin, sooty-looking coatings on ceilings and walls. Silicates occur as lenses, filaments, or cavity fillings in limestone; they weather or erode out in various forms, including geodes. Terra rossa (Italian, "red earth"), rich in iron minerals, an insoluble residue from impure limestone solutions, accumulates in joints and may cover cavern floors as an insoluble layer. Another insoluble residue, bauxite, is familiar in limestone caverns of tropical and mid-latitude regions and, commonly, in weathered igneous rocks.

On cave floors usually there are blocks that have fallen from roofs during the past. Some of these weigh hundreds of tons, but large rockfalls occur so rarely that as hazards they are insignificant. Caverns developed for sightseers are inspected regularly; no injury due to a rockfall has ever been reported from these caverns.

Ponds and lakes accumulate in lower portions of caves on insoluble rock strata or where drainage has been blocked by clay or other insoluble debris. Some ponds are dammed and enclosed by rims. These stone barriers, called rimstone, up to a few feet high, may have around their edges scallop forms known as "lily pads." Some cave ponds are habitats for fauna, such as blind fish, that evolved in darkness and never see daylight.

Some cave floors display scatterings of "cave pearls." These are grains of rock, sometimes sandstone, around which travertine has been deposited. They have been rounded by rolling in currents of water.

Two Limestone Caverns

Mammoth Cave

Best known among limestone caverns in the United States is Mammoth Cave, in Kentucky, site of a national park. This is a cavern of the horizontal type, more than 336 miles of which have been explored and 200 or more miles are still beckoning; only about 10 miles are open to the public. The cave system has been hollowed out in Mississippian limestone strata of the Cincinnati Arch, beneath a protective sandstone cap, and the system drains along the gentle northwestward dip of the strata to the Green River, some 300 feet below the cap. Among remarkable forms in the cave are impressive shafts (the tops are "domes") formed by vertical flows; tunnels and rooms, generally following the dip of the rock strata; profound canyons; and arrays of speleothems of almost every known kind. Indian artifacts found in the cave have been dated at more than 4,000 years.

The system is thought to have begun forming as long as 30 million years ago, possibly much later, with an acceleration of erosion during the Pleistocene, when streams of the region were deranged. Erosion continued during the pluvial period following.

A few of the major features in Mammoth Cave are Frozen Niagara, a flowstone mass 75 feet high and 4 feet wide; Mammoth Dome (one of many domes), an erosional cavity 192 feet high; and The Rotunda, the largest of the chambers, or "rooms," 139 feet wide and 40 feet high.

The Chester Upland, on the sandstone layer overlying Mammoth Cave, has numerous sinks resulting from collapse of cave roofs in limestone below. It also has haystack hills – residual limestone masses with resistant caps. South of Mammoth Cave is the Pennyroyal Plateau, sometimes called the Southern Sinkhole Plain, which offers a broad assortment of karst features, including innumerable swallow holes and collapse sinks, sinking streams, dry valleys, and a karst window.

"The Big Room": In Carlsbad Caverns' Big Room, about 4 million years old, some stalactites have grown to form "spearheads" tens of feet long, and others have joined stalagmites to make spectacular columns with conelike bases.

that much rock sculpturing was done by seepage rather than streamflow; hence the multitudes of speleothems. Also, much dissolution was done by water containing sulfuric acid (not carbonic acid) derived from the stratum of gypsum that formerly covered the cave site. The terrain above the caverns is arid and shows few karst features.

Radiometric dating of Carlsbad stalactites has indicated growth at the rate of 1 inch of thickness per 400 years – a rate that may apply in some other caverns, also. However, growth rates of speleothems depend on a multitude of factors and therefore vary widely.

Some of the cave floors are more or less covered with blocks of rock fallen from ceilings. The rockfalls occurred long ago as the climate became gradually dryer, the water table went down, and ceiling rock thus lost support. Among the many large fallen blocks is Giant's Coffin, with a weight estimated at 200,000 tons. Geologists believe no blocks of substantial size have fallen during the past few thousand years; thus the hazard for visitors is minimal.

Principal features in Carlsbad include Bat Cave, above the 200-foot level, which accommodates over a million wing-footed residents; the Big Room, about 750 feet below the surface, which is some 280 feet high and covers 14 acres; and the Lower Cave, at about 850 feet.

Carlsbad Caverns

In the 73 square miles of Carlsbad Caverns National Park, New Mexico, at least 65 caverns have been found. Most remain unexplored; few are open to visitors.

The caverns are in the Guadalupe Escarpment, a mass of Permian reef limestone formed in an arm of the sea here some 200 million years ago, then covered with sediments. Cave-forming started about 60 to 70 million years ago, after the sea had withdrawn but while the reef limestone was still below the water table. During the late Pliocene or early Pleistocene, regional uplift speeded groundwater circulation, solution accelerated, the climate grew dryer, the water table went down gradually, and more and more caverns were left high and dry. Today the water table is almost 600 feet below the deepest part of the caverns; below that level cavern formation may be continuing. The caverns are unusual in

Some Well-known Caverns (in order of measured length)				
Country	Cave Name	State or Region	Maximum Length (miles)	Maximum Depth (feet)
U.S.A.	Mammoth Cave	Kentucky	336	379
Ukraine	Optimisticeskaja	Ternopol	114	66
Switzerland	Holloch	Schwyz	91	2,845
France	Reseau de la Coumo d'Hyouerneo(e)	Haute-Garonne	56	3,294
Mexico	Sistema Purificacion	Tamaulipas	47	2,966
United Kingdom	Easegill System	Yorkshire Dales	41	692
Austria	Hirlatzhöhle	Oberösterreich	35	2,008
Spain	Red del Rio Silencio	Cantabria	33	1,614
Italy	Complesso Fighiera-Farolfi-Antro del Corchia	Toscana	31	3,885
New Zealand	Bulmer Caverns	South Island	22	2,559

Karst Cycle: (**A**) A limestone mass capped by sandstone (top layer) is uplifted. Streamwork accelerates, cutting valleys in the sandstone. (**B**) Water percolating through limestone dissolves out caverns and tunnels. Sinks develop at surface. (**C**) As tunnels and caverns enlarge, sinks and valleys coalesce. The sandstone layer has mostly disappeared. (**D**) On a lowered landscape, a sandstone-capped residual rises above a plain near base level on which streams meander and downcutting has ceased.

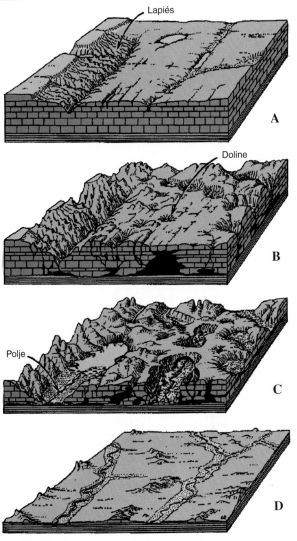

Karst Landscapes Through Time

Rates of land lowering for limestone terranes have been estimated at 15 to 20 inches per 1,000 years under wet and cold climates, 8 inches under periglacial conditions, 4 inches in temperate regions, and 3 inches in the tropics. These rates are, as one might expect, higher than rates for terranes on insoluble rocks.

Karst lands may evolve through more or less predictable stages. The first could begin when, under a humid climate, a well-jointed mass of hard limestone at least 400 feet thick and of regional extent starts a rise of 500 feet or more. As uplift proceeds, percolation of water through joints accelerates. Caves begin to form near the surface, with sinkholes leading down to them. Gradually extensive cave systems develop and sinkholes multiply. Then, as erosion thins the limestone and lowers the land surface to a level near the water table, cave formation ceases. Meanwhile, at the surface, accumulations of insoluble residues are slowing the solution process. The terrane is now patched with low, residual hills, many of them perhaps capped with insoluble crust. With erosion now occurring mainly at the surface, on insoluble rocks, the karst phase wanes and the terrane becomes an ordinary "old" erosional landscape.

This scenario is an attractive one, especially for those who like to think about landscapes in terms of time. It may well apply in some limestone regions despite those many conditions that can upset "orderly evolution" – especially interruptions by insoluble layers within the limestone, episodes of uplift or subsidence, a climate changing to more or less humid, or burial under layers of insoluble sediments. As with other landscapes, interruptions of one kind or another may be the rule rather than the exception.

A "Mature" Karst Landscape: East of Ploce, Croatia, at the edge of the Karst Plateau, erosion of the limestone terrane is far advanced. Note haystack shapes of the hills and the solution valley occupied by a lake. Presence of the lake suggests that the valley bottom is near regional base level or is on relatively insoluble rock, or both.

A Straight Coast: At Oregon Dunes National Recreation Area, Oregon, the beach is remarkably straight for miles, demonstrating the "Pacific-type" coast. Developed on weakly consolidated materials, the beach is wide and gently sloping. Waves come in on a broad wave-cut platform. The line of vegetation at left indicates the limit of ordinary high tides.

A Shore on Glacial Drift: Cape Cod, in Massachusetts, consists of reworked Pleistocene terminal-moraine materials. Along the eastern shore, plantings of beach grass have checked wind erosion, but layers of clay, sand, and gravel are being undercut rapidly by waves and currents. Thus the cliffs are kept steep and keep retreating.

16
Coasts and Shores

COASTS and shores, especially those fronting the sea, are full of meanings for humankind. These lands border an ever-restless, seemingly limitless realm, for thousands of years the domain of monsters and other terrors, and today, even in an age of sophistication, still awe-inspiring. Coasts and shores are scenes of great storms, and of great calms. They are sites of harbors for ships, fishermen, and boaters, for trade between far-flung regions, for ocean travelers going or coming, for hateful villains or innocent maidens to be thrown into the sea. In war they are a vital line for attack or defense. For vacationers they offer frolics on the sand and tanning in the sun. For all, there are walks along cliffs or beaches, sunrises and sunsets over the water, quiet moments for looking out to the edge of the world.

Coasts and shores are realms of obvious change. Storm waves crash against sea cliffs, undermining them and causing rockfalls and occasional landslides, and endangering roads, buildings, and people. Storm waves batter beaches and sometimes, especially at high tide, rush over and beyond them, sweeping away docks, fences, houses, pavements. Not only storm waves, but everyday waves and currents, transport sand alongshore, depleting some beaches and building up others, or perhaps simply moving sand out into deeper water. Meanwhile, wind blows beach sands hither and yon, piling some into dunes, transporting some inland and some out to sea. At the shore we come to appreciate the significance of that familiar phrase: "the restless sea."

A Realm of Change

Like all other terrains, coasts and shores are subject to the familiar processes of weathering, gravity movements, igneous and tectonic activity, and erosion and deposition. Unlike other terrains, shores are subject also to the daily, often energetic action of tides, waves, and currents, which at a relatively high rate erode rocky shores and transport erosion debris from place to place. In these processes, energy is exerted not in one direction, as with streams and glaciers, but alternately in opposite directions, so that erosion and deposition, with the transformation of landforms, can be quite rapid.

The work of tides, waves, and currents themselves is subject to change by a broad variety of events, with durations

Coasts on Volcanic Islands: The Virgin Islands, like others in the West Indies, are of volcanic origin, but much of the coastal areas is covered with coral. Therefore the coasts represent a combination of volcanic and organic types. This view is from Bordeaux Mountain, St. Johns.

ranging from minutes to millions of years. Storms, earthquakes, volcanic eruptions, landslides – all these can have sudden effects on coast and shore erosion. Over the long term, coastal and shore processes change as tectonic activity raises, lowers, and perhaps disrupts the terrain and the sea bottoms near shores, and as ocean currents take new routes as a result of tectonic and igneous activity and climate changes. Coast and shore processes are affected by changes in sea level due to tectonic movements, growth or melting of the world's glaciers, and changes in the density and temperatures of ocean water. Changes in climate due to plate drift, changes in land elevations (the higher, the cooler), and human activities, such as deforestation and the burning of fossil fuels, influence the full spectrum of processes acting to create and modify coasts and shores. These terrains seem subject to a wider and more complex variety of sculpturing events than any other category of landscapes.

"Shore" vs. "Coast"

"Coast" and "shore" are terms often used interchangeably, with little awareness of any difference between them. These features do differ.

Specifically, a shore is a zone, at the edge of an ocean or a lake, which is subject to the regular action of waves, currents, and tides. It is the area between the high-water mark and the low-water mark, and thus every part of it is at times under water. The fluctuating line along which the shore meets the water is the shoreline. An ocean shore extends seaward to the edge of the continental shelf (the submerged edge of the continental block) or the beginning of the continental slope, which extends down into deep water.

Landward from the shore, beyond the usual reach of high water, is the coast. Its boundary on the shore side, known as the coastline, may be either a cliff face or a line marking the inland limit of tidewater. Its boundary on the landward side is usually the edge of a highland or some other kind of terrain distinct from the shore; but some coastal borders are without clear distinction. Many coasts are sea bottoms uplifted to become dry land, and thus may exhibit features characteristic of shores even though the sea never reaches them now.

Origins of Coasts and Shores

Coasts can be classified in a score of ways. Here they are grouped according to their original nature and geomorphic events that have shaped them. Relatively new coasts are called "primary"; they have been little modified by marine processes. They include (1) subaerially eroded, then submerged terrains (such as the Maine coast); (2) terrains of material deposited by running water (Gulf Coast), glaciers (Cape Cod and Long Island), wind (desert shores), and mass movements (some cliffy shores); (3) lands on volcanic materials (Hawaii); and (4) terrains previously shaped by tectonic events (California and Adriatic coasts). "Secondary" coasts are (1) those already eroded by waves and currents (England's southeast coast), (2) lands formed by marine depo-

Sea Cliffs of Chalk: Southeastern England's coast is famous for chalk cliffs. The chalk is firm and lacks joints; little water can percolate through it; thus it can stand as cliffs. Chalk falling to the beach is soon decomposed by salt water and disintegrated by wave action; thus there is little talus to hinder undercutting, which keeps the cliffs steep. The scene is near Dover, on the Sussex Coast.

sition and uplift (Florida), and (3) terrains built by marine organisms (mangrove islands, atolls, barrier reefs).

Not to tax one's memory, but to suggest again the variety of conditions that influence coasts and shores, other modes of classification are worth mention. A coast-and-shore topography can be classified as to the amount of energy at work on it, its degree of stability, the materials that constitute it, its geometric pattern, whether it is a feature mainly of erosion or deposition, its status in an erosion cycle, whether it is emerging from the sea or submerging (and, if so, whether its present characteristics are mostly those of emergence or submergence), by what agencies it was formed, and – so on!

Whatever the classifications, when we look at a coast or shore some awareness of its "heredity" and its "experience" can add much to the enjoyment of the scenery.

Shore Configurations

Configurations of shores – their straightness or irregularities – depend on their origins and on the nature of the shaping processes that have been at work.

Bays along a rocky coast generally are where valleys meet the sea, and headlands are divides between valleys. Valley orientation is usually influenced by the direction of the structural "grain" (if any) of the coastal rock. (An example of such structure is the north-northeast to south-southwest direction of the fold axes of the Folded Appalachians.) If the grain is mostly parallel to the coastline, as along the Oregon coast, the coastline is likely to be relatively smooth, straight or gently curving, and indented by the mouths of few river valleys, because most valleys are subsequent and follow the grain. Such coastlines are said to be of the Pacific type. If, on the other hand, the grain is at an angle to the shore, as in Maine, Norway, or the Atlantic coast of Spain, many more valleys will reach the shore and the shore will have closely

spaced bays. These coastlines are said to be of the Atlantic, or "ria" type (Spanish *ria*, a long, narrow inlet).

The rock of some shores lacks grain, and coastline characteristics are determined by other conditions. Shores on lava flows, as in Hawaii, tend to be arcuate seaward, because fronts of flows usually are arcuate. Shores on birdfoot deltas, such as the Mississippi Delta, are much indented; those on alluvial fans are usually smooth and arcuate, as is the Nile Delta. Coastlines on glacial drift are determined by the forms of the drift; for example, shores on the drumlins of Boston,

A Coast on Marine Deposits: Florida is on a mass of limestone formed on the sea bottom and uplifted as a broad anticline, with a north-south axis, during the past few million years. This view north from over Miami shows the near-flatness of Florida and part of the barrier island along its eastern shore.

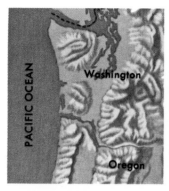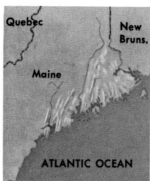

Pacific-type Coast   Atlantic-type Coast

Coast Profiles: Where the grain, or lineation, of the regional rock structure is approximately parallel to a coastline, the coastline is said to be of the Pacific type. Where the grain is at a wide angle and the coastline is indented by major valleys, the coastline is of the Atlantic type.

Massachusetts. On lands like former Yugoslavia's Adriatic Coast, coastline contours are determined by fault patterns. Coastlines on extensive, deep masses of sediments, as along the coastal plain from New Jersey to the Carolinas, are characterized by numerous islands of sand, most of them elongated parallel to the shore.

Ocean Water as a Land Sculptor
Tides

Tides are created by the gravitational pull of the Moon, mainly, and of the Sun. High tide and low tide each occur twice in about 12½ hours, high tide at a given location being when the Moon is overhead and low tide when it is at either horizon. Spring tide and neap tide, which are the highest and lowest tides, respectively, occur twice during the lunar month, which is about 27½ days. Spring tide comes when Moon and Earth are lined up with the Sun and thus the Moon's pull is reinforced by the Sun's pull. Neap tides occur when the Sun-to-Earth direction is at right angles to the Moon-to-Earth direction; then the Moon's pull is at minimum strength. Spring and neap tides at any given place have a range about 20 per cent greater or less, respectively, than the average high tide.

As tides rise and fall, seawater moves inland through tidal channels, then withdraws. The greater and more rapid the change of water level, the greater the velocity of the stream, the erosive effect, and the amount of material transported and deposited. Tidal action is noticeable mostly on coastal lowlands beyond the reach of breakers; there tidewater meanders and forms pools. The vertical range of tides varies according to the size, surface shape, and bottom topography of the basin in which tidal movement is occurring. In the spacious central Pacific area, the range is no more than about a foot;

A Rugged Shore: At Newton Head, County Waterford, Eire, the coast is on strongly deformed, diversified rocks. Because the foliation dips landward, the cliffs remain steep as undercutting by waves proceeds. This is a coast of submergence; waves come in over deep water without forming breakers.

High and Low Tide: At Sandy Cove, Digby Neck, Nova Scotia, the average tide range is about 18 feet. The small fishing fleet is piered in the cove where there is always, or nearly always, enough water under the keels. The cove's floor is on deposits of glacial drift.

in the relatively small, shallow North Sea, about 12 feet. In the Bay of Fundy, where the basin narrows in the direction of tidal movement, the range reaches as much as 45 feet.

Tidal ranges are affected by all or most of the many factors already indicated as influencing shore forms. Also they are affected by changes in atmospheric pressure and in the density and volume of seawater, and by earthquakes, variations in ocean-current velocities, and the growing or shrinking of the world's glaciers. Storm surges, such as the heaping up of ocean water by hurricane winds, also may affect tides. In fact, any of these factors in itself can somewhat alter sea level and thus affect erosional and depositional events.

Tsunamis

An ocean wave or series of waves produced by a sudden upheaval, slumping, sliding, or subsidence of a part of the sea bottom, due to a disturbance such as an earthquake, a volcanic eruption, or a fault movement, is a tsunami (tsoo-NAH-mee) – Japanese for "storm wave." Tsunamis, often incorrectly termed "tidal waves," occur continually, perhaps every day, somewhere – and a few times a year some coasts are hit by a really big one. At sea the wave can be up to 15 feet high, but is so extended that it can take hours to pass a given point, and thus may be hardly noticeable from ships.

As it approaches land, however, and begins to meet the resistance of the sea bottom, it can pile up to a height of 90 feet or so and rush against the shore at 30 miles per hour. Most common in the Pacific area, tsunamis such as the one produced by the great Krakatoa volcanic explosion of 1883 often sweep entirely over islands, causing great loss of life and property, and incidentally radically reshaping sandy shores. In October 1998 a quake-triggered tsunami, with waves averaging 30 feet in height, hit the coast of Papua New Guinea, carrying away an estimated 2,500 people.

A Stream of Tidewater: The erosive ability of the stream on this beach is suggested by the turbulence of the rushing water and the creation of the undercut *(left)* and slipoff slopes.

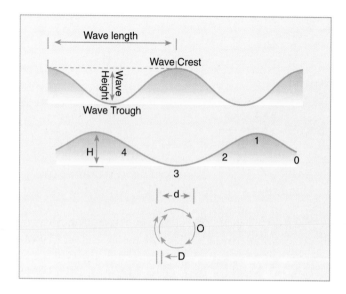

Waves Made by Wind

However great the destructive power of a tsunami, the land-sculpturing action of wind-made ocean waves is greater, simply because there are more of them and they are continuously at work. During hurricanes and other great storms, rock in cliffs is broken up, beaches are radically reshaped, and huge amounts of rock debris are swept along the shore. Waves of lesser storms and ordinary "windy days" also are destructive. Given time, ordinary waves are impressively efficient shore sculptors.

Ordinary ocean waves are described in terms of height

Wave Mechanics: Wave work on shores depends largely on the forward movement of water particles and on the breaking of waves as they enter shallows. **D** is the progress made by a water particle **O** while the wave form is traversing the distance **d**. **H** is the wave height.

(distance from trough to crest), wavelength (distance from crest to crest), period (time required to pass a given point), steepness, ratio of height to water depth, energy per unit of width per wavelength, and wave power (energy of flow). Generally these characteristics are determined by fetch (distance over which the wind is acting) and wind velocity. Over long ocean fetches, wavelengths may reach 2,000 feet and periods more than 20 seconds; over short fetches, corresponding figures are more like 20 feet and 2 seconds.

In deep water, the motion of wind-made waves is not a simple forward motion; rather, each particle in a wave moves in an orbit – forward, down, back, and up again. With each revolution the particle makes a slight forward gain. The wavelength is relatively long, the wave front not steep. However, as the wave nears shore and begins to "feel bottom," the orbit of the water particles becomes elliptical. The wave loses velocity and, because of the push of water behind it, its wavelength decreases and its height increases.

Advancing to the shore, the wave breaks. If its front is steep and the wind is onshore, water tends to spill down the front, forming a spilling breaker. If the wave is a swell (a far-traveled, relatively long wave without a crest), the water

Wind Power and Wave Power: Along 17-Mile Drive at Monterey, California, scenery is spectacular on a windy day. Breakers crash against the volcanic rocks with thundering sound and extraordinary erosive effect, and the air fills with spray.

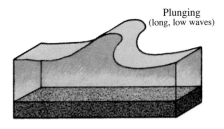

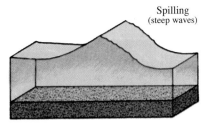

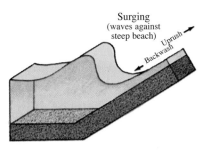

Plunging
(long, low waves)

Spilling
(steep waves)

Surging
(waves against
steep beach)
Uprush
Backwash

Kinds of Breakers: A plunging breaker (*left*) is produced by a long, low wave called a swell. A spilling breaker (*middle*) is made by a wave with a steep front, when wind is onshore. A surging breaker (*right*) is produced by a wave moving up a steep beach.

piles up and, pushed by water behind it, plunges forward, beyond the water beneath it, thus forming a plunging breaker. This type is common on days of offshore winds. In surging breakers, which are a third type, the base of the wave moves upbeach while the crest collapses over it. Surging breakers are common on beaches of steep gradient.

Shore Currents

Reflected, or turned back, by the beach slope, water from waves becomes backwash, flowing seaward. As it merges with incoming waves, some water spreads sideward and meets other sideward-moving water, and the combined waters form an elongated cell from which water flows seaward as a rip current. A rip current (also called sea puss or, inaccurately, "rip tide") extends to the so-called rip end, as much as a half mile offshore, where the water disperses in various directions.

Meanwhile, some water from backwash and incoming waves flows sideward, parallel to the shore, as longshore (littoral) currents. These are created in part by waves meeting the shore obliquely. Longshore currents dig out sediments, carry them away, and thus make longshore troughs. Longshore currents commonly feed into rip currents, mainly those on the downwind side.

Some swimmers may be glad to hear that the current called "undertow," waiting malevolently to pull swimmers under, probably does not exist. Some swimmers do get carried out by vigorous backwash or a rip current. The proper action then is to swim sideward, out of the current.

Sand and coarser sediments are transported upbeach by breakers and swash, downbeach by backwash, and parallel to the shore by longshore currents. When waves are at an angle to the shoreline, longshore currents are in the downwind direction, carrying sediments with them. Downwind movement of sediments, called beach drift, results in depletion of some beaches and the building up of others. To check drift, engineers build groins (sturdy rock walls perpendicular or at an angle to the shoreline), but although these help to save beaches on the upcurrent side of the groins they deplete beaches on the downcurrent side. Beach drift in the direction of prevailing winds is so persistent, and competition for sand

Changing Waves: South of Point Sur, California, swells form plunging breakers, which in turn become translational waves. These crumble and make swash on the beach. The cliffs are eroded by storm waves only, and the vertical portions are eroded most rapidly, by undercutting.

Waves at an Angle: At a beach on the east side of Cape Cod, swells are expending much energy along the shore as well as against it; thus sand is moved downwind (leftward) along the beach by longshore currents, and longshore bars are built up while the cliff is being cut back, about 3 feet per year.

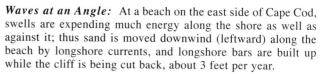

Shore Erosion by Marine Processes
Erosion of Rock Cliffs

The side of any highland can become a sea cliff when the ocean, advancing inland by prolonged erosional action of waves and currents, added to the usual processes of subaerial erosion, reaches the highland. Cliffs can be created when a shore rises and what was sea bottom becomes land subject to marine erosion, as along most of the California coast in recent time. Where land has subsided, as the coast of Maine did under Pleistocene ice, the sea can move in; thus the shoreline here is on river-valley walls. Sea cliffs are made also when glaciers erode coastal valleys below sea level, making them fiords. Lava pouring from volcanoes and rifts, as in Hawaii and New Zealand, builds up and makes cliffs along shores. Reefs constructed underwater by corals and other marine organisms become cliffy islands if sea level goes down, as happened when the growth of Pleistocene glaciers depleted the ocean.

Sea cliffs are subject to chemical and mechanical weathering, mass movements, slopewash, and glacial and wind erosion. Ordinary weathering effects are augmented by solution by wave splash, which incidentally, with its salt content, has a high capacity for dissolving limestone. Cliffs are subject

so intense among property owners, that coastal experts often recommend a do-nothing policy, allowing nature to take its course.

Waves coming up against a rock cliff that rises above deep water do not form breakers like those on a beach. They simply break against the cliff, causing more or less rock wear-and-tear, and the water falls back, forming rip and longshore currents. If the water is not too deep, these currents may form longshore bars and troughs, and bottom sediments may be moved alongshore in the downwind direction.

A Blocky Shore: Much of Mt. Desert Island's east shore is on granite, so resistant that marine erosion is slow to reduce it to sand. Blocks are parts of granite sheets disintegrated along joint planes. In heavy storms when waves are high, blocks are buoyed up by immersion and thus can be moved by the powerful waves. Because storm waves have more velocity than backwash does, blocks carried far up-beach tend to stay there.

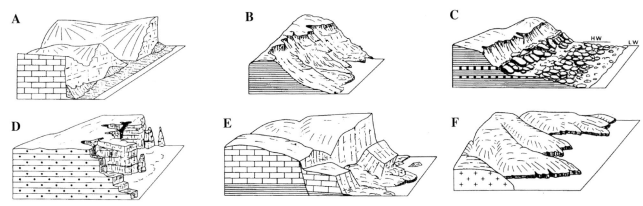

A B C

D E F

An Assortment of Cliffs: Cliff forms vary endlessly, but six kinds shown here, with different rock types and structures, can represent a wide variety: (A) and (B) – Chalk and horizontally bedded sandstone, respectively. (C) – Weak cliffs of shale and clay. (D) – Chalk overlying weak clay. (E) – Interbedded sandstones and shales. (F) – A complex of ancient rocks after prolonged marine and subaerial erosion and a change in sea level. (From Chorley et al., *Geomorphology*, © 1984 R.J. Chorley et al., Methuen & Co., used by permission of Methuen & Co.)

also to attack by boring organisms, such as oysters and clams, sponges, and barnacles. But all these together may not equal the erosional work done by waves.

Storm waves breaking against a cliff face exert pressures that have been measured at 1,241 pounds per square foot. Such pressure, called wave shock, loosens blocks of rock along joints and bedding planes, and blocks fall to the base of the cliff, there to be ground up by the action of other rock fragments as these are tossed about by waves.

Incoming storm waves lift fallen rock fragments and fling them against the cliff. Surprisingly large fragments can be lifted, because of the buoying effect of immersion; thus a block of granite one cubic foot in size, weighing perhaps 168 pounds, is buoyed up by a force equal to the weight of a cubic foot of seawater, some 63 pounds. But when the block is hurled against a cliff, the impact may involve the full 168 pounds. With waves reaching velocities as high as 20 miles per hour, such impacts, known as water hammer, can be impressively destructive.

A wave entering a joint suddenly compresses air ahead of it, creating high pressure on the sides of the joint. In a relatively large space, the compression produces a booming sound that may be heard more than a mile away. As the wave recedes in one of these "thunderholes," compression gives way to suction, and this, in the process called quarrying, or plucking, pulls out loosened blocks. Wave following wave, large masses of rock are loosened along joints and, fragment by fragment, fall from the cliff.

Sea cliffs are subject also to abrasion. Rock fragments are scraped against them by waves and currents much as a stream uses rock fragments to cut a channel in bedrock.

Assaults on sea cliffs by all erosional agents concentrate in the zone of wave action. Thus cliffs at or near the water level are subjected to continual undercutting. As the cliff base is cut back, overhangs develop, rockfalls occur, and the entire cliff gradually retreats.

Cliff Profiles

Sea cliffs on massive, relatively resistant rock tend to be steep because the effects of weathering, mass movements,

Influence of Jointing: Joints in the mostly granitic, resistant cliff rock at Otter Point, Mt. Desert Island, are predominantly almost vertical. As waves undercut the cliff, blocks loosen along joint planes and fall, keeping the cliff vertical.

Dismembering a Tuff Cone: On Isabela Island, in the Galápagos, marine action has cut away two-thirds of a tuff cone, exposing its layered interior structure. At far right, on the tuff, is the edge of a basalt lava flow.

and slopewash are subdued relative to undercutting. Relatively little material falls to the cliff base; thus this part of the cliff lacks protection from undercutting. Undercutting results in overhangs, and as these break off and fall, the cliff retreats and remains steep.

A sea cliff on loose material, such as glacial drift, is likely to have an even slope at about the angle of repose, around 35 degrees, which is maintained by mass movements as waves do their undercutting at the cliff base. A cliff on horizontal strata of weak rock such as shale will be somewhat steeper, with a profile more or less uneven because resistance of the strata will vary.

Predominantly vertical jointing in resistant rock favors a vertical cliff with a smooth front. If jointing and bedding planes are mostly horizontal, the cliff profile will be stepped. Inclined strata will yield a profile like a tilted stairway. Where rock structures are more or less chaotic, as in strongly folded and faulted rocks of different types, or jumbled lava masses, profiles likewise will be chaotic.

Cliff profiles vary partly according to climate. In the north temperate zone, cliff steepness is favored because of undercutting by waves driven by the relatively strong prevailing westerly winds. In the humid tropics, low-angle (gently sloping) cliffs are favored because they tend to be protected from erosion by dense vegetation, and because waves, driven by the weaker winds of low latitudes, have less energy for undercutting. Where desert cliffs front the sea they are likely to be steep because desert weathering is subdued and vegetation for slowing mass movements is lacking. Sea cliffs at high latitudes tend to be at a low angle because talus from the high rate of frost action protects the cliff base and sea ice limits wave action. Obviously, profile characteristics result from many causes, and exceptions to what one might expect are numerous.

Coast Erosion in Chalk: France's Normandy Coast is noted for coastal scenery of chalk. Arches here will eventually collapse, leaving stacks. This coast is part of the broad Wealden Dome, which underlies southeastern England, the Dover Straits (which bisect the dome), and northern France.

Undercutting by Marine Action: Along this shore at Corcovado, Costa Rica, Pacific waves and currents have undercut a terrane on weak, much-weathered volcanic rocks. With each tsunami or substantial storm, the shoreline retreats.

Sea Cliffs in Retreat

Rates of cliff retreat vary substantially. Cliffs on rock that is chemically or mechanically weak, such as a limestone or shale, are likely to retreat relatively fast. So may cliffs on hard rock, such as granite or basalt, if these are closely jointed and thus the more vulnerable to wave action. Joints and bedding planes inclining downward toward the cliff face favor slides and rockfalls. A high cliff may retreat slowly because, being high, it has a larger area from which rock fragments can fall, and thus there can be more talus protecting the cliff base against undercutting. Cliffs that face prevailing winds may retreat relatively fast because of more sustained wave action. In cold northern latitudes, cliff retreat may be rapid because tides are greater and swifter there and have a wider range, thus allowing wave action over a larger area; yet in the same regions retreat can be slowed because of large accumulations of talus. Retreat may be faster if waves approach the shore over a wave-cut platform which is inclined steeply seaward and thus causes waves to build higher.

Wave-cut Platforms

As a rock cliff retreats, it leaves behind a lengthening underwater rock platform which inclines gently seaward.

Variously called a wave-cut platform, wave-cut bench, wave-cut terrace, or marine terrace, it is the surface on which ocean waves become breakers and move up to the cliff. Usually not more than 30 feet beneath the ocean surface at its seaward end, it is covered more or less with erosional sediments from the coast and shore; these may be swept away by waves and currents during storms, and replaced by new sediments during quieter periods.

Wave-cut platforms usually are gently sloping and extend

A Chaotic Shore: Near Otter Cliffs, on Mt. Desert Island, rock structures are chaotic. The native granite and diorite masses are intricately mixed and strongly deformed; thus erosion of the shore has produced rugged, irregular forms.

far out under the water if the rock is weak; they are steep and narrow if the rock is strong. Some platforms are lower than normal because of shore subsidence or a rise in sea level, as has occurred with the world-wide melting of Pleistocene glaciers; other platforms are now high and dry because of shore uplift, as along much of the California coast. Because of wide fluctuations of sea level during the past few million years, especially as a result of Pleistocene events, the conditions for sea-cliff erosion have changed substantially, and wave-cut platforms as found today are not what they would be, in elevation or otherwise, if sea level had been constant.

Wave-cut platforms now above sea level, indicating uplifts in recent geologic time, are numerous. Examples include those on the Pacific Coast from Washington to southern California, ranging up to 1,700 feet above present sea level on Santa Catalina Island. The platform on Bainbridge Island, in Puget Sound, Washington, is known (by dating of wood and vegetation) to have been raised 21 feet in an earthquake 1,000 to 1,100 years ago. A rise of 10 feet has occurred in the crust beneath Bermuda, southern Florida, North Carolina, and Louisiana during the past 4,000 years. Along shores of the northwestern Mediterranean, platforms indicate uplifts at the rate of nearly 3 inches annually between 300 B.C. and A.D. 150. Wave-cut platforms on New Zealand's Southern Alps indicate an uplift of as much as 5,500 feet during the past 135,000 to 140,000 years – a result of the continuing collision between the Australian and Pacific plates.

Minor Features of Cliff Erosion

Erosional sculptures along cliffs reflect rock types and rock structures as well as various kinds of marine processes, and perhaps tectonic movements also. Generally speaking, the denser the jointing, the more acute the deformations; and the faster the cliffs retreat, the greater is the number and variety of sculptures.

Among the commonest cliff features is the wave-cut notch. This indentation at the base of a cliff is made within the tidal range. Usually it is best developed in cliffs of headlands, because these are subject to wave attack against the sides as well as the front. On shores uplifted in recent time,

An Uplifted Wave-cut Platform: The view south from Otter Crest, Oregon, includes an uplifted wave-cut platform. Buildings *(center)* are on parts of the platform; the top surface of the stack *(foreground)* is another platform remnant. Uplifted areas can be traced along the shore into the far background. The small ridges *(low center)* pointing toward the cliff base are portions of dikes exposed by marine erosion.

Exploiting a Weak Zone: Parts of the eastern shore of Mt. Desert Island, Maine, are in a "shatter zone," where masses of ancient metamorphic rocks and diorite were broken up and mixed with invading granite. Anemone Cave, shown here, and other caves have been hollowed out by marine erosion in this zone.

notches may be seen well above the present sea level and are evidence of the uplift. The Hawaiian Islands offer many examples.

Rocky shores frequently have the long, narrow indentations called reentrants, resulting from differential erosion. Some reentrants are mouths of valleys; some develop in joints that intersect the shoreline; others are made in masses of weak rock that likewise intersect the shoreline, such as basic dikes in granite, or crushed rock in a fault zone.

Marine erosion can produce hanging valleys. These are stream-cut valleys that end at the top of sea cliffs. A valley is kept hanging because undercutting by marine action makes the cliff retreat faster than the stream cuts down. This special type of hanging valley is sometimes called a valleuse – a French term for the features seen along some French coasts. Fine examples occur also along the coast of Canada's Cape Breton Island. Hanging valleys on some sea cliffs result not from marine processes but from glacial action, as in fiords; from faulting, as on the Adriatic Coasts; or from volcanic activity, as on Icelandic and Hawaiian shores.

Sea caves, or grottoes, are hollowed out by marine action at joint intersections and in zones or pockets of relatively weak rock. The noted grottoes of Capri, Italy, are in limestone; those of Hawaii are in lava rock of uneven texture and resistance; and those of Mt. Desert Island, Maine, are in masses of basic igneous rock occurring in granite, and in zones of rock shattered by tectonic activity.

A Coastal Hanging Valley: Wave and current action along the Na Pali coast of Kauai Island, Hawaii, undercuts old lava flows and beds of pyroclastics, creating vertical cliffs. Undercutting here has cut off the end of a small valley and left it hanging as a "valleuse." Sea caves are numerous along Hawaiian shores because of the relatively high erodibility of lava rock and the unevenness of its resistance.

Remnants of Dikes: At Seal Rocks State Park, Oregon, the linear arrangement of stacks suggests a dike, perhaps with one or two branches. The large, columned stack in the foreground has been strongly undercut. Rockfalls from it are occurring faster than waves are disintegrating and removing the debris. The human figure *(bottom center)* provides scale.

Beach Sculptures in Conglomerate: Sea cliffs of relatively weak rock are especially subject to undercutting, which may carve out arches and stacks. These examples are at The Rocks Provincial Park, New Brunswick, on the Bay of Fundy, where wave action is reinforced by high, powerful tides.

Sculptured in Pyroclastics: At Santiago Island, in the Galápagos, waves and currents have shaped layers of tuff into a sea arch, a wave-cut notch, and a wave-cut platform. Eventually erosion will probably reduce the arch to a stack or pair of stacks.

Now and then part of the roof of a sea cave collapses, and the resulting opening becomes a blowhole. As a wave rushes into the cave, it compresses the air, and this rushes out through the roof opening with spray and a booming or whooshing sound. As the wave withdraws, the sound may repeat. Blowholes develop often in shores on lava, as in Hawaii and the Galápagos Islands.

A headland, or promontory, is especially vulnerable to marine erosion because of being exposed on three sides. Narrow headlands are likely to be cut through by erosion in weak zones; thus an arch may be formed. Most such features are in small promontories and are limited to widths of 50 feet or so, but some are much larger. Scenic sea arches occur along the lava coast of Washington state, and along the limestone coasts of southeastern England and Normandy.

When the top of a sea arch collapses, its "legs" remain as columns or pillars called sea stacks. Stacks form also when marine attack cuts through a headland at several locations without making arches. Some stacks are resistant portions of wave-cut benches still being cut; others rise above uplifted benches. Still other stacks are remnants of basalt columns, as along Scotland's shores, and some consist of dikes that remain standing after surrounding pyroclastics have been eroded away, as along coasts of Oregon and Washington state.

A Reentrant in a Dike: Here a basaltic dike in a granite sea cliff at Schoodic Point, Maine, is being eroded out; thus a reentrant is being formed. Some reentrants are "thunder holes," in which invading waves produce a loud booming sound.

A Hawaiian Blowhole in Action: At Spouting Horn, Poipu Beach State Park, Kauai, waves and currents hollowed out a lava cliff to make a sea cave. Part of the cave roof collapsed, leaving a hole. Now, when a wave crashes into the cave, air is blown out through the hole with spray and a thundering sound.

On Deformed Strata - Hardly a Beach: At Ilfracombe in Devon, England, a wave-cut platform has been developed on strata of schist that dip landward. Ridges on the platform are edges of tilted layers that are relatively resistant. The schist is tough, and sand is produced at a relatively low rate.

An Evolving Beach: The island of Barbados, consisting mostly of coral limestone, is ringed by cliffs retreating before the onslaughts of the sea. Remnants of former cliffs rise as stacks above beaches-in-the-making. Here a stack encircled by a wave-cut notch stands on a portion of wave-cut platform. Beyond are other cliff remnants, with profiles that are rough and irregular probably because of the uneven composition and structure of the coral rock.

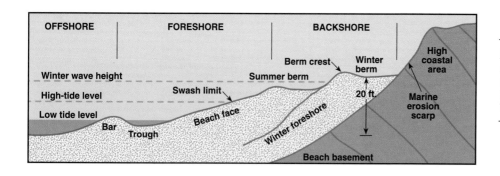

OFFSHORE			FORESHORE		BACKSHORE		

Profile of a Sand Beach: The zone represented here abuts a high coastal area and includes most of the features likely to be seen on a well-developed beach.

Sandy Shores

Some shores are on broad masses of sediments left by the erosion and retreat of sea cliffs, as along the coasts of California and Oregon. Sandy shores of the Chesapeake Bay area and southeastern England are on portions of sediment-covered continental shelf that have been uplifted; that is, they are on parts of coastal plains. Many sandy shores are on edges of deltas, such as that of the Mississippi River. Others originate on reworked deposits of glacial drift, notably those as on Long Island and Cape Cod, or originate on erosional debris from nearby highlands, or on masses of pyroclastics.

In many sectors, sediment consists of quartz and feldspar debris from nearby highlands. Sands of some shores, as on coasts of France and Italy, are of limestone. Sands may be of basalt, as on the Hawaiian and other volcanic islands; still others are of coral, as in the West Indies and islands of the southwestern Pacific.

Shores on loose material naturally are very vulnerable to marine attack. During the past century parts of the east shore of Cape Cod, exposed to this attack on all sides, have retreated as much as 50 feet in a single storm. Some of this sand has been added to the hook at the north end of the cape and much has been added in the south, below the elbow. From New Jersey to Florida, and all along the Gulf Coast, the sandy shores are subject to continuous change, being depleted of sand at some locations and built seaward at others.

The Anatomy of Beaches

Along rocky shores, beaches may be developed in indentations where sediments moving alongshore become trapped. Beaches develop mainly on shores consisting entirely or mostly of sediments. Beaches are divided into zones according to processes dominant in each.

On a fully developed beach the zone farthest to seaward is called the offshore area. This begins at the low-water mark or, according to some definitions, at the edge of the continental shelf. The offshore zone is always or almost always under water; its sediments are mostly of sand size.

The breaker zone is where incoming waves break over a longshore bar or bars. Such bars are ridges, usually of sand, parallel to the shore. They form usually where sand moving onshore meets sand being moved seaward by backwash; thus some sand moves sideward, forming the bars. Bar building commonly occurs during storms, and bars made during one storm may be erased by the next. The breaker zone is where movement and turbulence are greatest and finer sediments are carried away; thus in this zone sediments are coarse, usually including much gravel and perhaps some cobbles.

Landward from the breaker area is the surf zone. This includes a longshore trough ("runnel") or troughs bordering the longshore bars. Some water from breaking waves plunges into the trough and moves upbeach as translational waves, without rotation. Other water, meeting backwash from the upper beach, spreads sideward, flowing as a longshore current

Makings for a Beach: Sand Beach, on Maine's Mt. Desert Island, is a bayhead beach fronting on a small delta built by a stream from the nearby highlands. Large rock fragments eroded from the cliff (not visible) in the center foreground make the nearer part of the feature a boulder beach.

A Winter Berm: The winter berm, or storm beach, is not always of coarse sediments. This ridge at Chincoteague State Park, Virginia, on a barrier island, is of sand. Beachgrass plantings and a snow fence retard loss of sand by wind action.

in the trough. The materials in this zone tend to be coarse.

The swash zone is the area, usually steeply sloping, up which the translational water rushes, spreads, reaches to or near the high-water mark, and reverses direction, becoming backwash. Because translational waves slow as they climb the slope, the sediments they can carry are only of sand size. However, this kind of wave does have higher velocity than backwash, especially when the tide is rising; therefore some of the material the wave moves up beach is not carried back by backwash, but remains to build up the beach. In storms, waves are higher, move farther up beach, and deposit more material (including larger calibers); but as the storm wanes, backwash carries much of the material back again.

At the upper end of the swash zone there may be a beach scarp, a few inches to a few feet high, with a vertical front, cut into a steep slope by wave action. Whether or not a scarp will be cut depends mainly on the caliber of the sediments and the nature of the wave action.

Landward from the swash zone is the backshore: a narrow strip bounded on the upper side by the coastline. It is the zone reached only by the highest tides and waves, usually those of winter storms. Here is the berm, a ridge of material deposited by waves where they stop, before receding. Berms may be multiple, separated by beach scarps. Berms are like-

Sorted Sediments: At Hampton Beach, New Hampshire, eroded masses of schist crop out above the sand *(background).* Landward from the sand strip, seaweed left by swash shows the upper limit of the last high tide. Farther landward is a mix of seaweed and cobbles deposited probably during a recent storm. In the foreground is part of the beach ridge, consisting of shingle, marking the highest reach of storm waves.

A Boulder Beach: A beach on Cape Breton Island, Nova Scotia, is partly covered with boulders. Some have weathered off the low cliff in the background; others were deposited by melting glacier ice; most have been more or less rounded by tumbling in the waves.

ly to include very coarse materials – even boulders, which only very powerful storm waves can transport far up the beach. Often there is a summer berm, made by the highest waves of summer, with a winter berm above it.

The steepness of beaches commonly depends on the caliber of the sediments and on the vigor of wave action. The larger the caliber of sediments, the steeper the beach tends to be, because larger calibers are more difficult for waves to move far up the beach slope. Unusually powerful waves can move cobbles and boulders far up the beach and thus widen it.

On a relatively youthful beach being formed of sediments from cliff erosion there usually are many large, rugged rock fragments; time has not been sufficient to smooth them and reduce them in size. If the cliffs are of layered rock, the fragments tend to be flat. Flatness may occur also because rock fragments being worked by waves slide rather than roll; but, eventually, rolling may produce rounding anyway. Beaches of flattened rocks, and sometimes of rounded rocks larger than gravel, are called shingle.

Rock fragments that are blocky and crowded together become more or less rounded by tumbling; thus the term boulder beach may apply. Cobble beaches, with rounded stones smaller than boulders, are intermediate. Generally, as mentioned, beaches of large-caliber material are narrow and steep.

A Shingle Beach: On Espanola Island, in the Galápagos, sediment-laden storm waves washing over remnants of a basalt lava flow and causing them to slide up and down a steep beach have smoothed and flattened them, creating a rough pavement. The slabs average 4 to 6 feet in length, 1 to 2 feet in thickness. On most shingle beaches rock fragments are much smaller.

Made by Currents: On a Florida beach, these ridges and troughs were made by a current parallel to the shoreline. They are up to 5 feet long. Typically such marks are slightly flattened along crests and in troughs.

Preserved for the Future: Ripple marks can become fossilized by burial and lithification under accumulating sediments. These marks, on rocks in the Maligne Range of Canada's Rockies, were made by water currents on an ancient beach.

Minor Sedimentary Features on Beaches

Doubtless the commonest marks on beach faces are those left by swash. The swash mark is a low, thin, wavy or arc-shaped ridge of fine sand, usually less than 0.1 inch high, deposited at the upper limit of swash. Often detritus such as seaweed is deposited with it. As the tide goes down, rows of swash marks are left on the beach face; as the tide rises and waves reach higher, the marks are erased.

Much water from swash or seepage moves down the beach face as small streams called rills. These cut miniature channels known as rill marks, or runnels, up to about 0.5 inch wide, 1 to 2 feet long, and .04 inch deep. Like swash marks, rill marks are soon erased by new swash.

Back-and-forth wave action on sand often produces oscillation ripples, also called wave marks, usually but not always in shallows. The marks are sets of small, undulating, nearly parallel, somewhat irregular ridges perpendicular to the flow direction, with heights up to 2 inches and wavelengths usually of 8 inches or so, sometimes up to many feet. Sides of ridges have the same inclination. Troughs are rounded, crests sharp. Heavier sand grains settle in the troughs. Patterns can become complicated by changes in the rate or direction of flow.

Made by Waves: At low tide on a Florida beach, wave marks are exposed beside a tide pool. These marks are relatively fresh, unchanged by any later waves. Compare with adjacent photo.

Rill Marks and Wave Marks: On another Florida beach, rill marks lead downbeach (*toward left*) to small wave marks which were flattened and otherwise altered by later waves.

tween the horns. Exactly how cusps form is not well understood; perhaps an irregularity on the beach face or in the wave action triggers cusp formation. In any case, the seawater as it surges up the beach and then down again appears to shape the beach materials into forms – that is, the cusps – which tend to minimize turbulence in the flow and produce equilibrium.

On the landward side of the backshore a ridge of coarse material – gravel, cobbles, boulders - may be built up by especially powerful storms. Such features, called beach ridges, are beyond the present reach of tides and storm waves. On an advancing beach (one that is rising and widening relative to sea level), there may be several beach ridges, representing different stages of advance.

Barriers and Barrier Islands

Where sand is plentiful, as along edges of coastal plains, breaking waves and longshore currents often build ridges of sand offshore and nearly to high-tide levels. These sandbars, or sand barriers, are parallel to the shoreline. Commonly they are scores or hundreds of feet long; but in one strong storm they can be removed and rebuilt in another location.

Some sand barriers are built well above high tide and, with time and with the help of wind, may grow to widths of miles and lengths of tens of miles. These are barrier islands. Being permanent, or nearly so, they become covered with vegetation and are often developed for human use. They are subject to reshaping by continuing wave, current, and wind

Current ripples, made by sustained flow, also are perpendicular to the flow direction. They may somewhat resemble wave ripples but are asymmetrical; as in wind-made sand ripples, the lee sides of the ridges are steeper than the stoss (upstream) sides. Powerful beach currents can produce ripples much higher and with much greater wavelengths than occur in wave marks.

On a steep, straight or broadly curving beach, a row of crescentic indentations may form, with horns of the crescents extending seaward. These features, called beach cusps, are made by surging nonbreaking waves moving directly or almost directly onshore. Cusps vary in size, usually being spaced regularly at about 30- to 175-foot (occasionally much wider) intervals; the larger the waves, the larger the cusps, usually. Cusps can be of coarse to fine material, which is sorted by swash so that larger particles form the ridge be-

Sandbars and a Lagoon: At Estero Beach, south of Fort Myers, Florida, is this large bar, with a beach scarp showing the limit of wave action under ordinary conditions. At left is a smaller bar, which at high tide will be under water. At right is a lagoon, which at low tide will be a mudflat.

Mudflats: Near Yarmouth Harbor, Nova Scotia, extensive mudflats are exposed at low tide. In the foreground is a tide pool; beyond is a winding tidal stream. Seabirds enjoy rich hunting on the flats.

Patterns of Sediments: On the north shore of New York's Long Island, a spit has grown eastward across a bay, forming a baymouth bar with an inlet at the east end. At the south end of the bay a very small spit appears to be growing across a cove. A third spit, also small, seems to be growing out from a small cuspate foreland east of the bay.

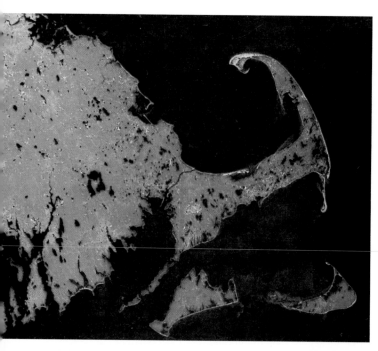

A Spit par Excellence: Cape Cod's original deposits of glacial drift are being reshaped by marine and wind action, sediments toward the northeast being swept north and west to form the spit. In the land area here, dark patches are kettle lakes. The color is artificial.

mainland of Long Island, New York. Usually shallow, a lagoon may gradually fill with sediments washed or blown off the barrier island, brought in by tidal currents, or deposited by streams and wind from the mainland. At low tide the bottoms of some lagoons become exposed as mudflats.

Spits, Tombolos, and Cuspate Forelands

Longshore currents sweeping around the end of a barrier may deposit enough sand there to connect the barrier to the mainland; thus a spit is formed. Most spits rise 10 feet or more above present sea level. The seaward end usually curves to form a hook; some spits have several hooks. Large spits, such as New Jersey's Sandy Hook, or Spurn Head, on the east coast of England, are relatively long-lasting, though subject to continual reshaping by waves, currents, and wind.

Any bar that grows across the entrance of a bay is known as a baymouth bar. Formed usually between two headlands, it may consist of a single elongated spit, or of two spits that meet, or of a longshore bar that has migrated shoreward. Most baymouth bars are cut through by an inlet which connects the bay with the ocean. A bar across a bay near its head (its shoreward side) is a bayhead bar.

A ridge of sediments connecting an island to the mainland is called a tombolo, or tiebar. Often it is built up where the longshore movement of sediments is interrupted or slowed by other currents or by the configuration of the coast; but other conditions may be involved. One well-known tombolo connects Morro Rock in Morro Bay, California, to the main land; another is the tie between St. Stefan Island, off the Adriatic Coast, and the mainland of Montenegro; and a third is the tie between Bar Island and the mainland of Maine at Bar Harbor.

Sediments moved along a coast may form a nearly triangular projection of land that has crescentic indentations on each side and ends in a point. Such features, often many miles in extent, are known as cuspate forelands. How they originate is not fully known; they may develop from spits,

action and may be breached during the heaviest storms. If sea level is rising, barrier islands are remade closer to the mainland by waves and currents.

The area of quiet salt water between a barrier island and the mainland is known as a lagoon, or bay. An example is Great South Bay, between Fire Island (a barrier island) and the

An Island "on Leash": A favorite tombolo, among those who "collect" them, is the one that ties the tiny island of St. Stefan to the coast of Montenegro on the Adriatic Sea.

Land in the Making: In shallow waters off the coast at Everglades City, Florida, mangrove forests flourish. Their roots stabilize sand and organic detritus, which accumulate and may eventually form land.

or where longshore currents are blocked by the shore configuration, or where opposing currents meet. Examples include Cape Hatteras and Cape Canaveral, in the United States, and Dungeness, on England's southern coast.

Organic Coasts and Shores

Long stretches of the world's coasts and shores are on materials produced by marine organisms, principally corals, calcareous algae, sponges, and marsh plants.

Organisms other than corals strongly determine the nature of some coasts, especially in lagoon environments. Salt marshes fringing the northern coasts of Canada and eastern England, and mangrove forests along the southwestern coast of Florida, are prominent examples.

Corals have built reefs which, having become exposed above sea level, are among our more massive and impressive landforms. Such reefs exist today in subtropical seas as huge, growing colonies; they exist also as fossil relics making up much of the world's limestones, some with ages of hundreds of millions of years. Among reef formations in the United States is the one that makes up New Mexico's Guadalupe Mountains, which contain Carlsbad Caverns. Fossil reefs make up parts of the Canadian Rockies, also.

Coral colonies grow upward from the ocean bottom in subtropical waters, mostly between latitudes 30 degrees N. and

S. Restricted to clear waters (they cannot live in delta and other muddy environments), they favor depths shallower than 75 feet, though some species tolerate 500 feet. Their calcareous parts, or "skeletons," accumulate by the billions to form the reefs, which may be several hundred feet thick, several miles wide, and hundreds of miles long. By compression, solution, and redeposition the skeletons become consolidated and cemented to form limestone, which, though well consolidated, still has some porosity – an important factor for geomorphic change.

A coral reef commonly has three facies (groups of genetically related sedimentary units): the reef itself, built by corals; the forereef, consisting of talus and other broken ma-

Raw Material for a Coral Island: On an island in the Galápagos a mass of brain coral, about 8 feet high, looms. The Galápagos were built up by a combination of coral growth and volcanic activity. Lowering of sea level has exposed masses of coral, originally formed under water, to the destructive effects of wave action and weathering.

Part of an Organic Reef: The Guadalupe Mountains, in west Texas, are an eroded portion of the Capitan Reef. Formed during the Permian Period, when a clear, shallow sea covered the region, this reef is one of the largest of its kind. Of its 300-mile length, only about 40 miles is exposed. Guadalupe Peak (El Capitan), 8,751 feet, is the highest summit. Behind the escarpment, movements of subsurface water have formed the Carlsbad Caverns.

terials on the reef sides, inclined about 30 degrees seaward and including various non-coral organisms; and the backreef, which is the landward side of the formation, between the reef proper and the mainland, with beds of limestone, reef detritus, and various non-coral organisms. The backreef may be separated from the mainland by a lagoon.

A reef attached to a mainland is a fringing reef. It has no permanent lagoon but may be more or less underwater except at low tide. Its seaward side slopes steeply down to the sea bottom. Fringing reefs are very common in subtropical seas, such as the Caribbean and the southwest Pacific.

A reef separated from the mainland by a lagoon is called a barrier reef. Small reefs, called reef knolls or patch reefs, may rise from the lagoon.

Barrier reefs are built up in relatively deep water and may reach enormous proportions. Tahiti, the Pacific island, is encircled by a barrier reef. The Great Barrier Reef of Australia (actually a composite of some 2,500 small reefs) is about 1,200 miles long, with a lagoon tens of miles wide.

A third kind of coral structure is the table reef, a flat-topped feature rising from a relatively shallow sea bottom. An example is the Bahama Platform, extending over some 60,000 square miles, in the Caribbean.

Atolls are the fourth major type of coral reef. Built up in deep water, they are roughly elliptical, with diameters as great as tens of miles and a central lagoon averaging about 40 feet in depth. The sides of atolls slope steeply down to depths as great as 1,500 feet. Of the approximately 330 atolls known, nearly all are in the southwest Pacific, where they have grown from the tops of submarine volcanic cones. Well-known examples are Bikini and Eniwetok.

Since the maximum depth for coral activity is around 500 feet, how can coral reefs extend down to 1,500 feet, as some do? One answer is sea-floor subsidence. Corals are able to maintain their building activity upward as the sea floor sinks. According to varied evidence, the sea floor in the vicinity of the Pacific atolls has sunk at rates averaging around 1 inch per 1,000 years. At various depths, atolls have terraces which indicate former sea levels.

Another question: If the sea bottom has been subsiding, and if corals can live and work only under water, how can tops of atolls be above water? One answer is that storm waves can break up coral rock at sea level and pile it up higher – as happened, for example, in 1972 on Funafuti Atoll, in the Ellice Islands, when a single storm built coral fragments into a ridge 29 miles long, with a mean height of 11 feet and a mean width of 120 feet. Clearly, storm work has done much to build atolls above sea level.

Another cause for the rise of atolls above sea level was perceived by Charles Darwin long ago: Before the Pleistocene, world climates were relatively mild and sea level relatively high; corals built atolls up to this level. During the Pleisto-

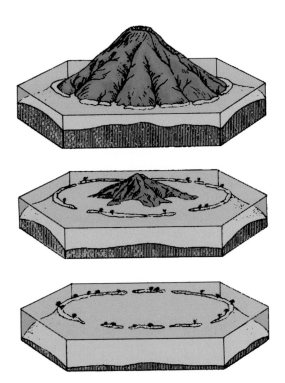

Development of an Atoll: Atolls may develop as a result of the submergence of a volcano or a rise in sea level, growth of a fringing reef around the volcano's sides, and, finally, lowering of sea level enough to expose the reef. Also, storm waves can break up underwater coral and pile it high enough to form land.

cene sea level fell hundreds of feet, and atoll tops were exposed to wave and current action, which covered them with erosional debris and cut terraces in their sides. Today world temperatures are still substantially lower than before the Pleistocene, and sea level has not risen far enough for the atolls to be covered as they were before the ice age.

Lakeshores

Because even the largest lakes are very small compared to oceans, they have much smaller waves and currents, and these cannot compare in effectiveness with waves and cur-

Built by Corals: Swain's Island, one of the American Samoa group in the southwest Pacific, is an atoll of classic form. Approximately elliptical in shape, a square mile in area, it is bounded by a beach of white coral sand. Landward from the beach is the backreef area, with coral detritus and other organic remains, and in the center is the lagoon. Note refraction of waves, caused by drag against the bottom along the sides of the atoll.

rents that work upon seashores. Further, lakes are naturally short-lived – they tend to fill with sediments and to be emptied by streams downcutting at their edges; thus lake waves and currents have insufficient time for creating large landforms. Finally, lake tides where they do exist are generally too slight to provide waves and currents with a wide vertical range to work in.

Although astronomical lake tides (those caused by Moon-Sun gravitational attraction) are relatively insignificant, other movements of lake water can be substantial. In the event called a seiche, wind pushes water up against one shore and the water then flows back to the opposite shore, like water in a washtub being rocked. Seiches can be important in the transportation of sediments. In Lake Erie seiches have been measured up to 8 feet in height; those in Great Salt Lake, made by north or south winds, up to 2 feet. In the south arm of this lake the oscillation period (time covered by one back-and-forth movement) is about 6 hours. At the southwest end

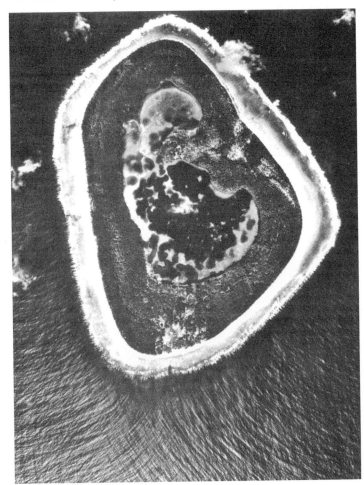

of Lake Geneva, Switzerland, seiches rise to dangerous levels because of the funnel-like contour of the shore.

Seiches can be produced also by earthquakes, sudden changes in atmospheric pressure, heavy rain, surges of glacial meltwater from nearby mountains, and variations in water density.

Waves on large bodies of water such as the Great Lakes shape shores of loose material, building sand barriers (including spits) as well as beaches with scarps, berms, and beach ridges. Storm waves attacking weak bedrocks fringing lakeshores can undercut them, creating cliffs. In cold regions in winter, expanding lake ice pushes sand and rock fragments up the beach, creating ridges. Waves on the largest lakes, such as Lake Bonneville during the pluvial period, have cut terraces on adjacent highland slopes.

Lakeshores also have their exhibits of rill marks, swash marks, and ripple marks. On the landward side of beaches there may be enough sand for dunes to form; for example, the spectacular Indiana Dunes, of glacial drift, at the southeastern edge of Lake Michigan.

Wind works effectively as a land sculptor wherever substantial amounts of sand are available. This condition tends to rule out sea cliffs as sites for much wind work, but coastal areas and beaches bordering the ocean and large lakes qualify. These areas may be on the edges of deserts, coastal plains, floodplains, deltas, or any other sand-rich terranes. Sands may consist of erosion sediments from nearby highlands, eroded lavas, glacial drift, fragmented coral or other sea-bottom sediments (on a recently uplifted ocean floor), or simply sand blown in from an inland location.

As in deserts, the numbers and sizes of dunes depend largely on the extent of the sandy terrane, sand thickness, prevailing temperatures and humidity (hot, dry conditions favor dune-building), amounts of rainfall and vegetation, and strength, direction, and persistence of winds. Although rain may be fairly frequent, warm winds quickly dry the sand, so that it is readily movable. Windy conditions are common because air tends to move onshore during the day and offshore at night. On shores that face prevailing winds (such as the shores of California, Oregon, and Washington), much sand tends to blow inland, and beaches widen, but on shores fac-

Barriers Against the Sea: At Hampton Beach, New Hampshire, a sea wall has been built to protect a highway and a residential area. Storm waves have built a beach ridge of cobbles against the wall. A groin of large stone blocks, partly buried by the cobbles, extends seaward from the wall.

ing oppositely (such as shores from Long Island to Florida), much sand tends to blow seaward). Dunes include the smaller types: mainly domes, barchans, and the barchanoid variety; beaches are not extensive enough, and commonly lack sand enough, for development of dunes to match the seifs, whalebacks, and other giants of the Arabian and Sahara deserts.

Many coasts and beaches are only semiarid, and vegetation there has succeeded in establishing itself in substantial patches. These tend to anchor some areas of sand and distort dune formation. For example, when vegetation anchors the horns of a barchan but the wind continues to blow freely between them, the nose (forward end) of the barchan continues to migrate windward, passing the horns, and thus a sort of reverse barchan – a parabolic dune – takes form. On such terrains, distortions of basic dune types are common. Among them are compound dunes, consisting of two or more coalesced dunes of the same type (such as two or three barchans), and complex dunes, formed by two or more dunes of different types. Some coasts and shores have enough vegetation to prevent dune formation altogether; and, in fact, that is a common objective for those who want sand to stay where it is. Where the natural vegetation is insufficient, the addition of beach grass and snow fencing often gets results.

In the coterminous United States, notable areas of wind work on seacoasts and shores occur intermittently from Cape Cod (which is on glacial drift) to Florida, and along the Gulf Coast of Texas. Northward from Tenmile Creek, on the Oregon coast, are impressive dune chains, especially of the linear and transverse types, migrating before the prevailing winds and inexorably invading coastal forests. Also in the Northwest are the amazing fields of barchans in the Moses Lake area of Washington. On the southern shores of Lake Michigan are heaped-up dunes of glacial drift spectacular enough to be preserved as the Indiana Dunes.

Windwork on sandy coasts and shores continues on portions of all the continents. Among the more interesting areas

Marching Dunes: At Oregon Dunes National Recreation Area, Oregon, prevailing winds from the Pacific, often very strong, keep building dunes and moving them eastward. In this forested area near the shore, the number of dead saplings suggests that the invasion by dunes has been relatively rapid; trees are killed before they grow large.

To Save a Resort: At Eastbourne in Kent, England, on the chalk coast, closely spaced groins protect the promenade. The beach is of cobbles, suggesting that it is relatively steep and backwash is strong enough to carry away much of the sand.

are the shores of Libya and Egypt, where the northeast trade winds keep the land hot and dry, and keep sand steadily migrating inland. Also spreading inland are sandhill concentrations on the coasts of Holland, northern Germany, and southwest Africa. In England similar sand migrations are occurring on the northwest coast of Cornwall, and in Scotland on the southern shore of Moray Firth.

Coasts and Shores in Evolution

Formerly most geologists thought seacoasts and shores undergo an evolution from youth to maturity and old age, much as inland landscapes were supposed to do. A newly uplifted land fronting the ocean would become subject to lowering by subaerial erosion. Waves and currents would undercut the seaward side, creating cliffs. Weak and strong zones would be selectively eroded to make bays and headlands. At the same time, the cliffs would be retreating because of undercutting, the beach would be widening landward, and the wave-cut platform would be lengthening. Meanwhile headlands, especially vulnerable to erosion, would be shrinking, and bays, protected from much marine erosion, would be filling with sediments. As erosion continued to lower the land generally, and as the beach and the wave-cut terrace kept widening, a low, nearly flat terrain would be developed with a very broad beach, a nearly straight shoreline, and a wave-cut platform extending far seaward.

Today it is understood that coasts and shores are not "standard" landforms in various stages of predictable evolution. Rather, these forms reflect the fact that natural forces and the conditions under which they operate tend toward a state of equilibrium, or balance. The age of a coast or shore – the length of time it has existed – is but one factor in the overall complex of processes and conditions that shape it.

Coasts are undergoing changes which seem to deny anything like a regular and predictable evolution. All coasts now are subject to a eustatic (world-wide) rising of sea level due primarily to world-wide melting of glaciers, the melting being due, at least partly, to global warming. Rising sea level is steadily altering the dynamics of coastal change; and, incidentally, it threatens to inundate some of the world's great port cities – New York, London, Stockholm, Shanghai, and many more. Yet many coasts, such as those of Scandinavia and New England, are rebounding after the Pleistocene glaciation, and others, as in the Chesapeake Bay area, are sinking as a result of tectonic forces unrelated to glaciation. Coasts of Hawaii, Iceland, and other regions are being altered frequently by volcanic eruptions. All coasts are changing as, in the course of erosion, different kinds of rocks are becoming exposed to subaerial erosion and to wave and current action. Shore-shaping currents are changing in force and direction as global warming raises ocean temperatures, as tectonic forces reshape ocean-bottom topography, as submarine volcanic activity spreads more lava over ocean floors. In this immense, complex scenario human activities, too, have their roles – in deforestation, well-drilling, groining, transfers of sand for beach maintenance, road-building, draining or filling of marshes, and – on the larger scene – burning of fossil fuels.

Shore Change: Controllable?

From early antiquity, perhaps even before the days of King Canute, human ingenuity has been challenged to stop or modify shore processes. Seawalls of concrete or ton-size stone blocks, not always invincible, are built to protect the base of cliffs and expanses of beach from wave attack. Breakwaters, those long walls of concrete or stone, are constructed parallel to the shore to shelter harbors, docks, and beaches. Groins, which are shorter walls extending seaward from shore, are used to restrict movement of sand alongshore. Fences are built and beach grass is planted to retard blowing sand. Where sand can be spared, it is picked up and trucked to places where sand is needed. And so engineers and earth-moving machinery find plenty to do.

As with some other measures for protection against natural processes – measures such as flood control and soil stabilization – schemes to resist tides, waves, and currents are often in vain. Waves can have enormous power; currents are implacably persistent. Very massive walls can be undermined and breached in minutes, groins save sand for one beach by stealing it from the next one, and wind does blow through beach grass and fences. Protection is not always in vain, but it is expensive and must be sustained. Along ocean shores property is continually being lost to waves and wind, and courts are kept busy with suits to prevent loss or to be reimbursed for it.

Even if human sagacity could master tides, waves, and currents, in the long run the ocean with its changes in level, over which human control seems highly unlikely, would win the day. The ocean is greater than all the lands. Its level has been changing continuously for billions of years, and it will continue to do so as a result of continuing tectonic events, climatic change, and igneous activity.

Sea level is rising. Coasts around the world with less than a few hundred feet of elevation are threatened with inundation during the next few thousand years, conceivably within the next few centuries. Those coasts include not only the world's great port cities but millions of square miles of agricultural lands upon which hundreds of millions of people depend for survival. But inundations around the world are only one of several possibilities. Perhaps within the next 12,000 years, according to some glaciologists, sea level could go down again because of the onset of a new ice age. And at some time in the future it will certainly be going up or going down because of tectonic activity, which never ceases.

Truly, we live on a busy Earth.

The Enduring Sea: The world ocean surrounds all lands. With its waves, currents, and changing temperatures it strongly affects world weather and climates, breaks down and builds up shores, and provides water for streams and glaciers that shape and reshape the land. As landforms change and change again, the ocean, changing in its own multitudinous ways, endures.

Acknowledgments

This book owes much to the influence and efforts of many persons. Long ago, Louis Moyd and Howard Jaffe introduced me to the excitements of field geology. George Gaylord Simpson, J. Tuzo Wilson, Kirtley Mather, and Arthur Strahler, among others, encouraged me to write about geology for the layman. (Dr. Strahler has donated some of his superlative drawings for this book.) Over many years, George F. Adams guided me over hundreds of miles of geological trails, and joined me in producing the Golden Science Guide *Landforms*. Gladys Rosalsky and the late Maurice Rosalsky hosted us for many a show of Maurice's fine photographs of geological features around the world. (Gladys has generously provided a selection of Maurice's photographs for these pages.) Betty Randall suggested the book's title. In recent years Donald R. Coates has lent his expertise and wisdom to the project with countless hours of effort and sustained encouragement, expressing his desire that the public know, and enjoy, more about Earth scenery.

When the time came for production, Maja Britton, of Adastra West, advised on the typesetting, performed most of the illustration processing (with an assist from Vincent Gargiulo and masterly troubleshooting by Tonianne Schlamp), created the cover and title-page designs, and provided vital advice and management in other phases of production. Alex Rainer, also of Adastra West, with versatile expertise assisted in the typographical work, assembled materials into pages, and read proof. Roland Meyer provided knowhow in arranging for the printing, Thomas Grissom contributed invaluable production advice, and Robert Etchells generously shared with us his computer and internet expertise. All associated with Adastra West participated most helpfully in planning.

Thanks are owed to the many persons, government agencies, business firms, and other sources that provided copyrighted illustrations.

Most of all I am grateful to Elaine, my wife, for her everready wisdom, her belief in the project, and her stalwart support during so many years of living with an author.

- J.W.

Credits

All photographs and drawings in this book are © Jerome Wyckoff unless otherwise credited.

Adams, G.F., 296*t*; American Geophysical Union, 13; Atlantic Oceanographic Laboratory, 57*t*; Australian News and Information Bureau, 269*b*; Barbados Tourism Authority, 321*b*; Bureau of Reclamation, 219*b*; Cahayla-Wynne, Richard, 302*tr*; Carlsbad Caverns National Park, 304*t*; Coates, D.R., 77*b*, 79*b*, 91, 282*t*; Cold Regions & Engineering Laboratory (A.E. Corte), 57*m*; Crandall, Dwight, 79*t*; Craters of the Moon National Park, 153*tlr*; Department of Energy, Mines, and Resources, Ottawa, Canada, 259*b*; El Baz, Farouk, 276*tr*; Finnish National Tourist Office, 260*t*; French Government Tourist Office, 133, 316*b*; Geological Survey of Canada (Baird, D.M.), 199; Hawaii Natural History Association, 138*tb*, 142*t*, 151*b*; Hayes, Miles, 130*b*; Hong Kong Government Information Services, 65; Information Service of South Africa, 221*b*; Jacobson, C.E., 121*b*; Japan Air Lines, 144*b*; Kentucky Geological Survey, 303; Landsat (images processed by Earth Satellite Corporation), 232*t*, 284*tlr*, 328*t*; Lehman Caves National Park, 301*t, br*; Lobeck, Armin K., 174*t*, 262*t*; Luray Caverns, Virginia, 298; Morrison, Jack, 201*b*; NASA, 9; NASA–U.S. Department of Agriculture, 129*t*; National Air Photo Library, by permission of Natural Resources Canada, 245*t*; National Park Service, 153*tlr*, 301*bm*, 308; National Speleological Society (courtesy Bob Gulden), 304*b*; Nicholson, Sharon, 284*b*; Northern Pacific Railway, 157*b*; Oregon State Highway Department, 147*br*; Pirsson and Schuchert, 130*t*, 185*b*; Rosalsky, Gladys, 67*t*, 139*t*, 146*tl*, 207*tb*, 221*t*, 295*b*, 328*b*; Royal Canadian Air Force, 254*b*; Sevon, William, 81*t*; Sharpe, C.F.S., 71*t*; Spence Air Photos, 226*b*; State of Colorado, 194*t*; Strahler, Arthur, 160*t*, 169, 181*b*, 187*b*, 188, 227*b*, 305*t*; Swiss National Tourist Office, 70*t*; Texas Highway Department, 330; Thornbury, W.D., 217, 233*t*; U.S. Air Force, 17, 143*t*; U.S. Coast and Geodetic Survey, 131*b*, 331*b*; U.S. Forest Service, 103, 115*t*, 182*b*, 211, 242*t*, 301*bl*; U.S. Geological Survey, 11, 12*b*, 57*b*, 78*tb*, 134*b*, 135, 137, 140, 143*t*, 151*tr*, 168, 174*b*, 186*tl*, 194*b*, 213*m*, 218*b*, 234, 244, 245*b*, 246, 247, 275, 278, 283, 295*tr*, 299*b*; U.S. Navy, 57*b*, 138*b*; von Engeln, O.D., 185*t*; Western Publishing Co., 49*b*, 99*b*, 104*t*, 107*t*, 109*t*, 112*b*, 113*t*, 123*b*, 125*b*, 128*b*, 131*t*, 144*t*, 161, 172*t*, 177*tl*, 179*m*, 181*t*, 183*m*, 186*tr*, 210, 240*b*, 310*t*, 322*t*; Wyckoff, E.V., 20*t*, 58*b*, 67*b*, 84*b*, 105*b*, 114*b*, 154*t*, 164*br*, 173*b*, 230*t*

Additional credits appear with illustrations in the text.

Further Information

The publications listed below can be helpful to a general reader who wants to know more about landforms. Books that cover landforms comprehensively for general readers are few; therefore the list includes some items designed primarily for students and professionals. Some of the books are presently out of print but may be reprinted in the future. Some are available from bookstores or via book hunters who advertise in magazines or on the Internet.

Books and pamphlets on specific localities, such as states or national parks, and on specific kinds of landforms, such as mountains, glaciers, or volcanoes, are relatively numerous. They may be available at bookstores, via the Internet, at outlets such as visitor centers at parks and museums, or directly from the publishers. Some outlets offer color slides, CD-roms, and videotapes containing geological materials for teaching or other uses.

American Association of Petroleum Geologists, Tulsa, OK: Regional Geological Highway Maps. Showing geological provinces and kinds, structures, and ages of rocks.

American Geological Institute, Alexandria, VA: *Geotimes*. A journal of geological news and research for both professional geologists and ordinary readers. Also, *Dictionary of Geological Terms* (1976), Anchor Press-Doubleday, Garden City, NY. Pocket size.

Bird, J.B.: *The Natural Landscapes of Canada* (1980), John Wiley and Sons, Toronto, Canada.

Chesterman, C.W.: *Audubon Society Field Guide to North American Rocks and Minerals* (1978), Alfred A. Knopf, New York.

Chorley, R.J., and others: *Geomorphology* (1985), Methuen and Co., New York.

Coates, D.R., and Vitek, J.D.: *Thresholds in Geomorphology* (1980), Allen and Unwin, London.

Cvancara, A.M.: *Field Manual for the Amateur Geologist* (1995), John Wiley and Sons, New York. Highly detailed, well illustrated, many information sources listed.

Fairbridge, R.W.: *Encyclopedia of Geomorphology* (1968), Reinhold Book Corporation, New York. International scope; for all readers.

Faul, H., and Faul, C.: *It Began with a Stone* (1983), John Wiley and Sons, New York. History of the development of geological science, for all readers.

Geological Society of America, Boulder, CO: *Centennial Field Guide* (to six regions of U.S.A. and parts of southern Canada). Very comprehensive.

Goudie, A.: *Landforms of England and Wales* (1990), B. Blackwell – Cambridge University Press, New York.

Harris, A.G., and Tuttle, E.: *Geology of National Parks* (1990), Kendall-Hunt Publishing Co., Dubuque, IA.

King, L.C.: *Morphology of the Earth* (1967), Oliver and Boyd, Edinburgh, Scotland. Mainly for professional readers.

Lobeck, A.K.: *Geomorphology* (1939), McGraw-Hill Book Co., New York. Still a classic, especially for illustrations.

Mountain Press Publishing Co., Missoula, MT.: *Roadside Geology* Series. Detailed, well-illustrated guides for U.S.A.

Penrose, C.: *Introduction to the British Landscape* (1973), Macmillan, London.

Pough, F.H.: *Peterson First Guide to Rocks and Minerals* (1991), Houghton Mifflin Co., Boston, MA. Standard.

Ritter, D.F., Kochel, R.C., and Miller, J.R.: *Process Geomorphology* (1995), Wm. C. Brown Publishers, Dubuque, IA. For students and professionals.

Shelton, J.T.: *Geology Illustrated* (1966), W.H. Freeman, San Francisco, CA. Superb photographs.

Skinner, B.J., and Porter, S.C.: *The Dynamic Earth* (1995), John Wiley and Sons, New York. Standard; mainly for students and professionals.

Soons, J.M., and Selby, M.J., *Landforms of New Zealand* (1992), Longman Paul, Auckland, N.Z.

Stone, Carolyn R.: *Australian Landforms* (1974), Wren, Melbourne, Australia.

Strahler, A.N.: *Introduction to Physical Geography* (1965), John Wiley and Sons, New York. A classic.

Thornbury, W.D.: *Regional Geomorphology of the United States* (1965), John Wiley and Sons, New York. Still a massive source of information, useful for all readers.

Trueman, A.E.: *Geology and Scenery in England and Wales* (1971), Penguin Books, New York. Pocket size.

Tuttle, S.D.: *Landforms and Landscapes* (1980), Wm. C. Brown Publishers, Dubuque, IA.

Whittow, J.B.: *Geology and Scenery in Ireland* (1975), Penguin Books, New York. Pocket size.

Whittow, J.B.: *Geology and Scenery in Scotland* (1977), Penguin/Harmondsworth, New York. Pocket size.

Whittow, J.B.: *Geology and Scenery in Britain* (1992), Chapman & Hall, London and New York.

Zim, H.S., and Shaffer, P.R.: *Rocks and Minerals* (1957), Golden Press, New York. Simplified; in color; pocket size.

Numerous books, pamphlets, and leaflets providing geological information – international, national, regional, local – are published by national and state or provincial geological surveys, and by geological societies. Present addresses are generally available from libraries and the Internet. Lists of publications may be obtained on request. Some major sources are listed here.

British Geological Survey
Canadian Provincial geological surveys
National Park Service, Washington, D.C.
National Park associations
State geological societies
State geological surveys of the United States
U.S. Geological Survey, Washington, D.C.

Index

M